Nanotechnology Toward the Sustainocene

Nanotechnology Toward the Sustainocene

edited by

Thomas A. Faunce

Pan Stanford Publishing

Published by

Pan Stanford Publishing Pte. Ltd.
Penthouse Level, Suntec Tower 3
8 Temasek Boulevard
Singapore 038988

Email: editorial@panstanford.com
Web: www.panstanford.com

British Library Cataloguing-in-Publication Data
A catalogue record for this book is available from the British Library.

Nanotechnology Toward the Sustainocene

ISBN 978-981-4463-76-8 (Hardcover)
ISBN 978-981-4463-77-5 (eBook)

Printed in the USA

This book is dedicated to Dr. Bryan Furnass,
Canberra ecophysician and developer
of the concept of the Sustainocene and to ANU
photosynthesis researcher Warwick Hillier who
attended the 2011 Lord Howe Island 'Towards Global
Artificial Photosynthesis' conference, but tragically
departed this life in 2014.

Contents

Foreword

Analogical Precedents: Informatics, Biotechnology, and Energy

The Hon. Michael Kirby AC CMG

Former Justice of the High Court of Australia (1996–2009). One-time chairman of the Expert Group of the OECD on Transborder Data Flows and the Protection of Privacy. Formerly a Member of the World Health Organisation Global Commission on AIDS and of the UNESCO International Bioethics Committee. Commissioner of the UNDP Global Commission on HIV and the Law (2010–2011).

Introduction

The "Towards Global Artificial Photosynthesis: Energy, Nanochemistry and Governance" conference I attended at Lord Howe Island in August 2011 was concerned with radical changes in energy services that will be necessary as the finite sources of energy from fossil fuels are depleted (Faunce, 2012). Whichever way our species turns, there will be challenges. Some of these will be technological and scientific. Some will be economic, environmental, and social. But others will be governmental and legal, necessarily, at least so far as artificial photosynthesis' global deployment.

In medieval times, a revolution occurred in the communication of ideas following two technological changes, perfection of spectacle glasses and invention of the printing press. The former allowed the monks, who spent their years inscribing religious texts, to extend their working lives beyond presbyopia. The printing press then released words (and hence the ideas represented by words) from the calligraphy of the monks.

For holy men, the words were written to be said or sung. But after Caxton, printed words took on a life of their own. Their meaning could be delivered without mouthing the sounds they conjured up. In a forerunner to the urgencies of the present-day email, text messages, and social networks, ideas could be communicated four times faster than they could be said. A revolution in communication had begun. It continues to expand.

In addressing energy, science, and technology this conference opened a new dimension for me. My past encounters with my chosen issue had been with information technology and biotechnology. These are major features of the contemporary world. From the viewpoint of law, they present a common difficulty—that no sooner is a relevant law or agreed guideline developed to address some of their features and to regulate those deemed necessary for regulation by reference to community standards and the technology itself has raced ahead. The law in the books is then in great danger of becoming irrelevant, in whole or part. Because of fast-moving science and technology, laws, and institutions conceived at one time may have little or no relevance to events that happen very soon thereafter.

My encounters with information technology and biotechnology suggest possible directions in which laws and institutions may emerge to respond to the remarkable developments that are occurring in the fields of energy technology. Everyone knows that fossil fuels as a source of energy now have a limited life. Most experts also believe that fossil fuels have serious environmental disadvantages, despite their enormous efficiency for use, transmission, and storage. It is these features of fossil fuels that have led to urgent endeavors to find viable alternatives. At first, it was believed that nuclear technology would be the cheap, safe, and plentiful energy source for the future. However, experiences in Chernobyl and more recently in Japan have demonstrated the vulnerability of this source to accidental and natural disasters.

As well, the risks of nuclear terrorism have combined with accidental events to turn attention urgently to other energy sources and the technologies that will deliver them efficiently and economically for worldwide human use. It is here that solar energy and wind energy (often in symbiotic combination) present themselves to complement hydro sources as safe and powerful energy sources for the future. These and the developments of global artificial photosynthesis (if it can be perfected) are presented as the energy alternatives to replace fossil fuels that have dominated the past century (Faunce et al., 2013).

Whether by default, or by design, many issues presented to the law by contemporary technology are neglected or ignored because one suspects that, in many instances, it is too complicated or seen as too sensitive. I now will identify a number of paradoxes, or at least curiosities, that emerge about regulating technology generally but that are likely to provide valuable precedents for governing the technologies that eventually make our world sustainable.

Five Paradoxes

Doing the Best without Experts

The first of the curiosities is a reflection not only on my own limited competence but also on the limited competence of every lawyer and every would-be regulator.

There are no real experts on the generic subject of regulating technologies. They do not exist in the United Kingdom, the United States, Australia, or elsewhere. It is much easier to find an expert on the intellectual property implications of biotechnology and information technology than it is to find someone skilled in considering what new law or institutional arrangement, if any, should be created to deal with a particular issue presented by technology and how it should be devised. It is easier by far to find an expert on income tax or unjust enrichment or international human rights law than to find governmental officials, judges, or even legislative drafters who can claim to be experts on subject matters such as those of safely globalizing artificial photosynthesis, as presented at the Lord Howe Island conference.

Each new area of technology will have its own features, relevant to the regulators and institutions appointed to the task of responding to its national and international challenges. Take, for example, the provision of intellectual property protection for computer programs—a comparatively new technology. Lawrence Lessig was the founder of Stanford School's Center for Internet and Society. His book *Code and Other Laws of Cyberspace* (now updated by *Code V2*) blazed a distinctive trail. He is something of a guru on the interface of cyberspace and the law. He challenges lawyers and technologists to think freshly. His thesis is that "Code," or the architecture of technological systems, will sometimes incorporate regulatory imperatives into the information technology itself, thereby obviating or negating any real choice on the part of the user as to whether or not to conform to a particular law.

In the High Court of Australia, the judges came face to face with this reality in *Stevens v. Sony Computer Entertainment* ([2005] 224 CLR 193). The case was about a claim by Sony Corporation of a breach of a "technological protection measure" installed by it in the program of its computer games. Sony asserted that the measure was protected under the Australian Copyright Act 1968. Sony argued that Stevens had unlawfully sought to circumvent the device-incorporated computer games that it produced and sold on CD-ROMs for use in its PlayStation consoles.

Applying a strict interpretation to the expression "technological protection measure," the court held that Sony's device did not fall within the statute. I agreed in this analysis (Stevens [2005] 224 CLR 193 at 246). The case was a vivid illustration of the way in which, for copyright, contractual, and other legal purposes, attempts are now often made to incorporate regulatory provisions in the relevant technological codes (Brennan, 2006). It is a new development, although I suppose one might see primitive attempts directed at the same purpose in the safety provisions incorporated in the engineering designs of houses, bridges, and airplanes. Computer PlayStations simply take this to a higher level of sophistication and technological capability. Professor Lessig identified this new development with particularity. Inevitably, his expertise did not include all of the current major technologies such as energy technology, still less the way in which law can regulate them all.

We may complain about the absence of law concerned with new cutting-edge technology. We may acknowledge our own imperfections for addressing the gap. We may recognize, with Professor Lessig, that regulation in the future may not necessarily come in the form of instruments made by or under the legislature and published in the *Government Gazette*. It may take new and different forms and sometimes be grafted onto the technological setup itself.

Nevertheless, the issue of regulating technology, including sustainable energy technology, is undoubtedly one having the greatest importance for the future of the rule of law in investigating governance for the globalisation of artificial photosynthesis in every society.

The conference at Lord Howe Island in 2011 surveyed what is substantially a blank page. Increasingly, the content of law, like the content of life, will be concerned with renewable energy technologies like artificial photosynthesis and with their many consequences for society. The importance of energy technology regulation therefore belies the comparatively little that is written, said, and thought about it. Paradoxically, those who first lay claim to expertise may thereby participate in a self-fulfilling prophecy.

Too Much/Too Little Law

The second paradox is that the failure to provide a framework for legal regulation to deal with the consequences of a particular technology, such as global artificial photosynthesis, is not socially neutral. Effectively, to do nothing often amounts to making a decision.

In so far as law expresses prohibitions supported by sanctions that uphold the command of a relevant governmental or intergovernmental power, silence may, for once, imply consent or at least nonprohibition. Nothing then exists to restrain the governmental or intergovernmental power except its own ethical principles, any institutional ethics requirements, the availability of funding, and the prospects of a market.

The recognition that inaction in the face of significant technologies may amount to making a decision coexists with our appreciation, as observers of the law, that premature, overreaching, or excessive lawmaking may, in some cases, be an option worse than doing nothing. It may place a needless impediment upon local scientists and technologists, obliging them to take their laboratories and experiments offshore.

In a big world with diverse cultures, religions, and moral beliefs, it is generally possible to find a place offering a regulation-free zone in exchange for investment dollars. Just as bad is the possibility that laws will be solemnly made and then ignored or found to be ineffective, as was temporarily the case with the "technological protection measure" considered in the Australian Sony litigation. Following the decision of the High Court of Australia in that case, and under pressure from the US government under the United States-Australia Free Trade Agreement, Australian law was changed. The amended law represented an attempt to overcome the High Court's decision, although in a somewhat different way (deZwart, 2007).

Most of participants at the "Towards Global Artificial Photosynthesis: Energy, Nanochemistry and Governance" conference at Lord Howe Island would recognize that, in their own fast-moving field of energy technology, premature, overreaching, and ill-targeted laws might diminish experimentation, burden innovation, and cause economic and other inefficiencies. It is this attribute that reacts with anxiety about the dangers of the so-called precautionary principle (Andorno, 2004), which initially emerged in an environmental context.

While the precautionary principle appears to be gaining increasingly widespread acceptance in the international community, particularly in respect of protection of the global environment, it carries risks of its own. If taken too far, it could instill a negative attitude toward science and technology and encourage excessive regulation in the attempt to avoid *any* risks. Life is risky. Most technological innovations carry at least some risk. An undue emphasis on precaution, for fear of *any* risks, would not be good for science or technology or for the global economy, energy sources, or innovation, in thought as well as action. On the other hand, energy technologies clearly involve risks, as the accidents in connection with nuclear power plants demonstrate and as the vulnerability of large-scale power grids and environmental dangers of wind, solar, and hydro technology may present.

The second paradox is thus more of a contradiction or tension, difficult to resolve. At the one time we must accept that doing nothing to regulate technologies involves making a societal decision. Yet we must also recognize that sometimes doing nothing will be a better option than making laws that unduly impede innovation and burden efficiency.

First Amendment and Copyright Law

The third paradox concerns the proposition that while the law generally recognizes that while "free" expression and access to a "free" media constitute important human rights, they are not unlimited. They have to be harmonized with other fundamental human rights. These include the right to individual honor and reputation and to protection of privacy and family relationships (International Covenant on Civil and Political Rights [1976], articles 17.1, 17.2, and 19.3). They also include protection of the legitimate rights of inventors (Universal Declaration of Human Rights [1948], article 27.1; International Covenant on Economic, Social and Cultural Rights [1976], articles 15.1[b] and [c]).

In the field of energy technology we are inevitably also influenced by US regulatory models and traditions, simply because of the very significant role of US scientists, technologists, and energy companies operating in this field. However, now there are other players, particularly China and India, with their insatiable demands for energy to sustain their fast-expanding economies. The influence of the regulatory traditions and economic needs of these other countries is likely to emerge as a major consideration in this field. And in so far as international trade and investment treaties are needed and negotiated, they should necessarily reflect the interests of developing countries with their huge potential energy needs rather than those of multinational corporations.

Technology's Democratic Deficit

A fourth paradox derives from the way in which contemporary technology at once enhances and diminishes our facilities of democratic governance. When it comes to legal regulation few participants, at least at a conference on sustainable energy technologies, such as that on Lord Howe Island, will question the desirability of rendering laws and regulation generally available and accountable to the people from whom authority to govern society is ultimately derived. However, on balance, does technology enhance or reduce or encourage or discourage democratic accountability for the state of the resulting regulations?

There can be little doubt that information technology has improved communication that is essential to converting the requirements of electoral democracy into the realities of genuine accountability of the governors to the governed. Radio, television, worldwide satellite communications, the Internet, podcasts, blogs, and so forth have revolutionized the distribution of information about those persons and institutions whose decisions affect the regulation of our daily lives. In this sense, democratic governance has moved from small town-hall assemblies of earlier times into huge national and international forums, both public and private.

Paradoxically, however, the very quantity of technological information today has resulted in its manipulation and presentation in ways that are often antithetical to real democratic accountability. The technology stimulates a demand for the simplification and visualization of messages, personalization of issues, trivialization of conflict, confusion between fact and opinion, and centralization and "management" of news. The so-called spin and infotainment are characteristics of the present age. They tend to concentrate power in a way that even George Orwell could not have imagined. This energy technology conference was mercifully free of these deficiencies. However, another deficiency was certainly present in abundance. I refer to the complexity of the basic science and technological controversies addressed by the scientists and technologists. The participants could speak with each other. Even then, those whose training was in physics confessed to understanding only part of the discourse of the energy chemists. In relation to such exchanges, the intelligent layperson was often left out in the cold.

Moreover, the effective incorporation of regulation in the technology itself, and the way it is rolled out, going beyond what is strictly required by local law, denies effective opportunities for those affected to challenge the regulation so imposed. Who can, or would, challenge the overinclusive software designed to bar access to Internet sites selected as "harmful to minors" but sometimes operating in an overinclusive way?

Once, when I was serving in the High Court of Australia, I found that the website of the archbishop of Canterbury was barred from use! My staff was unable to procure one of the archbishop's addresses. This was because an Internet filter, instituted to deny access to websites deemed undesirable, had erected a bar. Presumably, this was because, in the manner of these times, one or more of the archbishop's addresses dealt with issues of sex, specifically homosexuality. In fact, that was exactly why I wanted the nominated speech. I was surprised to find that at the same time the Vatican website was accessible without any restriction. This may say something either about the prudence of the Pope's choice of language, his discomfort with candid discussion of sex, the power of the Roman Catholic Church in such matters, or the religion of the filter programmer. I gave directions that led to the filter being overridden and duly secured a copy of the desired speech. Others might not have been so lucky.

Given the importance of technology generally to the current age, how do we render those who design, install, and enforce such programs accountable to the democratic values of our society? As information technology, such as "Code," enlarges and replaces the old style of legal regulation of technology, how do we render its architects answerable to the majority views of the people? How, if at all, are transnational corporations, like Sony or Esso or BP or Shell, for instance, rendered responsible to the democratic values of the nations in which their products are used? How do we render the inescapably complex discourse of energy scientist and technologists—and their large implications for peace, justice, economic advancement, and human rights—understandable to democratic legislations, the officials who advise them, the judges who supervise them, and, especially, the citizens who elect them?

These are legitimate questions because the fourth paradox involves the coincidence, at the one time of history, of technologies that took notions of freedom over the Berlin Wall, and the energy technologies that may promise an end to poverty and inequality in our world. Yet both these fields of technological revolution also sometimes diminish genuine debate, enlarge unreviewable "technological" and corporate decisions, and expand the potential of a relatively few decision makers to "manage" news in a way inconsistent with real transparency and accountability of decision making to the people most affected.

Vital but Neglected Topics

I reach my fifth, and final, paradox. This is one of the most important issues for the future health of the rule of law in every country. Because of the elusiveness of much contemporary technology to effective regulation, large and increasing areas of activity in society find themselves beyond the traditional reach of law as we have hitherto known it. When regulation is attempted, as I have shown, it will often be quickly rendered ineffective because the target of the law has already shifted or because corporate lobby groups can use narrow international investment arbitration rules to attempt to shut down regulation perceived as undercutting their profits. Typically, in the past, the drawing up of laws has been a slow and painstaking process. Consulting governments and those primarily affected, not to say the people more generally, takes much time. In that time, the technology may itself change, as I have demonstrated from my experience in the design of human tissue transplantation and privacy laws. Now new forms of regulation are being developed in the form of what Professor Lessig calls "Code." Yet this form of regulation is not so readily susceptible, if susceptible at all, as conventional laws have been, to democratic values and to the participation (or even appreciation) of most of those affected in the moral choices that determine the point at which the regulation is pitched.

It sometimes falls to small groups, particularly in professions, to lead the way and to bring enlightenment to the many. This, then, is the fifth paradox—at least it is an oddity. Such an important topic as the regulation of burgeoning technologies such as globalised artificial photosynthesis with the capacity to shift modern society towards sustainability should engage the interest and attention of all who claim to be lawyers, sociologists, and philosophers and express an interest in the health of the rule of law. Yet, for the moment, and for most such observers, this is *terra incognita*. The contributions at the Lord Howe Island conference suggest that it will, and should, not be so for long.

Seven Lessons

Recognize a Basic Dilemma

Certain general lessons about how new technologies may help us achieve a sustainable world stand out from the presentations at the Lord Howe Island conference. Some of them have already been touched on.

The first is that the regulation of technology faces a fundamental dilemma, hitherto uncommon in the law. This is that, of its character, technology is normally global. Law, on the other hand, comprising the enforceable, consistent, and predictable commands of an organized community, on traditional "positivist" interpretations has been tied to the governmental and judicial apparatus of a particular geographical jurisdiction. In recent years the need for extraterritorial operation of municipal law has been recognized, and upheld (Martinez, 2003, *Re Aird; Ex parte Alpert* [2004] 220 CLR 308 at 344-350 [114]-[133]; [2004] HCA 44, referring to the case of the *SS "Lotus"* [1927] Permanent Court of International Justice, Series A, No 10, Judgment No 9, pp. 18–19). Nevertheless, the fact remains that the focus of most national law remains upon acts that occur within the territory of that nation. By way of contrast, the focus of regulating technology must be the technology itself wherever it is produced or utilized across the globe (Dow Jones [2002] 210 CLR 575 at 615-619 [78]-[92]). Sometimes, that feature of the technology will make effective regulation by national law difficult, or even impossible.

It is into this context that direct enforcement of certain laws by "Code," written into software programs or otherwise imposed, adds a new dimension to global technology. The values and objectives of transnational corporations may be even more unresponsive to national regulation than the rules of the municipal legal system are. Moreover, "Code" of this kind may opt for caution and overinclusion so as to avoid dangers to markets in the least right-respecting countries.

The contractual arrangements entered by the government of the People's Republic of China and the corporations selling access to Yahoo! and Google in China illustrated the willingness of the latter to succumb to the demands of the former so as to avoid endangering a lucrative economic market for their products. In this way the provider and also the users are subjected to forms of censorship that might not be tolerated in other societies. A smaller country, with a smaller market, is unlikely to exert the same clout. Considerations of economics rather than of legal principle, ethical rules, or democratic values may come to predominate in such cases, and especially in respect of energy technology, with its distinctive transnational features and needs.

Recognize That Inaction Is a Decision

In the past, proponents of technological innovation have often favored containment of law and a generally "libertarian" approach to developments of technology. Yet most lawyers recognize that there are limits. Unless such limits are clearly expressed, and upheld in an effective way, the absence of regulation will mean, effectively, that the society in question has effectively made a decision to permit the technological advances to occur, without democratic and legal impediments.

Those who are hesitant about adopting any form of the precautionary principle may nonetheless recognize the need for some restraints. Thus, unlimited access to child pornography will probably promote crime and sustain the need for regulation of the Internet to prohibit, or restrict, access to such sites. However, that will still leave room for debate about the detailed content of the regulation: the age of the subjects depicted, any permissible (computer graphic rather than human) images, the means of enforcing the law, and the provision of effective sanctions (*Bounds v. The Queen*, 2006). Cases on these issues, and on any constitutional questions that they present, are now quite common (*The Queen v. Fellows and Arnold*, 1997; *The Queen v. Oliver*, 2003; *cf Lawrence v. Texas*, 2003).

Likewise biotechnology—views may differ over whether regulation is necessary, or even desirable, to prohibit therapeutic cloning, reproductive cloning, or the use of human embryonic stem cells. Yet nonbinding prohibitory resolutions and declarations have been adopted in the organs of the United Nations on this subject (Macintosh, 2005). Even those nations, like the United Kingdom, that have not favored prohibitions or moratoriums on experiments with human cloning for therapeutic purposes might well accept the need to prohibit, or restrict, some biotechnological experiments. Hybridization and xenotransplantation of tissue across species clearly require, at the very least, restrictions and safeguards so as to prevent cross-species transmission of endogenous viruses. To do nothing is therefore effectively to decide that nothing should be done (Robertson, 2001). It does not necessarily amount to a decision to "wait and see."

Nuclear energy obviously requires strict regulation because of the risks inherent in the technology itself and its vulnerability to nature and man-made disasters, detrimental to human life and well-being. A hydroelectrical development may have significant associated environmental and cultural hazards. A huge oil spill, occasioned by the search for even more remote sources of fossil fuels, will necessitate national and international regulation.

This is why the regulation of a new globally distributed technology, such as artificial photosynthesis, is such an important topic in the context of creating a sustainable world. It is not a subject that can be ignored, simply because the subject matter, and the available regulatory techniques, is difficult, uncertain, and controversial.

Recognize the Limited Power to Regulate

A third lesson, derived from the first two, is that the normal organs of legal regulation often appear powerless in the face of the pace, complex content and significant social and environmental impacts of a new technology. This is clear in the case of attempts to regulate new information technology. So far as the Internet is concerned, the regulatory values of the United States inevitably exert the greatest influence on the way the Internet operates and what it may include. This means that both the First Amendment and copyright protection values, established by the law of the United States, profoundly influence the Internet's present design and operation. An attempt by another nation's laws (such as those of France) to prohibit transnational publication offensive to that country's values (such as advertising Nazi memorabilia) may face difficulties of acceptance and enforcement in the Internet (*League against Racism and Anti-Semitism [LICRA], French Union of Jewish Students, v. Yahoo! Inc. [USA], Yahoo France* [2–1], 2000).

The same is true of biotechnology. The Australian Parliament initially enacted the Prohibition of Human Cloning Act 2002 [Cth] and the Research Involving Human Embryos Act 2002 (Cth). These were part of a package of laws aimed at the consistent prohibition in Australia of human cloning and other practices deemed unacceptable at the time. Both acts were adopted on the basis of the promise of an independent review two years after the enactment. Such a review was duly established. It was chaired by a retired federal judge, the Honorable John Lockhart. The review presented its report on December 2005. It recommended an end to the strict prohibitions of the 2002 legislation, the redefinition for legal purposes of the "human embryo," and the introduction of a system of licensing for the creation of embryos for use for therapeutic purposes (Australian Government Legislation Review: the Prohibition of Human Cloning Act 2002 and the Research Involving Human Embryos Act 2002, Report, Canberra, December 2005).

Initially, the Australian government rejected the recommendations of the Lockhart review. However, following political, scientific, and media reaction, a conscience vote on an amending act, introduced by a previous health minister, was allowed. In the outcome, the amendments were enacted. They passed the senate with only a tiny majority (Commonwealth Parliamentary Debates [House of Representatives], 6 December 2006, 127. Commonwealth Parliamentary Debates [Senate], 7 November 2006, 48).

The main arguments that promoted this regulatory response to a new technology were the recognition of the pluralistic nature of the society, the availability of widespread reports on the potential utility of the research and experimentation, and the expressed conviction that experimentation would proceed in overseas countries with results that, if they proved successful, would necessarily be adopted and utilized (Finkel and Cannold, 2006; Skene, 2006; Carr, 2006). Interestingly, both the then prime minister and the then leader of the federal opposition voted against the amending act (Commonwealth Parliamentary Debates [House of Representatives], 6 December 2006, 117 119).

The global debates on the regulation of experiments using embryonic stem cells have often been driven by countries that, to put it politely, are not at the cutting edge of the applicable technology (Macintosh, 2005). On the other hand, the United States has also often adopted a conservative position on these topics in UN forums. As happened in Australia, this may change in time as the potential benefits of the new technology become more widely understood.

Recognize Differentiating Technologies

So far as regulation of technologies is concerned, the 2011 Lord Howe Island conference established the need to differentiate emerging sustainability-focused technologies for the purpose of regulation. The consensus was that it could not be a case of one response fits all. Self-evidently, some forms of technology are highly risk sensitive and urgently in need of regulation. Thus, unless the proliferation of nuclear weapons is effectively regulated, the massive destructive power that they present has the potential to render all other topics, including of regulation, hypothetical. Similarly, some aspects of the regulation of biotechnology are highly risk sensitive, including the use of embryonic stem cells and germ line modification. For some, the risk sensitivity derives from the technology's perceived threat to deep religious or other beliefs concerning the starting point of human existence. For others, it arises out of fears for humanity's health or moral worth if it sanctions irreversible experiments that go wrong.

Somewhat less risk sensitive is the regulation of information technology. Yet, this technology too presents questions about values concerning which people may have strong differences of opinion. To outsiders, Americans seem to imbibe First Amendment values with their mother's milk. US lawyers sometimes have to be reminded that their balance between free speech and other human rights is viewed in most of the world as extreme and disproportionate.

The regulation of energy technology at first sight may be less controversial in terms of such moral conflicts. However, as regulations are adopted, they will exert influence on national and regional tendencies to shift from one energy source to another. In such shifts, it cannot be expected that corporations selling current forms of energy will be altruistic over decisions that may profoundly affect national, corporate, local, and individual interests. Such debates are the stuff of politics and cannot be expected to follow a course of plain sailing.

Recognize Different Cultures

Most of the participants in the Lord Howe Island conference came from the developed world. They therefore reflected general attitudes of optimism and confidence about the outcome of rational dialogue, scientific research, technological inventiveness, and the capacity of human beings ultimately to arrive at reasonable responses to regulating technologies on the basis of calm debate. However, as present energy sources are depleted and change, it cannot be expected that experts, investors, and other citizens from developing countries will be quite so dispassionate.

Basing Regulation on Good Science

In the early days of the HIV pandemic, I served on the Global Commission on AIDS of the World Health Organization. One of the members, June Osborn, then a professor of public health at the University of Michigan, taught us all the importance of basing all regulatory responses to the epidemic upon good science. The danger of responses based on assumptions, religious dogmas, intuitive beliefs, perceived national self-interest, or popular opinion was such that they would not address the target of regulation effectively.

The intervening decades have suggested that the countries that have been most successful in responding to HIV/AIDS have been those that have observed June Osborn's dictum (Plummer and Irwin, 2006). The same is true of the subjects of biotechnology, information technology, and energy technology examined at the Lord Howe Island conference. All too often, science and technology shatter earlier assumptions and intuitions.

For example, the long-held judicial assumption that jurors, and judges themselves, may safely rest conclusions concerning the truth of witness testimony on the basis of the appearance of witnesses and courtroom demeanor has gradually evaporated because scientific experiments shatter this illusion (*Fox v. Percy*, 2003). One day, by subjecting witnesses to brain scans, it may be possible to demonstrate objectively the truthfulness or falsity of their evidence. If, and when, the day arrives, other issues will doubtless be presented for regulators. We are not there yet. But any regulation must recognize the need to remain constantly abreast of scientific knowledge and technological advances.

Addressing the Democratic Deficit

This brings me to the last, and most pervasive, of the lessons of the Lord Howe Island conference on governance issues confronting the globalisation of artificial photosynthesis. Technology races ahead of our predilections and predictions. Its innovations quickly become out of date, both in terms of social usefulness and safety. Laws addressed to a particular technology are rapidly overtaken and rendered irrelevant or even obstructive. Nowadays scientific knowledge, technological inventions, and how they are perceived to mesh with community values change radically in a very short space of time.

In such an environment, there is an obvious danger for the rule of law. It is impossible to expect of legislatures, with their many responsibilities, that they will address all of the technological developments for regulatory purposes. The average legislator finds such issues complex and impenetrable. Such issues are rarely political vote winners. They struggle to find a place in the entertainment and personality politics of the present age as well as with the many other competing questions awaiting political decision making. This leaves a gap in democratic involvement in this sphere of regulation. It is a gap that is being filled, in part, by "Code" that incorporates regulations designed by inventors of information systems themselves in the structure of such systems but without a democratic input or the necessity of individual moral judgment.

The democratic deficit presented by contemporary technology is thus the largest potential lesson for the global deployment of artificial photosynthesis from the 2011 Lord Howe Island conference. In an age when technology is so important to society, yet so complex and fast moving that it often defies lay understanding, how can we adapt our accountable lawmaking institutions to keep pace with such changes? One means is by the use of consultative mechanisms such as law reform commissions (Chalmers, 2005) or independent inquiries (Cooper, 2006; Stobbs, 2006; Karpin, 2006). In such cases, the very process of consultation and public debate promotes a broad community understanding of the issues, an appreciation of the competing viewpoints, and an acceptance of any regulations adopted, even when they may give effect to conclusions different from one's own.

Conclusion

Adapting the legislative timetable and machinery of regulating energy sources to the challenges of modern governance is a subject that has engaged law reform bodies and executive government for decades. Often they lie unattended for years, or indefinitely, not because of any real objections to the proposals but because of the legislative logjam (Mason, 1971; Kirby, 2006).

In the face of radically changing sustainability-focused technologies such as globalised artificial photosynthesis and the danger of a growing democratic deficit if their roll out is impeded by corporate interests in say the fossil fuel or global agribusiness industries, it will obviously be necessary to adapt and supplement the lawmaking processes we have hitherto followed in most countries. Different types of delegated legislation may need to be considered. So may the enactment of overarching laws, expressed in general terms, which will not be quickly reduced to irrelevancy by further technological change (*R v. Quintaralle [on behalf of Reproductive Ethics] v. Human Fertilisation and Embryology Authority*, 2005). So adopting international principles (such as the OECD Guidelines on Privacy) or preventing corporate fraud, evading tax or unduly influencing governments may prove so influential in helping individual countries to act on the needs for regulation. And to do so in a way that is broadly consistent with a common approach chosen internationally.

Addressing the weaknesses in the democratic accountability of the large and complex modern government is an important challenge to legal and political theory. The 2011 Lord Howe Island conference demonstrated once again the particular ingredients and substantial urgency of the problem. It will take more conferences to provide the solutions appropriate to the differing systems of government operating in different countries. But the urgent need is for leadership through bodies such as the United Nations to bring together all those working in the complementary and competing sources of future energy so that a global approach to the challenges may be adopted for the whole world. That approach should seek to minimize the risks to the biosphere and to the human species. It should seek to harmonize the national and transnational economic interests at stake. And it should help maximize the contribution that future energy sources may make toward reducing the toll of poverty on humanity, increasing the sustainable pace of economic and social development, and protecting the universal human rights of all human beings for all time into the future.

References

Andorno, R. (2004). The precautionary principle: a new legal standard for a technological age. *J. Int. Biotechnol. Law*, **1**, 11–19.

Bounds v. The Queen, 228 ALR 190 at 197 [26], 211 [94]; (2006) HCA 39.

Brennan, D. J. (2006). What can it mean 'to prevent or inhibit the infringement of copyright'? A critique of Stevens v Sony. *Aust. Intellectual Property J.*, **17**, 186.

Chalmers, D. (2005). Science, medicine and health in the work of the Australian law reform commission. In *The Promise of Law Reform* (Weisbrot, D., and Opeskin, B., ed.). Sidney, Australia: The Federation Press, 374–387.

Cooper, D. (2006). The lockhart review: where now for Australia? *J. Law Med.*, **14**, 27.

deZwart, M. (2007). Technological enclosure of copyright: the end of fair dealing. *Aust. Intellectual Property J.*, **18**(1), 7–38.

Faunce, T. A. (2012). Towards global artificial photosynthesis (global solar fuels): energy, nanochemistry and governance, *Aust. J. Chem.*, 65(6), 557–563.

Faunce, T. A., Styring, S., Wasielewski, M. R., Brudvig, G. W., Rutherford, A. W., Messinger, J., Lee, A. F., Hill, C. L., deGroot, H., Fontecave, M., MacFarlane, D. R., Hankamer, B., Nocera, D. G., Tiede, D. M., Dau, H., Hillier, W., Wang, A. R. (2013). Artificial photosynthesis as a frontier technology for energy sustainability. *Energy Environ. Sci.*, **6**, 1074–1076.

Finkel, B., and Cannold, L. (2006, August 7). Day for stem cells and the hope of finding cures. *Sydney Morning Herald*, 9.

Fox v. Percy, 214 CLR 118 at 129 [31]; (2003) HCA 22.

Karpin, I. (2006). The uncanny embryos: legal limits to human reproduction without women. *Sydney Law Rev.*, **28**, 599.

Kirby, M. (2006). Law reform and human rights: Scarman's great legacy. *Legal Stud.*, **26**(4), 449–474.

Lawrence v. Texas, 539 US 558 at 590 (2003).

League against Racism and Anti-Semitism [LICRA], French Union of Jewish Students, v. Yahoo! Inc. [USA], Yahoo France [2–1] (2000). *Electronic Business Law Reports*, **1**(3) 110–120 (The County Court of Paris).

Macintosh, K. L. (2005). Human clones and international human rights university of technology. *Sydney Rev.*, **7**, 135–136.

Martinez, J. (2003). Towards an international judicial system. *Stanford Law Rev.*, **56**, 429.

Mason, A. F. (1971). Law reform in Australia. *Fed. Law Rev.*, **4**, 197.

Plummer, D., and Irwin, L. (2006). Grassroots activities, national initiatives and HIV prevention: clues to explain Australia's dramatic early success in controlling the HIV epidemic. *Int. J. STD AIDS*, **17**, 1.

R v. Quintaralle [on behalf of Reproductive Ethics] v. Human Fertilisation and Embryology Authority, (2005) UKHL 28 at [25].

Robertson, J. A. (2001). Why human reproductive cloning should not in all cases be prohibited. *Legis. Public Policy*, **4**, 35.

Stevens v. Sony Computer Entertainment, 224 CLR 193 (2005).

Stobbs, N. (2006). Lockhart review into human cloning and research involving human embryo: closing the gap. *Queensland Lawyer*, **26**, 247–251.

The Queen v. Fellows and Arnold, 2 All ER 548 (1997).

The Queen v. Oliver, 1 Cr App R 28 at 466-467 (10) (2003).

Preface

Our world is now in a perilous situation; its sustainability is being endangered or destroyed by the misguided activities of artificial human entities (corporations) we've allowed to dominate the earth's governance. This is occurring just as real people begin to expand their moral sympathies sufficiently to prioritize protecting our world's interests. The immediate threats to sustainability of humanity and its surrounding biosphere include not only extreme weather events associated with climate change but also the extent and toxicity of pollution, loss of biodiversity, and the disrupted integrity of land use, oceans, and atmosphere.

At this crucial moment, our species has developed a new technology—nanotechnology. Yet, governance of nanotechnology is currently polarized as either a threat to human safety or a source of great profit in new industries. A theme linking the chapters in this book is that they envision nanotechnology as fulfilling a destiny to provide the solution to that most intriguing and important of riddles—how this generation can shape the conditions for our race's secure and sustainable future on the earth over a span of millions of years.

The contributions in this volume are drawn from researchers at the Australian National University in a variety of fields related to nanotechnology and to its implications for human security and environmental sustainability. One unifying theme of these papers is that nanotechnology (if properly used) has the capacity to usher in a many-million-year period of human flourishing and stewardship within the earth's ecosystems, a period that for public policy purposes has been termed the Sustainocene. The concept of the Sustainocene was developed by Canberra ecophysician Dr. Bryan Furnass.

The Sustainocene is a concept drawing directly upon utopian literature but also coherent with modern neuropsychology and brain plasticity studies. The latter emphasize that to enhance human performance, visualization of ideal performance can be critical. Such capacity seems, however, to be denied to the citizens of modern human civilization. Part of the reason is that they can be corralled into shopping malls and housing estates; mollified with mass media television, videos, and videogames; and subjected to increasingly sophisticated surveillance with their freedom of choice eroded down to purchasing decisions. The mass centralized production and global distribution of energy and food facilitates such corporate control. A new decentralized energy and food technology such as artificial photosynthesis (capable of being employed in every building) cuts across this. It promotes a world where individual families and small communities address their basic energy and food needs locally with attendant local responsibility environmentally and socially. Thus are established the basic preconditions for that ecocentric transformation of human consciousness sustained by contemplative traditions as our collective destiny. Such is the vision at the core of the Sustainocene. This book explores some ways in which various applications of nanotechnology may contribute to this transition.

The foreword was written by the Honorable Michael Kirby, AC CMG, in the context of his attending the first international conference dedicated to the creation of a global artificial photosynthesis project at Lord Howe Island in August 2011 (http://150.203.86.5/coast/tgap/conf.htm). The conference was coordinated by the author under the auspices of the UNESCO Natural Sciences sector and was an official event of the UNESCO 2011 International Year of Chemistry. Many of the eminent nanotechnology researchers in this volume attended that meeting and/or have had their research focus altered by it. The papers were edited by the author in a special open source edition of the *Australian Journal of Chemistry* published in 2012, and he also coordinated a subsequent conference at Chicheley Hall, in the United Kingdom, focused on developing those initial ideas into a practical framework to establish a global artificial photosynthesis project with the assistance of the Royal Society in 2014.

The author has elsewhere promoted the idea that global artificial photosynthesis is nanotechnology's moral culmination (Faunce, Wilton Park, 2012; Faunce, BBC, 2012; Faunce, Energy Futures Lab, 2012; Faunce, ANU Public Lecture, 2012; Faunce, *Nanotechnology for a Sustainable World*, 2012). In many ways, the chapters in this book, by a range of nanotechnology experts at the Australian National University, are arranged and conceived as case studies or thought experiments to test that hypothesis.

Thomas A. Faunce
Canberra, Australia
Autumn 2014

Chapter 1

Nanotechnology Toward the Sustainocene

Thomas A. Faunce
ANU College of Law, Building 6, North Wing, Australian National University, Canberra, ACT 0200, Australia
ANU College of Medicine, Biology and the Environment, Building 131, John Curtin School of Medical Research, Australian National University, Canberra, ACT 0200, Australia
thomas.faunce@anu.edu.au

The hypothesis explored in this book is that the development of a Sustainocene, of a human civilization capable of sustaining over millions of years the flourishing of both its people and the ecosystems of which it is steward (Furnass, 2012) and companion, is a worthwhile policy goal that requires the utilization of nanotechnology, particularly for energy security and climate change mitigation. One particularly important (and perhaps unifying) aspect of this investigation involves nanotechnology's role in engineering an improved, nonbiological ("artificial"), photosynthetic component into every road, building, and vehicle we build and utilize on the surface of the earth (Faunce, 2013a, 2013b).

Yet, as will be shown, there are many other ways by which nanotechnology may help facilitate such a Sustainocene period. This

Nanotechnology Toward the Sustainocene
Edited by Thomas A. Faunce

ISBN 978-981-4463-76-8 (Hardcover), 978-981-4463-77-5 (eBook)
www.panstanford.com

volume showcases some of the excellent research being undertaken at the Australian National University in which nanotechnology is being utilized to not merely address the great public health and environmental challenges of our times but to pave humanity's way toward the Sustainocene. This introductory chapter aims to set that research in a broader policy and governance context.

1.1 Governance Recognition of Our Need for New Energy and Environmental Technologies

Humanity is in desperate need of a technology capable of effectively satisfying its increasing demand for energy, whilst protecting (and improving or rectifying) the integrity and resilience of the natural environment in which it dwells. What the world needs, in other words, is for humanity to create a planetary therapeutic. One hypothesis, explored here, is that nanotechnology is destined to play an important role in such development. The development of human civilization in its current form relied heavily on technologies utilizing archived photosynthetic energy in the form of what are euphemistically (and with considerable geopolitical intent) called "fossil" fuels. Coal, for example, powered the industrial revolution of the 19th century, while oil was a fuel central to the system and policies of corporate globalization that dominated throughout the 20th century and continue to shape our social fabric in the 21st.

The big profits of corporate multinationals and laws requiring them to proritise shareholder profit create powerful vested interests that aim to ensure the continuation of human economic reliance on environmentally damaging, carbon-intensive fuels despite the adverse consequences in terms of environmental pollution and accelerated anthropogenic (or perhaps more accurately "corporatogenic") climate change. Around two-thirds of the world's electricity comes from coal, oil, and natural gas, as well as the overwhelming proportion of its transportation fuels (International Energy Agency [IEA], 2013). With increasing populations and the enhanced standards of living in many developing nations, the global demand for energy has doubled since 1980, sharply increasing not only greenhouse gas emissions and associated severe weather events but also the political and economic power of the corporations controlling the global extraction, refining, and distribution of carbon-intensive fuels (IEA, 2013). In specific terms, the projected increase of the global human population is to around 10 billion by 2050, with associated energy consumption rising from ~400 EJ/yr to more than 500 EJ/yr (IEA, 2013).

The 2009 *Copenhagen Accord* is characteristic of the effectively impotent early 21st-century international agreements that tried to reassure humanity that governments of nation states were capable of responding to scientific evidence where that evidence predicted potentially catastrophic social and environmental consequences stemming from a multinational corporate-influenced failure to deal with the impacts of population growth and from fossil fuel–driven climate change (United Nations Framework Convention on Climate Change [UNFCCC], 2009). For the first time, all major CO_2-emitting countries agreed to a target of keeping global warming to less than 2°C above preindustrial levels, made hopeful-sounding undertakings concerning mitigation (including the Copenhagen Green Climate Fund), and established mechanisms to accelerate the development and deployment (including transfer to developing nations) of new renewable energy technologies many of which are critically based on nanotechnology. None of these commitments involved obligations linked to viable enforcement mechanisms (such as trade sanctions or loss of corporate privileges).

Idealistic aspirations to redirect new technology (such as nanotechnology) toward critical global challenges, including reducing poverty and addressing the lack of necessary fuel, food, water, housing, and access to health services, while encouraging environmental sustainability, were also expressed in the United Nations Millennium Development Goals (UNMDGs) (United Nations Industrial Development Organization [UNIDO], 2009). Africa's energy consumption per capita is low in global terms (10 and 5 times less, respectively, than that of the United States or Europe [IEA, 2013]), and its primary energy supply is 5% of the global total—mostly from traditional biomass for cooking energy (IEA, 2013). Yet the UNMDGs and this disparity highlights the extent to which that continent lacks access to energy, particularly in the form of fuel. There is no clear and universally accepted definition of "energy access," but the term is often referred to in policy documents as the number of households supplied by an electricity grid system (IEA, 2013).

Africa, particularly in its rural areas, is grossly underserved by large-scale electricity grid infrastructure. Its installed generation capacity is 122.6 GW, or just 2.6% of the world total (IEA, 2013). In sub-Saharan Africa, only about 30% of the population has access to electricity (with 14% electrification in the rural areas) (IEA, 2013). Further, those connected to the grid experience highly unstable and unreliable electricity supplies and relatively high costs per kilowatt hour (IEA, 2013). Yet the global need for an energy transition from high carbon emission–intensive fuels may provide an ideal driver for a technological "bypass" of such fossil fuels in developing countries and regional areas. It has been claimed that the days of the power grid are numbered, even in developed nations, owing to the development of new forms of "home grown" green energy (Martin, Chediak, and Wells, 2013; MacKay, 2009).

In 2012, the United Nations (2012a) ran a program "Sustainable Energy for All," declaring that ". . . access to modern affordable energy services in developing countries is essential for the achievement of the internationally agreed development goals, including the MDGs, and sustainable development, which would help to reduce poverty and to improve the conditions and standard of living for the majority of the world's population." Charles Holliday, chairman of Bank of America, and Kandeh Yumkella, chair of UN-Energy and director general of UNIDO, co-chaired the secretary general's High-Level Group on the topic designed to "mobilize commitments from governments, the private sector, and civil society partners to take actions that will make sustainable energy a reality for all over the next two decades" (United Nations, 2012). Nevertheless, it is difficult to determine what practical and beneficial outcomes have arisen from that initiative. One suspects that the impressive range of corporate stakeholders advising the secretary general would have been disinclined to excessively promote technologies disruptive to the interests of their organizations, which have heavily invested in and are drawing substantial profits from carbon-intensive industries.

The vision behind this text on the other hand is of governance changes driving a global transformation in the use of nanotechnology to address the critical public health and environmental problems of our time. The corpus of international law as one example of a relevant global governance system, does contain incipient obligations supporting the development and deployment of new technologies that address critical contemporary issues, such as energy security and climate change, and their connection to the interests of the poor, future generations, and ecosystems. These arise, for example, under the international right to health (set out in Article 12 of the United Nations' *International Covenant on Civil and Political Rights* and elsewhere) and obligations to share the benefits of scientific progress and its applications (set out in Article 15 of the same convention). Further statements of the human right to enjoy the benefits of scientific progress and its applications (REBSPA) are found in Article 27 of the United Nations' *Universal Declaration of Human Rights.* Article 12(2) on freedom from hunger refers to making full use of scientific knowledge, and Articles 2(1) and 23 of the *International Covenant on Economic, Social and Cultural Rights* (*ICESCR*) refer to scientific assistance in achieving the specified rights. Principles supporting the development of new technologies for similar goals (and directed to individuals, communities, and private corporations, as well as states [Article 1]) appear in UNESCO's *Universal Declaration on Bioethics and Human Rights* (particularly the social responsibility principle in Article 14(b)—"access to adequate nutrition and water"; 14(c)—"improvement in living conditions and the environment"; and 14(e)—"reduction in poverty and illiteracy") (Faunce and Nasu, 2008).

UNESCO's *Declaration on the Responsibilities of the Present Generations towards Future Generations* issues, for example, a similar call to action for those engaged in areas such as nanotechnology research by declaring in Article 4:

> The present generations have the responsibility to bequeath to future generations an Earth which will not one day be irreversibly damaged by human activity. Each generation inheriting the Earth temporarily should take care to use natural resources reasonably and ensure that life is not prejudiced by harmful modifications of the ecosystems and that scientific and technological progress in all fields does not harm life on Earth (UNESCO, 1997).

Yet such altruistic "soft law" norms are in many cases directly undermined by obligations arising under trade and investment agreements that are protective of the profits of corporate multinationals whose business model is focused on centralised, carbon-intensive fuel and food production and backed strongly by the threat of trade sanctions or damages awards under arbitral proceedings. Part of the task of focusing the nanotechnology research endeavour (such as that exemplified in these chapters) on a vision of environmental sustainability, involves overcoming such corporate influence upon the democratic process by developing alternate jurisprudential approaches from enhanced understanding of how our conscience interacts with fundamental levels of reality. To this task we now turn.

1.2 Powering Toward the Sustainocene

1.2.1 Philosophic Foundations of the Sustainocene

In the 17th century, the philosopher Benedict de Spinoza wrote in his *Ethics* (Bk II, Prop. XLIV) that it is the nature of reason properly applied to perceive things truly, that is, as they are in themselves, not as contingently existing and revealed to us by variable sensory experience. This pronouncement and its implications have often been ignored or dismissed as a peculiar type of idealist rhetoric. Yet Spinoza's realization—that there could be true statements about reality that did not appear to correlate with common sense—paved the way for major breakthroughs in both scientific and jurisprudential thinking likely to underpin the use of nanotechnology to support a world dedicated to and exemplifying the virtue of environmental sustainability.

In the 18th century, for example, Immanuel Kant similarly and influentially contended that the capacity to form ethical concepts in the form of goals or end points for future actions based on universally applicable principles arises (like our understanding of time and space) a priori as one of the necessary preconditions for accurate sensory experience (rather than being determined by it). Kant made these claims respectively in his *Groundwork of the Metaphysics of Morals* and his *Critique of Pure Reason* (though philosophers tend not to grasp the linkage between his conceptual positions on ethics on the one hand and time and space on the other). The freedom of individuals to set ethical goals, Kant reasoned, presupposes a capacity to reject them, and laws (backed up by official enforcement) provide an external constraint upon persons whose selected ethical end points would otherwise unduly interfere with the capacity of other rational beings to choose their own goals. Yet the more, he held, people acted from a concept of duty (often against the opposition of their own sensual inclinations) to consistently apply such universally-applicable principles, the more humanity was morally flourishing toward a type of collective enlightenment. Kant summarized this by stating (in his *Introduction to the Doctrine of Virtue*) that virtue for individuals and societies arises from consistent voluntary decisions to act (despite internal or external obstacles) upon principles capable and worthy of application by all rational humans. Einstein read Kant in his youth and was certainly influenced by these ideas, which perhaps manifested in his acceptance of scientific anomalies that seemed contrary to common sense but which could be explained by reference to universal physical laws. Fitting quantum mechanics within such physical laws was a major difficulty for Einstein.

Well-educated people now readily accept that reality is made up of physical laws that underpin and explain the forces of nature, but are not readily discernable through sensory experience and reason based upon it. Such physical laws include quantum mechanics and electrodynamics, which describe electromagnetism and some nuclear interactions, or relativity, which describes time and gravity. There is even a beautiful and alluring mathematical elegance in ideas of supersymmetry, multiverses, and multiple additional dimensions associated with string theory. This approach to scientific discovery is explored seriously by scientists including those involved with nanotechnology irrespective of the fact that aspects of these laws do not conform to common sense. Similarly, well-educated people endorse the concept of universal human rights as an important true statement about our condition, regardless of whether they are expressed as such in a constitution, legislation, or any international convention to which their state is a party. Global public policy is now, it is posited here, at the point of acknowledging another such leap toward a reality that basic principles suggest is true despite apparent incoherence with our sense perceptions—our transition with the assistance of nanotechnology toward a multimillion year Sustainocene period of human stewardship over the earth.

The Holocene ("recent whole") period was the term given to the postglacial geological epoch by the International Geological Congress in Bologna in 1885. It began 10,000 years ago. From that time till about 1800 CE, humanity's activities were not enough to alter the basic components of the world's natural systems. Atmospheric carbon dioxide concentration, for example, did not rise during that period above natural variability. Since 1800, with the onset of the coal- and oil-driven Industrial Revolution; the development of the capacity to fix atmospheric nitrogen as a fertilizer (albeit with the use of carbon-intensive fuels); improved sanitation, healthcare, and transport; the dominance of governance systems by corporations; and the human population explosion, our impact on the earth's ecosystems has dramatically increased. Land ecosystems, for instance, were globally converted from "mostly wild" to "mostly anthropogenic" by the mid-20th century (Steffen, Persson, Deutsch et al., 2011).

Indeed, it has been argued that human activity has pushed this planet from the Holocene into what has been termed the Anthropocene period (Crutzen and Stoermer, 2000). The term "Anthropocene" was coined by Crutzen (2002). It refers to an epoch when human interference with earth systems (particularly in the form of influences on land use and land cover; coastal and maritime ecosystems; atmospheric composition; riverine flow; nitrogen, carbon, and phosphorus cycles; physical climate; food chains; biological diversity; and natural resources) has become so pervasive and profound that they are not only becoming the main drivers of natural processes on earth but also threatening their capacity to sustain life (Steffen, Crutzen, and McNeill, 2007). The author's contention is that the locus of power behind these detrimental decisions resides chiefly with mutinational corporations prioritising shareholder profit, to the extent that the epoch should more accurately be termed the "Corporatocene."

The term "Sustainocene" was coined by the Canberra-based Australian physician Bryan Furnass (2012). It refers to a period where governance structures and scientific endeavor coordinate to achieve the social virtues of ecological sustainability and environmental integrity as influentially propounded by eco-economists such as E. F. Schumacher (with his concept of "small (and local) is beautiful") and Kenneth Boulding (with his idea of "Spaceship Earth" as a closed economy requiring recycling of resources), as well as Herman Daly with his notion of "steady state" economies, drawing upon the laws of thermodynamics and the tendency of the universe to greater entropy (a system characterized by dispersal of energy) (Faunce, 2012b, 2012c, 2012d, 2013a). Five characteristic features of the Anthropocene epoch likely to be reduced in the Sustainocene are population, poverty, preparation for war, profits, and pollution (Furnass, 2012).

One area of academic research and policy development that fits well with "Sustainocene" thinking is that centered on the idea that this planet should be treated not just as a distinct living entity (James Lovelock's Gaia hypothesis [1991]) but also as a patient (Faunce, 2012a). "Planetary medicine," as this field has become known, has become a symbolic rubric, focusing public and governmental attention on the interaction between human health, technological development, and sustainability of the biosphere (McMichael, 2002). In this emerging discipline, characteristic features of the Corporatocene epoch, such as anthropogenic climate change and environmental degradation, as well as gross societal imbalances in poverty; the lack of necessary fuel, food, medicines, and security; and access to nature, are targeted as intrinsically global pathologies, the resolution of which requires concerted efforts to implement a wide range of not just renewable energy technologies (such as those using nanotechnology) but also bioethical principles, including those related to protecting the interests of future generations and preservation of biodiversity (Vines, Bruce, and Faunce, 2013; Faunce, 2012e; Faunce et al., 2013a, 2013b).

1.2.2 Environmental Sustainability

Environmental sustainability, whether conceived as a social virtue or an ethical principle, will probably be at the foundation of governance systems in the Sustainocene policy period. On one likely formulation it requires consideration of the greatest good for the greatest possible number of stakeholders (human and nonhuman) now and into the future, coherent with patterns of symmetry underlying nature and fostering continuing peaceful relationships between people.

Environmental sustainability as a primary social virtue (emerging alongside the more traditional justice, equity, and respect for human dignity) can be linked with the so-called ecocentric or biocentric ethics as expressed in Gaia theory, deep ecology, "The Earth Charter," or "The Earth Manifesto." It involves two key ethical principles. The first is that the flourishing and diversity of nonhuman life forms have intrinsic value, requiring their protection by laws and policies. The second holds that human flourishing itself requires a deepening respect for right relations with ecosystems that should be reflected in the choices our species makes about the use of new technologies such as nanotechnology.

It is difficult to discern environmental sustainability as a clearly defined ethical virtue or principle in contemporary media and policy debates between those with Christian, Islamic, Buddhist, or secular perspectives and those enmeshed in securing the embellishments of corporate financial power while influencing the global research agenda in fields such as nanotechnology.

Attempts have been made by some economists to frame the ethics of sustainability chiefly in terms of the fictional notion of perpetual growth in the gross domestic product (GDP). Such formulations often pay obeisance to the fictional power of deregulated markets and the "invisible hand" of entrepreneurial self-interest to ethically regulate demand upon the earth's resources. There appears to be a belief among such advocates (cultivated in many cases by corporate funding of their institutions) that the free market inevitably carries an implicit and sufficient ethics of competition. A logical application of this Corporatocene approach involved banks and other global financial institutions selling and reselling money and property in increasingly complex forms and ways (futures, hedge funds, bundled debt, etc.) till the bankers ended up knowingly and rapidly selling things that didn't really exist in a global game of "stop the music." Companies hid away their true debts, and the governments provided bailout loans to economies that had been almost bankrupted by this process, mostly to repay investors and ensure salaries and bonuses for those who officiated over the process. The "Occupy Wall Street" protestors, among other claims, are asserting that by facilitating this economic mess, the so-called free market has actually begun to undermine its claim to be ethical in any of its recognized conceptions. The hypothesis explored here is that such an understanding must spur the use of nanotechnology toward applications that undercut the corporatised centralisation of money, employment, fuel and food, to support more globally distributed, democratic and environmentally sustainable governance systems.

As mentioned, other economists, however, have striven to factor our moral responsibilities concerning the finite and fragile resources of the biosphere much more centrally into their economic calculations. The virtues of ecological sustainability and environmental integrity, for instance, were influentially propounded by eco-economists, such as Schumacher and Boulding. In doing this, the former drew upon Buddhist ethical principles and virtues, while the latter relied upon those resonating with the Quaker tradition.

Outcomes from related approaches include measures such as a global financial transactions tax (Buckley, 2012) or a global anti-false claims whistleblower financial incentive law, or corporate 'marriage' laws to reign in corporate governance toward a greater balance with public and environmental good.

1.2.3 Governance Transitions for Emerging Technologies and the Sustainocene

One of the main issues facing policymakers that seek to drive the shift from Corporatocene to Sustainocene will be determining the principles that should govern how emerging technologies (such as nanotechnology) assist in achieving the hydrogen economy. In the "hydrogen economy," hydrogen will be used ubiquitously as a carbon-neutral energy vector (e.g., as a source of electricity via fuel cells or as a fuel itself being stored safely, e.g., through complexing with metal hydrides in building materials or as ammonia fuel after complexing with absorbed atmospheric nitrogen) and a source of freshwater (when combusted). Major policy documents have outlined the case for such an economy (European Hydrogen and Fuel Cell [EHFC], n.d.; US Department of Energy [USDOE], n.d.; National Research Council and National Academy of Engineering [NRC and NAE], 2004; E4tech, 2004).

Some of the governance challenges associated with this transition include the need to remove taxpayer subsidies on fossil fuels so that the cost of hydrogen or ammonia fuel production becomes competitive with that of petrol. Another is the need to develop regulations and safety standards at national and international levels, as well as the need to develop stable incentive systems for large-scale investment that will not fluctuate with oil prices. One of the main problems at present with moving to a global hydrogen economy is the carbon-intensive energy required to produce hydrogen in large quantities by steam reformation of hydrocarbons, generally methane (Sartbaeva, Kuznetsov, Wells, and Edwards, 2008; Olah, Goeppert, and Surya Prakash, 2009). Indeed (as is set out in Chapter 13) one of the most exciting ways that nanotechnology can assist with the creation of a Sustainocene may be through mechanisms that allow hydrogen to be created by solar-driven water splitting.

Table 1.1 Comparison of Corporatocene and Sustainocene

	Corporatocene	Sustainocene
Powered by old photosynthesis fuels	Yes	No
Powered by new photosynthesis fuels	No	Yes
Corporate-led governance (through trade and investment agreements, mass media, indebtedness, lobbying, and infiltration of governments)	Yes	No
Community-led governance (through local control over power, food, and employment)	No	Yes
Governance protecting future generations	No	Yes
Governance giving rights to ecosystems	No	Yes
Governance giving rights to biodiversity	No	Yes

1.3 Nanotechnology for the Sustainocene

We now turn to how the chapters in this edited volume fit the theme of nanotechnology research oriented to assisting a secure and sustainable future; in more specific terms a Sustainocene epoch in which human interests are viewed in the context of the foundational virtue of environmental sustainability. They represent a cross section of the excellent research involving nanotechnology undertaken at the Australian National University (ANU). In this respect, the ANU can be considered representative of most of the great universities in the world, all of whom have considerable research programs involved in nanotechnology, though perhaps not in all cases with the same disciplinary breadth.

Nanotechnology is the science of making things from components that are not much bigger than a few atoms, less than 100 nm (a nanometer is a billionth of a meter). The chief policy interest with nanotechnology to date has been concerned with ensuring its safety (Faunce, 2008). Interest is growing, however, in linking fundamental nanotechnology research to foundational social virtues such as equity and environmental sustainability (Faunce, White, and Matthaei, 2008). Experts have encouraged nanotechnology researchers to systematically contribute to achievement of the UNMDGs, particularly energy storage, production, and conversion; agricultural productivity enhancement; water treatment; and remediation (Salamanca-Buentello, Persad, Court, Martin, Daar, and Singer, 2005). The idea that nanotechnology should be viewed as having its moral culmination in how it addresses the significant public health and environmental problems of this era (Faunce, 2012b) provides a major theme for this book.

1.4 Overview of Chapters

This chapter, Chapter 1, is designed to present an overview of how the chapters in this volume test the idea that nanotechnology has a critical role to play in resolving the most important public health and environmental challenges of our time. Nanotechnology experts from a variety of disciplines were encouraged to reflect upon how their work might systematically contribute to achievement of the UNMDGs and to the development of a secure and sustainable future for humanity and its ecosystems. They were encouraged, in other words, to describe how they viewed their nanotechnology research through the lens of conscience when applied to their social and environmental responsibilities.

In Chapter 2, "The Cosmic Context of the Millennium Development Goals: Maximum Entropy and Sustainability," Lineweaver and O'Brien set the contemporary challenges for nanotechnology in about as broad and profound a scientific and governance context as is possible. Their starting point for discussing nanotechnology and sustainability involves the first and second laws of thermodynamics: (1) energy is always conserved (so that "saving energy" fundamentally means merely to delay the dispersal of a specific kind of concentrated or low-entropy energy called "free energy") and (2) all systems (including our universe) run from order to disorder—from low entropy to high entropy—humans, other life forms like bacteria, galaxies, stars, convection cells, hurricanes, and bushfires being examples of "far from equilibrium dissipative systems" (Far FEDS) that convert low-entropy energy to high-entropy energy. They examine an extension of the second such law, the maximum entropy production principle (MEPP), by which life evolved on the terrestrial biosphere to maximize the long-term production of entropy.

Lineweaver and O'Brien boldly take sustainability back to quantum fluctuations at the Planck scale (10^{-35} m) and baryon nonconservation that occurred within the first nanosecond after the universe's postulated origin in what scientists refer to as the "big bang." Life evolved as constructs of nanometer-sized biomolecular machines continuously manipulating atoms and molecules to metabolize free energy (provided by cyanobacteria and plants converting photons into higher-energy states by dissociating water and carbon dioxide to produce carbohydrates and free oxygen), maximizing entropy through dispersible waste heat.

Lineweaver and O'Brien point out that given the above context, the achievement of MDGs 1–6 (eradicate extreme poverty and hunger; achieve universal primary education; promote greater gender equality and empower women; reduce child mortality; improve maternal health; and combat HIV/AIDS, malaria, and other diseases) can be facilitated by economic growth but not goal 7 (environmental sustainability). This is chiefly, they argue, because of the attendant pollution, including climate-altering impacts of atmospheric carbon dioxide emissions in a high-population and carbon-based energy world that exceeds the earth's capacity to renew, replenish, and restore. They argue for the development of a reliable "fuel gauge" for the planet using the laws of thermodynamics and the concept of entropy. They conclude that with proper management, hydrogen fusion in the sun and the temperature gradient between the hot interior and the relatively cold surface of the earth could be harnessed to sustain our energy needs for about a billion years (the mooted length of the Sustainocene).

In Chapter 3, "Nanophotonics for Light Trapping," Mokkapati, Beck, Wilson, Wang, and Catchpole detail how their research assists nanophotonics (particularly with plasmonics and grating structures) to significantly increase the photon absorption in thin-film solar cells. They point out that nanophotonics in light trapping could have a significant positive impact on sustaining world energy needs—given that the earth receives more solar energy in one day than its entire human population requires in one year—but that such an approach can be used to produce electricity or hydrogen (e.g., by splitting water in artificial photosynthetic systems) in a power-conserving decentralized manner.

In Chapter 4, "Growth and Characterization of GaAs Nanowires," Gao, Joyce, and Jagadish review the fundamental growth behavior of GaAs nanowires by metal–organic chemical vapor deposition via a vapor-liquid-solid (VLS) mechanism. The authors make the case that such nanotechnology enhances the capacity to deliver precise control of nanowire dimensions, orientation, structure, phase purity, and chemical composition. Applications relevant to sustainability include improved quantum computing and solar energy–harvesting devices.

In Chapter 5, "The Synthesis, Structure, and Properties of Titania-Coated Silica Nanowires," Shalav and Elliman discuss how nanostructured titanium dioxide (TiO_2) has many valuable applications likely to be of importance in achieving various UNMDGs, including environmental sustainability. Such nanowires also have applications for various aspects of artificial photosynthesis, including electrochemical dye-sensitized solar cells, conversion of CO_2 to gaseous hydrocarbons, and H_2 production via water splitting. Additional benefits include self-cleaning surfaces, disinfection, the degradation and elimination of offensive odors, and the oxidation of contaminants. The authors show how the achievement of these applications can be facilitated by stoichiometric silica glass (SiO_x) nanowire layers, grown via the oxidation of a Si substrate that can be conformally coated with TiO_2 thin films via different techniques to produce large-surface-area nanostructures.

In Chapter 6, "Global Health and Environmental Implications of Mimicking Biological Ion Channels Using Nanotubes," Hilder presents research on how synthetic ion-selective nanotubes can be used to perform functions of biological ion channels. This could result in a range of products with particular beneficial applications to developing a sustainable global civilization, particularly with respect to desalination, demineralization, decontamination, and purification of water and the detection of explosives, chemicals, and toxins at low concentrations, as well as the potential for delivering more effective antibiotics and anticancer agents. Hilder notes that significant advances need to occur in nanofabrication technology to reduce costs of nanotube-based devices so that they can be more readily accessible to local communities, particularly in what is now termed the "developing" world.

In Chapter 7, "Nanostructured Materials: Implications for Information Technology," Neshev discusses how nanotechnology-based fishnet metamaterials infiltrated with liquid crystal can enhance the effectiveness of optical information transfer systems. Potential beneficial applications relevant to sustainability include improving light capture for energy production, electrical routers on the Internet, and domestic security. Such nanotechnology-based metamaterials could also play an important role in "smart grid" systems facilitating distributed (household and community-sourced), zero-carbon renewable energy.

In Chapter 8, "Laser Trapping of Nanoparticle Agglomerates in Air," Rode, Shvedov, Hnatovsky, and Krolikowski present a new technique of trapping and moving nanoparticles in the air using pressure exerted by electromagnetic radiation ("optical tweezers"). They point out that this effect of light waves was deduced by Johannes Kepler, who observed that the tail of a comet always points away from the sun, was predicted mathematically by Maxwell, and was proven experimentally by Lebedev, Nickols, and Hull in 1901. An optical trap that they developed for airborne nanoparticles using a counterpropagating vortex beam is said to have applications not only in pollution control and workplace safety but also in touch-free transport of biological samples, interstellar dusty plasmas, and atmospheric physics. Such advances are likely to become critical to the safe global use of nanotechnology in the Sustainocene period.

In Chapter 9, "The Bhopal Disaster and Peroxide Bombs: Nanoscale Aspects of Oscillatory Thermal Instability," Ball studies nanoscale aspects of thermal oscillatory runaway, a chemical process here exemplified in a case study involving the venting of methyl isocyanate (MIC) by a Union Carbide plant at Bhopal, India, on December 2–3, 1984. Ball points out that the sustainability implications of such research (and its capacity to facilitate achievement of the UNMDGs) include better detection of terrorists through nanocalorimetry exothermic behaviors in trace quantities of liquid peroxide explosives.

In Chapter 10, "Fusion Power and Nanoscience Challenges for Extreme Materials," Hole and Cormac make the case for human energy requirements to be sustainably met by one of the main challengers to global artificial photosynthesis—the International Thermonuclear Experimental Reactor (ITER) fusion project aimed at harnessing the fusion energy that powers the sun. Unlike global artificial photosynthesis, this alternative for sustainable energy emphasizes centralized power generation. The authors claim that related issues surrounding radioactive neutron waste, health, safety, and weapons proliferation can be satisfactorily resolved. They discuss how nanotechnology is very useful in this context through the creation of materials capable of functioning properly in the extreme conditions involved in fusion reactors.

In Chapter 11, "Nanotechnology, Plasma, Hydrogen from Artificial Photosynthesis, and Fuel Cells: Powering the Developing World to the Sustainocene," Faunce and Charles argue such fuel calls offer attractive mass market applications to assist in the global deployment of artificial photosynthetic technology. A main focus initially may be on personal transport aimed at reducing urban pollution. As mentioned, such fuel cell efficiency strongly depends on a membrane–electrode assembly, and one of the challenges in its commercialization is the high cost of the noble metals used as catalysts, for example, platinum in the electrodes. Such efficiency will also require funding from governments and business that is congruent with achieving objectives such as the MDGs.

In Chapter 12, "Nanotechnology-Based Artificial Photosynthesis: Promoting Animal and Ecosystem Rights in the Sustainocene," Bruce and Faunce make the case that competition and citizen-consumer law could form a critical part of the global governance mechanisms that facilitate rapid deployment of artificial photosynthetic technologies. They explore some of the governance implications of widespread use of such technologies for other species, especially those currently used for human food production.

In Chapter 13, "Toward the Sustainocene with Global Artificial Photosynthesis," Faunce, Bruce, and Donohoo consider the proposition that a world powered by global artificial photosynthesis will be better able to implement rights of ecosystems and will be likely to support more democratic models of governance.

1.5 Global Artificial Photosynthesis as Nanotechnology's Moral Culmination

One important governance issue for nanotechnology is whether the field is converging on a single, elegant approach to resolving the critical energy, public health, and environmental problems of our times. Should we risk testing in our daily lives the hypothesis in other words that there is a moral destiny for nanotechnology in our contemporary world, or a purpose it is meant to be used for, if the flourishing of all life is to be maximized.

Is it not an intriguing hypothesis that if the destiny of the world is toward greater coherence and harmony at all levels, then the traditional foundational social principles of justice, equity, and respect for human dignity are soon to be supplemented as a basis for global governance arrangements by the less anthropocentric "respect for environmental sustainability"? The author's hypothesis is that this transition is being driven by our collective conscience—that part of our being that tunes in to the greater forces of harmony, symmetry, and mathematical elegance in the world; that part of our consciousness that (to make a physics analogy) identifies with an altruistic wave function rather than an egocentric particle orientation of interests. (Faunce, 2012b).

One vision underpinning this book (one that not all the contributors would necessarily agree with) is that nanotechnology's moral culmination involves artificial photosynthesis rolled out globally as a product driven by consumer and competition law to produce "off-grid" local and domestic energy, fertilizer, and food products. Nanotechnology-based global artificial photosynthesis could replace policy models of corporate globalization and ever-increasing economic growth predicated on preparation for war and use of nonrenewable and polluting energy sources (Faunce, 2011; 2012c).

Researchers (many using nanotechnological approaches) now are striving to improve photosynthesis to achieve low-cost conversion of sunlight, water, and carbon dioxide into fuel for heating and cooking (Hurst, 2010). Some nanotechnological innovations for what is generally known as artificial photosynthesis focus on improved light capture. Such systems use mesoporous thin-film dye-sensitive solar cells of semiconductor nanoparticles (Kalyanasundaram and Graëtzel, 2010) and carbon nanotubes harvesting and conducting the resultant electricity (Sgobba and Guldi, 2009). Nanomaterials and hybrid organic-inorganic nanostructures are improving the solar energy conversion efficiency of existing photovoltaic units that could be used in artificial photosynthetic light capture (Guter, Chone, Philipps et al., 2009; Konstantatos and Sargent, 2010).

Photosystem II (PS II) is central to the great invention whereby plants discovered how to use photons to extract electrons from water molecules; it has a core reaction centre where oxygen atoms cluster round manganese in a distorted Van Gogh chair shape [the Mn_4CaO_5 cluster now resolved at a resolution of 1.9 Å (Umena, Kawakami, Shen and Kamiya, 2011)]. Researchers are working on making a nanotechnological mimic of this protein (maquette) that is simpler (Koder, Anderson, Solomon, Reddy, Moser, and Dutton, 2009) and that incorporates designer molecules that prolong charge separation (Carmieli, Mi, Ricks, Giacobbe, Mickley, and Wasielewski, 2009). Nanotechnology is facilitating the construction of artificial photosynthetic electron pathways to the core water-splitting reaction center, which perform a single quantum computation, sensing many states simultaneously and thus enhancing the efficiency of the energy capture and transfer at physiological temperatures (Lee, Cheng, and Fleming, 2007; Engel, Calhoun, Read et al., 2007; Ball, 2010).

The most globally widespread water catalytic system will probably involve inexpensive and self-repairing components that operate at neutral pH with nonpure (salty or bacterially and chemically contaminated) water (Kanan and Nocera, 2008) and are stable to a variety of exposure conditions in air, water, and heat. Major scientific challenges for developing the water-splitting component of an efficient artificial leaf include designing a stable oxygen-evolving photoanode less expensive and more efficient than the geo-politically problematic iridium (Rao, Cai, Liu et al., 2014; Hu, Shaner, Beardslee et al., 2014) potentially involving cobalt oxide nanocrystal (Ullman and Nocera, 2013) and a photocathode for hydrogen evolution (Messinger, Lubitz, and Shen, 2014) within a system that is cost effective with good global life cycle analysis (Thapper, Styring, Saracco, Rutherford, Robert et al., 2013). Multiwalled carbon nanotubes and single-walled carbon nanotubes may produce the critical breakthrough here, but life cycle analysis will be critical.

New catalysts for H_2 production and methods for efficient H_2 usage (in a fuel cell to make electricity) or storage (as a fuel after cooling and concentrating) will need to be built (Magnuson, Anderlund, Johansson et al., 2009). It is unlikely methanol will turn out to be, at least in the short term, the most viable fuel produced from this side of the artificial photosynthesis process (Olah, Goeppert, and Surya Prakash, 2009). It is more probable that atmospheric carbon dioxide and nitrogen absorption will be used to make basic starches and fertilisers for food and materials production. In the artificial photosynthesis version of the "dark reaction," adenosine triphosphate (ATP) and reduced nicotinamide adenine dinucleotide phosphate (NADPH), as well as carbon dioxide (CO_2), will be used in an enhanced version of the Calvin–Benson cycle to make locally usable food or fuel (for domestic, heating, cooking, light, and transport) in the form of carbohydrates via the enzyme RuBisCO (Gray, 2009). Bioinspired self-repair strategies will ensure that this aspect survives damage from repeated cycles of thermodynamically demanding reactions (Wasielewski, 2006).

Numerous competitively funded research teams have dedicated artificial photosynthesis–related projects already underway in many developed nations (Sanderson, 2008; Faunce et al., 2013a, 2013b). As mentioned in the foreword and the preface, an international conference at Lord Howe Island in August 2011 and a subsequent meeting hosted by the UK Royal Society at Chicheley Hall have linked senior artificial photosynthesis and global governance experts, purportedly as a precursor to a macroscience Global Artificial Photosynthesis (GAP) project (Faunce et al., 2013a; Faunce, 2013b; Royal Society, 2014).

A world powered by artificial photosynthesis, in which an increasing proportion of human-engineered structures operate like trees (in using sunlight to split water and generate hydrogen fuel and oxygen while absorbing atmospheric nitrogen and carbon dioxide to make basic food and fertilizer), will be much closer to being ecologically sustainable. It might even mark a shift from the hopefully short-lived Corporatocene to the multimillion year Sustainocene, wherein humanity controls ecosystems more as a compassionate steward, exhibiting due respect for it and paying back not only its genetic but also its moral debt to the flourishing of all life on this fragile earth.

References

Allen, B. L., Kichambare, P. D., Gou, P., et al. (2008). Biodegradation of single-walled carbon nanotubes through enzymatic catalysis. *Nano Lett.*, **8**, 3899–3903.

Ball, P. (2010). Material witness: quantum leaves in fact and fiction. *Nat. Mater.*, **9**, 614–621.

Balzani, V., Credi, A., and Venturi. M. (2008). Photochemical conversion of solar energy. *Chemsuschem*, **1**(1–2), 26–33.

Buckley, R. (2012). A financial transactions tax. *Interecon.: Rev. Eur. Econ. Policy*, **47**(2), 99–103.

Carmieli, R., Mi, Q., Ricks, A. B., Giacobbe, E. M., Mickley, S. M., and Wasielewski, M. R. (2009). Direct measurement of photoinduced charge separation distances in donor-acceptor system for artificial photosynthesis using OOP-ESEEM. *J. Am. Chem. Soc.*, **131**, 8372–8378.

Committee on Alternatives and Strategies for Future Hydrogen Production and Use, National Research Council, National Academy of Engineering (NRC & NAE) (2004). *The Hydrogen Economy: Opportunities, Costs, Barriers and R&D Needs (2004)*. National Research Council and National Academy of Engineering report, http://www.nap.edu/catalog/10922.html.

Crutzen, P. J. (2002). Geology of mankind: the Anthropocene. *Nature*, **415**, 23.

Crutzen, P. J., and Stoermer. (2000). The "Anthropocene." *Global Change Newslett.*, **41**, 17–18.

E4tech, Element Energy, Eoin Lees Energy (2004). *A Strategic Framework for Hydrogen Energy in the UK*, http://www.berr.gov.uk/files/file26737.pdf.

Engel, G. S., Calhoun, T. R., Read, E. L., et al. (2007). Evidence for wavelike energy transfer through quantum coherence in photosynthetic systems. *Nature*, **446**, 782–793.

Faunce, T. A. (2008). Toxicological and public good considerations for the regulation of nanomaterial-containing medical products. *Expert Opin. Drug Saf.*, **7**(2), 103–106.

Faunce, T. A. (2012a). *Artificial Photosynthesis* (BBC, Frontiers program), http://www.bbc.co.uk/iplayer/episode/b01jrknf/Frontiers_Artificial_Photosynthesis/.

Faunce, T. A. (2012b). *Nanotechnology for a Sustainable World. Global Artificial Photosynthesis as the Moral Culmination of Nanotechnology.* Edward Elgar.

Faunce, T. A. (2012c). *Towards Global Artificial Photosynthesis* (ANU Public Lecture), http://www.youtube.com/watch?v=dZ8qFS0kcbU.

Faunce, T. A. (ed.) (2012d). Open-source edition of the *Australian Journal of Chemistry*, **6**, 557–693, http://www.publish.csiro.au/nid/52/issue/5915.htm.

Faunce, T. A. (2012e) Governing planetary nanomedicine: Environmental sustainability and a UNESCO universal declaration on the bioethics and human rights of natural and artificial photosynthesis (global solar fuels and foods). *Nanoethics* **6**(1), 15–27.

Faunce, T. A. (2013a). *Wilton Park Conference on Future Trends for Regions and State* (podcast), https://www.wiltonpark.org.uk/podcast/futures-wp1218/.

Faunce, T. A. (2013b). Powering the world with artificial photosynthesis. *The Futurist*, http://www.wfs.org/futurist/2013-issues-futurist/may-june-2013-vol-47-no-3/powering-world-artificial-photosynthesis.

Faunce, T. A., and Nasu, H. (2008). Three proposals for rewarding novel health technologies benefitting people living in poverty. *Public Health Ethics*, **1**(2), 146.

Faunce, T. A., Lubitz, W., Rutherford, A. W., et al. (2013). Energy and environment policy case for a global project on artificial photosynthesis. *Energy Environ. Sci.*, **6**(3), 695–698.

Faunce, T. A., Styring, S., Wasielewski, M. R., et al. (2013a). Artificial photosynthesis as a frontier technology for energy sustainability. *Energy Environ. Sci.*, **6**, 1074–1076.

Faunce, T. A., White, J., and Matthaei, K. (2013b). Integrated research into the nanoparticle-protein corona: a new focus for safe, sustainable and equitable development of nanomedicines. *Nanomedicine*, **3**(6), 859–865.

Furnass, B. (2012, March 21). *From Anthropocene to Sustainocene. Challenges and Opportunities* (public lecture). Australian National University.

Gray, H. B. (2009). Powering the planet with solar fuel. *Nat. Chem.*, **1**(1), 7–12.

Gujba, H., Thorne, S., Mulugetta, Y., Rai, K., and Sokona, Y. (2012). Financing low carbon energy access in Africa. *Energy Policy*, **47**(1), 71–78

Guter, W., Chone, J. S., Philipps, S. P., et al. (2009). Current-matched Triple-junction solar cell reaching 41.1% conversion efficiency under concentrated sunlight. *Appl. Phys. Lett.*, **94**, 223504.

Hilder, T. A., Gordon, D., and Chung, S. H. (2009). Boron nitride nanotubes selectively permeable to cations or anions. *Small*, **5**, 2870–2875.

Hilder, T. A., Gordon, D., and Chung, S. H. (2010). Synthetic chloride-selective carbon nanotubes examined using molecular and stochastic dynamics. *Biophys. J.*, **99**, 1734–1742.

Hurst, J. K. (2010). In pursuit of water oxidation catalysts for solar fuel oxidation. *Science*, **328**, 315–317.

Hu, S., Shaner, M.R., Beardslee, J.A., Lichterman, M., Brunschwig, B.S., Lewis, N.S., (2014) Amorphous coatings stabliize Si, GaAs, and GaP photoanodes for efficient water oxidation. *Science* **344**(6187): 1005–9.

Intergovernmental Panel on Climate Change (2007). *Climate Change 2007: Synthesis Report. Contribution of Working Groups I, II and III to the Fourth Assessment*. Report of the Intergovernmental Panel on Climate Change (core writing team, Pachauri R. K., and Reisinger, A. [ed.]). Geneva, Switzerland.

Kalyanasundaram, K., and Graëtzel, M. (2010). Artificial photosynthesis: biomimetic approaches to solar energy conversion and storage. *Curr. Opin. Biotech.*, **21,** 298–302.

Kanan, M. W., and Nocera, D. G. (2008). In situ formation of an oxygen-evolving catalyst in neutral water containing phosphate and carbon dioxide. *Science*, **321**, 1072–1078.

Konstantatos, G., and Sargent, E. H. (2010). Nanostructured materials for photon detection. *Nat. Nanotechnol.*, **5**, 391–398.

Lee, H., Cheng, Y.-C., and Fleming, G. R. (2007). Coherence dynamics in photosynthesis: protein protection of excitonic coherence. *Science*, **316**, 1462–1468.

Lovelock, J. E. (1991). *Gaia, the Practical Science of Planetary Medicine*. London: Gaia Books.

MacKay, D. J. C. (2009). *Sustainable Energy: Without the Hot Air*. Cambridge: UIT, 204.

Magnuson, A., Anderlund, M., Johansson, O., et al. (2009). Biomimetic and microbial approaches to solar fuel generation. *Acc. Chem. Res.*, **42**(12), 1899–1912.

Martin, C., Chediak, M., and Wells, K. (2013, August 23). Homegrown green energy is making power utilities irrelevant. *Bloomsberg Business Week*, http://www.businessweek.com/articles/2013-08-22/homegrown-green-energy-is-making-power-utilities-irrelevant#r=rss.

McMichael, T. (2002). The biosphere, health and "sustainability." *Science*, **297**(5584), 1093–1096.

Messinger, J., Lubitz, W., Shen, J-R., Photosynthesis: from natural to artificial. *Phys. Chem. Phys.*, **16**, 11810–11811

Michalsky, R., Parman, B. J., Amanor-Boadu,V., and Pfromm P.H. (2012) Solar thermochemical production of ammonia from water, air and sunlight: thermodynamic and economic analyses. *Energy* **42**, 251–260.

Olah, G. A., Goeppert, A., and Surya Prakash, G. K. (2009). *Beyond Oil and Gas: The Methanol Economy*. Weinheim: Wiley VCH.

Rao, P. M., Lili Cai, L., Chong Liu, C., Cho, I. S., Lee, C. H., Weisse, J. M., Yang, P., and Zheng, X., (2014) Simultaneously efficient light absorption and charge separation in $WO_3/BiVO_4$ core/shell nanowire photoanode for photoelectrochemical water oxidation. *Nano Lett.*, **14**(2), pp 1099–1105

Rogner, H. H. (2004). Energy Resources (Chapter 5) in *World Energy Assessment: Energy and the Challenge for Sustainability* New York, NY: United Nations Development Programme, 135–166.

Royal Society. (2014). *Do We Need a Global Project on Artificial Photosynthesis: Solar Chemicals and Fuels?*, http://royalsociety.org/events/2014/artificial-photosynthesis-global-project/.

Salamanca-Buentello, F., Persad, D. L., Court, E. B., Martin, D. K., Daar, A. S., and Singer, P. A. (2005). Nanotechnology and the developing world. *PLOS Med.*, **2,** e97.

Sanderson, K. (2008). The photon trap. *Nature*, **452**, 400–402.

Sartbaeva, A., Kuznetsov, V. L., Wells, S. A., and Edwards, P. P. (2008). Hydrogen nexus in a sustainable energy future. *Energy Environ. Sci.*, **1**, 79–85.

Sgobba, V., and Guldi, D. M. (2009). Carbon nanotubes—electronic/electrochemical properties and application for nanoelectronics and photonics. *Chem. Soc. Rev.*, **38**, 165–172.

Steffen, W., Crutzen, P. J., and McNeill, J. R. (2007). The Anthropocene: are humans now overwhelming the great forces of nature? *AMBIO: J. Hum. Environ.* **36**(8), 614–621.

Steffen, W., Persson, A., Deutsch, L., et al. (2011). The Anthropocene: from global change to planetary stewardship. *AMBIO: J. Hum. Environ.*, **40**(7), 739–761.

Stern, N. (2007). *The Economics of Climate Change: The Stern Review*. Cabinet Office HM: Treasury. Cambridge, UK: Cambridge University Press.

Thapper, A., Styring, S., Saracco, G., Rutherford, A., Robert, B. et al. (2013). Artificial Photosynthesis for Solar Fuels: an Evolving Research Field within AMPEA, a Joint Programme of the European Energy Research Alliance. *Green*, vol. **3**, ss. 43–57

Toma, F. M., Sartorel, A., Iurlo, M., et al. (2010). Efficient water oxidation at carbon nanotube/polyoxometalate electrocatalytic interfaces. *Nat. Chem.*, **2**, 826–832.

Towards Global Artificial Photosynthesis: Energy, Nanochemistry & Governance (conference proceedings), http://law.anu.edu.au/coast/tgap/conf.htm.

Tsikata, F. S., Brew-Hammond, A., and Osafo, Y. B. (2008). Increasing access to clean energy in Africa: challenges and initiatives. In *Moving Beyond Carbon: Energy Law in Transition* (Zillman D. N., Redgwell, C., Omorogbe, Y. O., and Barrera-Haernandez, L., ed.). Oxford University Press, 163–179.

Ullman, A. M., Nocera, D. G., (2013) Mechanism of cobalt self-exchange in electron transfer *J. Am. Chem. Soc.*, **135**(40), 15053–15061

Umena, Y., Kawakami, K., Shen, J-R., and Kamiya N. (2011). Crystal structure of oxygen-evolving photosystem II at a resolution of 1.9 Å. *Nature* **473**, 55–60.

United Nations (2012a). *Sustainable Energy for All*, http://www.sustainableenergyforall.org/.

United Nations (2012b). *We Can End Poverty: Millennium Development Goals and Beyond 2015*, http://www.un.org/millenniumgoals/.

United Nations. *International Covenant on Economic, Social and Cultural Rights* (ICESCR). Adopted and opened for signature, ratification, and accession by General Assembly Resolution 2200A (XXI) of 16 December 1966, http://www2.ohchr.org/english/law/cescr.htm entry into force 3 January 1976.

United Nations Educational, Scientific, and Cultural Organization (UNESCO) (1997). *Declaration on the Responsibilities of the Present Generations Towards Future Generations*, http://portal.unesco.org/en/ev.php-URL_ID=13178&URL_DO=DO_TOPIC&URL_SECTION=201.html.

United Nations Framework Convention on Climate Change (UNFCCC) (7–18 December 2009). Draft decision-/CP.15 Conference of the Parties. Fifteenth Session, Copenhagen, FCCC/CP/2009/L.7, 18 December 2009.

United Nations Industrial Development Organization (UNIDO) (2009). *Scaling up Renewable Energy in Africa*. 12th Ordinary Session of Heads of State and Governments of the African Union, Addis Ababa, Ethiopia, http://www.uncclearn.org/sites/www.uncclearn.org/files/unido11.pdf.

US Department of Energy (USDOE). *Hydrogen Posture Plan*, http://www.hydrogen.energy.gov/.

Vines, T., Bruce, A., and Faunce, T. A. (2013). Planetary medicine and the Waitangi Tribunal Whanganui River report: global health law embracing ecosystems as patients. *J. Law Med.*, **20**, 528–541.

Wamukonya, N. (2007). Solar home system electrification as a viable technology option for Africa's development. *Energy Policy*, **35**(1), 6–14.

Wasielewski, M. (2006). Energy charge and spin transport in molecules and self-assembled nanostructures inspired by photosynthesis. *J. Org. Chem.*, **71**, 5051–5058.

World Bank (2006). *An Investment Framework for Clean Energy and Development: A Progress Report*. Washington, DC: The World Bank.

Chapter 2

The Cosmic Context of the Millennium Development Goals: Maximum Entropy and Sustainability

Charles H. Lineweaver[a] **and Molly Townes O'Brien**[b]

[a]*Planetary Science Institute, Research School of Astronomy and Astrophysics, Research School of Earth Sciences, Australian National University, Building 113, Mills Road, Canberra, ACT 0200, Australia*

[b]*ANU College of Law, Australian National University, Building 5, Fellows Road, Canberra, ACT 0200, Australia*

charley.lineweaver@anu.edu.au

The nations of the world are converging in health and wealth as the world grows more polluted. Navigating a path away from this unsustainable development toward sustainable development requires an understanding of the relationships between development, energy consumption, and entropy. We explore these relationships and describe the nanocosmological processes of the big bang, which are the ultimate source of the free energy that we consume. We show that the biomolecular nanotechnology of animal muscles is more efficient than internal combustion engines. We also hypothesize that an extension of the second law of thermodynamics, the maximum

Nanotechnology Toward the Sustainocene
Edited by Thomas A. Faunce

ISBN 978-981-4463-76-8 (Hardcover), 978-981-4463-77-5 (eBook)
www.panstanford.com

entropy production principle, is consistent with sustainable values for the rate of entropy production.

To see a world in a grain of sand, and a heaven in a wild flower,
Hold infinity in the palm of your hand, and eternity in an hour.
—William Blake, *Auguries of Innocence*

2.1 The Millennium Development Goals: Sustainability vs. the Other Goals

The millennium development goals (MDGs) for the year 2015, adopted by the United Nations in the year 2000 (http://mdgs.un.org), are

1. eradicate extreme poverty and hunger;
2. achieve universal primary education;
3. promote gender equality and empower women;
4. reduce child mortality;
5. improve maternal health;
6. combat HIV/AIDS, malaria, and other diseases;
7. ensure environmental sustainability; and
8. develop a global partnership for development.

Substantial but uneven progress is being made toward these goals (United Nations, *Millennium Development Goals Report*, 2013). Economic growth has been the most effective path toward meeting the MDGs. For example, the growth of the economies of China, India, and other increasingly wealthy countries has reduced poverty and hunger for millions of people. As poverty and hunger are reduced, maternal and child health improves, female literacy increases, and this tends to stabilize population (Wardatul, 2002). However, as the wealth of this stable population increases, energy consumption and pollution increase. Thus, economic development helps achieve MDGs 1 through 6 but makes environmental sustainability (MDG 7) harder (Moran, Wachernagel, Kitzes, Goldfinger, and Boutaud, 2008; Togtokh, 2011).

As we celebrate (or mourn) the birth of the seven billionth human inhabitant of our planet (Tollefson, 2011), our most important challenge is how to promote development to avoid poverty while modifying development to avoid global pollution. We have no examples of increasing economic development without increasing energy consumption and CO_2 emissions (Rosling, 2009, 2010, 2011; Emerson, Levy, Esty, et al., 2010), so the challenge before us is an unprecedented and difficult one (Wilson, 2002). Where is the safest passage to sustainable development in a high-population world, where the use of our oceans and atmosphere as common waste sinks (Fig. 2.1) can no longer be taken for granted (Hardin, 1968, 1974; Daly, 1996, 2005)?

> The factors that have historically underpinned population health gains are now, by dint of their much increased scale, scope, and intensity, undermining sustainable good health as we exceed Earth's capacity to renew, replenish, provide, and restore. (McMichael and Butler, 2011)

Figure 2.1 Two ecospheres of different sizes (left: $\sim 10^7$ m; right: $\sim 10^{-1}$ m). Both are powered by sunlight but are otherwise self-sustaining—you never have to feed them. The larger one, on the left, is thought to be less susceptible to ecological collapse because of having more diversity in life forms. However, these life forms are constrained to live in the relatively thin surface layer, one-tenth as thick as the green line. The ecosphere on the right contains only purified seawater, algae, bacteria, and marine shrimp and has been known to last ~18 years (www.eco-sphere.com/about.html). An intermediate-sized ecosphere is described in Sagan (1990). Left image: NASA, Noon in Mozambique, 7 December, 1972.

An important policy debate is going on between neoclassical economists and ecological economists that explores whether economic life on our planet is limited. Neoclassical opinion is that "there are no . . . limits to the carrying capacity of the Earth that are likely to bind any time in the foreseeable future . . . . The idea that we should put limits on growth because of some natural limit, is a profound error . . ." (Summers, 1991; Solow, 1974; Stiglitz, 1979). Ecological economists, on the other hand, are ambitiously trying to recognize and measure the environmental overheads and weigh the trade-offs between the good and bad products of economic growth (Rees, 1992; Wackernagel and Rees, 1996; Daly, 1997a, 1997b). The argument centers around two points—which aspects of the economy are knowledge based and have no identifiable limits (or limits we haven't reached yet [Johnson, 2000]), and which aspects have thresholds beyond which growth is uneconomic and if continued will lead to ecological collapse (Rockström, Steffen, Noone et al., 2011; Diamond, 2004). Fishing is an example of the latter:

> The annual fish catch is now limited by the natural capital of fish populations in the sea and no longer by the man-made capital of fishing boats. Weak sustainability would suggest that the lack of fish can be dealt with by building more fishing boats. Strong sustainability recognizes that more fishing boats are useless if there are too few fish in the ocean and insists that catches must be limited to ensure maintenance of adequate fish populations for tomorrow's fishers. (Daly, 2005)

Forty years ago, Georgescu-Roegen (1971, 1975) introduced the concept of entropy into economics (Schneider and Sagan, 2005). There has been controversy ever since about what kinds of goods are subject to the second law of thermodynamics (Daly, 1997a, 1997b). Ecological economists, such as Daly (2005), invoke entropy as the ultimate limit on sustainability:

> [L]ack of sustainability is predicted by the first two laws of thermodynamics, namely that energy is conserved (finite) and that systems naturally go from order to disorder (from low to high entropy). Humans survive and make things by sucking useful (low-entropy) resources-fossil fuels and concentrated minerals--from the environment and converting them into useless (high-entropy) wastes. The mass of wastes continuously increases (second law) until at some point all the fuel is converted to useless detritus. (Daly, 2005)

To understand the limits of economic growth and to chart a path between the Scylla of poverty and the Charybdis of pollution, we need to understand what sets the limits on the earth's capacity to "renew, replenish, provide and restore." At what point will all the fuel be "converted to useless detritus"? It is easy to run out of fuel if you don't have a fuel gauge.

A good place to begin the task of devising a reliable fuel gauge for the planet is with the laws of thermodynamics. Daly is correct when he asserts that "at some point," all fuel will be converted to useless detritus. That is the inevitable ultimate result of the second law: $dS \geq 0$. However, the "some point" is rather far in the future. The universe will reach a heat death ~10,000 googol years (10^{104} years) from now, when there will be no more stars to shine (Egan and Lineweaver, 2010). A "lack of sustainability" is only "predicted by the first two laws of thermodynamics" on time scales longer than a billion years. There are two sources of the earth's capacity to "renew, replenish, provide and restore." For the life of the biosphere (estimated to be another billion years [Caldeira and Kasting, 1992; Lenton and von Bloh, 2001; Lovelock and Whitfield, 1982]), we can count on the fusion of hydrogen in the sun and the temperature gradient between the hot interior and the cold surface of the earth (Korenaga, 2008) to supply the earth with low-entropy energy to power winds, rain, and the biosphere and naturally recycle wastes that life forms produce. The sun provides ~300 W/m^2, while the heat of the earth's interior provides ~0.1 W/m^2 at the surface. The earth's surface and the biosphere will continue to be replenished by the supply of low-entropy free energy from these two sources—driving plate tectonics that build mountains and the hydrological cycle that erodes them down and driving volcanism that replenishes the nutrients in the soils and rains that leach the nutrients out, while providing freshwater at a given rate. That rate sets the rate of sustainable extraction. Thus, for the next billion years on the earth, the second law, $dS \geq 0$, is not the problem. The problem is much more immediate—the current rate of entropy increase is larger than a sustainable rate:

$$(dS/dt)_{\text{current}} > (dS/dt)_{\text{sustainable}} \tag{2.1}$$

We are digging up and burning fossil fuels faster than nature is burying them. We are drinking and irrigating with freshwater faster than the clouds, rivers, and aquifers can supply it (Trenberth, Smith,

Qian, Dai, and Fasullo, 2007; Wada, van Beek, van Kempen, 2010), and we are mining minerals faster than plate tectonics can create new deposits.

The amount of freshwater that the earth can produce is limited by the input of free energy from the sun, which evaporates surface water and drives convection cells and winds, which carry the clouds over the land, where freshwater falls as rain, recharging the rivers, ponds, aquifers, and plants (Kleidon, 2010; Lineweaver, 2010). As is the case for the fish in the sea, the highest rate at which water can be sustainably extracted is the natural rate at which the hydrological cycle, driven by the sun, can supply it. At faster rates, aquifer water levels get lower and wells get deeper. Much of civilization (farms, desalinization plants, oil refineries, modern fisheries, and mining) is based on speeding up the natural production of food, water, and almost anything that can be made with electricity.

2.2 Energy Conservation, Entropy Increase

Understanding the role played by the first and second laws of thermodynamics can help us measure the carrying capacity of natural recycling and the price of speeding it up or overloading it (Emerson, Levy, Esty, et al., 2010). Understanding energy and entropy can help resolve the tension between development and global pollution—or at least help us think less myopically about the trade-offs. Energy conservation (first law) and entropy increase (second law) are the unifying concepts that connect gravitational collapse to nuclear fusion, fusion to sunlight, and sunlight to food, to the carrying capacity of the earth and to sustainable development. First, let's review the sources of energy.

Figure 2.2 shows the most familiar sources of energy in the universe. As mass falls into a gravitational well (Fig. 2.2A), its gravitational potential energy can be used to do work (e.g., hydroelectric power from dammed rivers and geothermal energy left over from accretion of the earth). As protons and neutrons (Fig. 2.2B) fall deeper into a nuclear potential, they release energy in the form of gamma ray photons, which emerge as visible photons from the photosphere of the sun. These photons power the hydrological cycle, ocean currents, solar cells, windmills, and phototrophic life forms. For example, in cyanobacteria and plants, solar photons excite electrons into higher-energy states, dissociating water and CO_2 to produce carbohydrates and free oxygen. We aerobic animals breathe oxygen and oxidize these high-energy electrons down into lower-energy states (ΔE in Fig. 2.2C). We live off this ΔE.

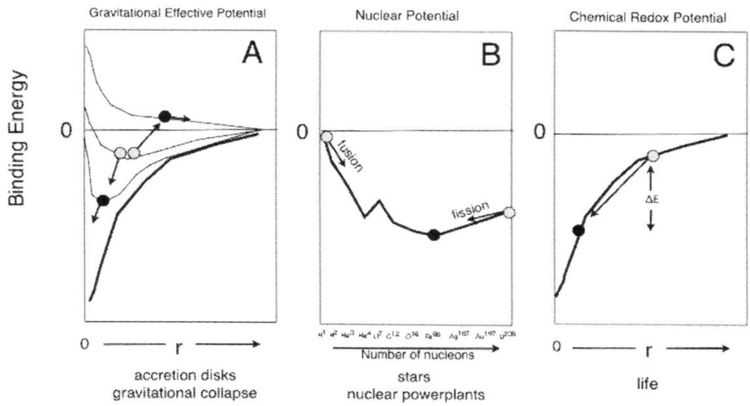

Figure 2.2 Sources of energy associated with three forces: gravity (A), the strong nuclear force (B), and electromagnetism (C). In A, gravitational binding energy is released when a particle falls deeper into the potential well. In B, nuclear potential energy is released when nuclei become more tightly bound through fusion (stars) or fission (nuclear reactors). In C, electrons release energy as they become more tightly bound into atoms and molecules (biological redox reactions).

Free energy can be extracted from the binding energies in Fig. 2.2 because the universe did not start out in a maximally bound ground state. The universe started out with potential energy. Unbound things have been able to fall into the three types of potential wells and release potential energy. For example, matter started out unclumped. As it falls and clumps into gravitational potential wells, it releases energy. Also, the hot big bang did not fuse all elements into iron. Rather, it left us with hydrogen, which can fall (fuse) into helium and eventually into iron, producing starlight.

Since energy is always conserved (first law of thermodynamics), "consuming" energy, "wasting" energy, or "saving" energy has nothing to do with the amount of energy. It has to do with consuming, wasting, and saving a specific kind of useful, low-entropy energy that is called free energy (F in Fig. 2.3). This is energy, such as the gravitational potential energy of unclumped matter or the nuclear potential energy of unfused hydrogen or the electrostatic energy of excited electrons (Fig. 2.2A–C), that can do work and has not yet been converted into waste heat (TS in Fig. 2.3).

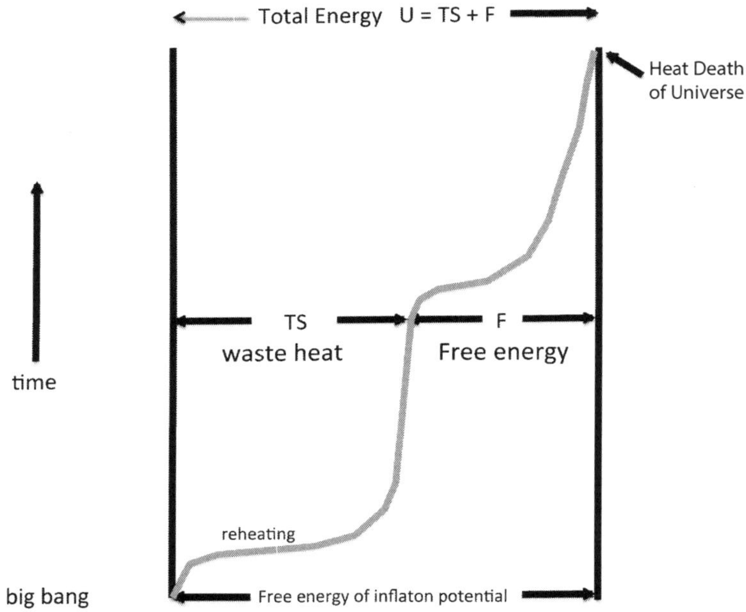

Figure 2.3 The big bang produced a low-entropy universe full of free energy (bottom of the plot). The total energy U is the sum of the waste heat TS and the free energy F. The second law ensures that with time, all of the free energy is converted into waste heat. This will occur at the heat death of the universe ~10,000 googol years from now, when there is no longer any free energy to sustain any life. Figure modified from Lineweaver and Egan, 2012.

Figures 2.3 and 2.4 show how free energy is, with the passage of time, inevitably converted into waste heat or high-entropy energy. This is the unavoidable second law of thermodynamics (entropy increase $dS \geq 0$) in action. The free energy available at one level comes from the level below it. Starting at the top of Fig. 2.4, we heterotrophs (e.g., humans and pigs) depend on phototrophs (plants) for our free energy. Phototrophs get their free energy from the photons produced by the nuclear potential energy of fusion in the sun. The nuclear potential energy was made available only because of the gravitational potential energy of unclumped matter (Fig. 2.2A), which clumped and formed stars whose cores were hot enough to access the free energy of unfused hydrogen left over from the incomplete fusion of the hot big bang (Fig. 2.2B).

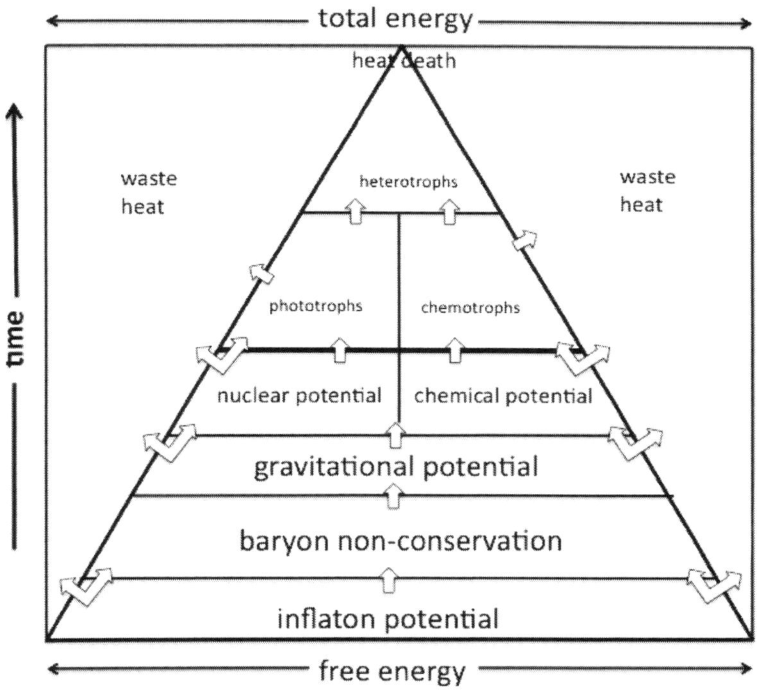

Figure 2.4 Universal trophic pyramid. The initial sources of low-entropy free energy at the bottom ("inflaton potential" and "baryon non-conservation") appeared within the first nanosecond after the big bang. The sources above them are still active today and are continually getting converted into high-entropy waste heat as their free energy drives more processes, spreading into smaller scales as waste heat is produced by dissipative structures (white arrows). The amount of free energy available narrows and disappears at the top (see also Fig. 2.3). As time goes by, free energy is converted into waste heat until the universe reaches a heat death.

The amount of free energy in unclumped matter (Fig. 2.2A) sets the amount of free energy available in gravity, represented by the width of the "gravitational potential" band of the trophic pyramid in Fig. 2.4. But what is the origin of this unclumped matter? Matter exists because of the process of baryon nonconservation that occurred within the first nanosecond after the big bang. Baryon nonconservation produced a one-part-in-a-billion excess of matter over antimatter (Sakharov, 1967). All the antimatter combined with matter and produced photons (which became the cosmic background radiation) and left a one-part-in-a-billion excess of matter (baryons). This excess is all the matter around us today. Without baryon nonconservation, equal amounts of matter and antimatter would have been produced. Their mutual annihilation would have left only photons. Photons do not clump, and therefore there would be no "gravitational potential." The photons would have been maximum-entropy energy, with zero free energy, and could not have produced galaxies, stars, planets, or life (Lineweaver and Egan, 2008). Thus, baryon nonconservation is responsible for the excess of matter and the free energy associated with this excess.

But what is responsible for there being any matter or antimatter in the first place? Our best ideas about the origin of matter and antimatter involve an epoch of rapid expansion during the early universe that happened $\sim 10^{-43}$ seconds or $\sim 10^{-35}$ seconds after the big bang (Kolb and Turner, 1990; Lyth and Liddle, 2009). Inflation took the tiny, irreducible virtual fluctuations of the vacuum and expanded them by many orders of magnitude. This is represented by the arrow on the right side of Fig. 2.5 labeled "inflation." Inflation lasted less than a nanosecond and came to an end during a process called reheating, when the energy of these inflated, formerly virtual quantum fluctuations were dumped relatively homogeneously into the universe in the form of radiation, matter, and antimatter. The level of the inflaton potential above the ground state of the vacuum determined the amount of energy that was dumped into the universe, but we have very little knowledge about what set that level. Thus, we don't know what is beneath the inflaton potential in the trophic pyramid of Fig. 2.4.

Some energy can be used to do work, while other energy cannot. The capacity of energy to do work has to do with the number of degrees of freedom over which that energy is distributed. The energy in light is distributed over the number of photons. The chemical

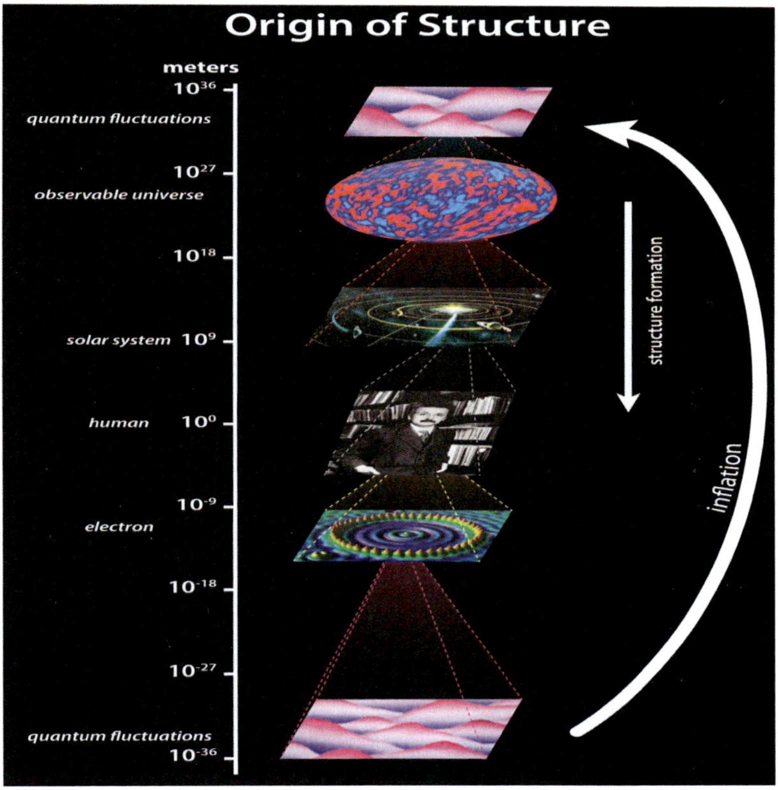

Figure 2.5 The origin of structure. The inflation of quantum fluctuations is responsible for all the structures in the universe. The tick marks on the vertical size axis are separated from each other by 9 orders of magnitude. During a brief period 10^{-43} seconds or 10^{-35} seconds after the big bang, quantum fluctuations at the Planck scale (10^{-35} m) inflated into the largest over- and underdensities of matter, currently observable as temperature fluctuations in the cosmic microwave background. The overdensities gravitationally collapsed to form large-scale structures, galaxies, stars, planetary systems, and life forms (Lineweaver and Egan, 2008). Thus, when we discovered the temperature fluctuations in the cosmic microwave background radiation (Smoot, Bennett, Kogut et al., 1992; second image from the top), we discovered Planck-scale quantum fluctuations that had been magnified by inflation to scales larger than the observable universe. Thus, we simultaneously discovered the largest and the smallest structures ever observed. This is probably the most profound and direct connection between the largest and the smallest scales in the universe.

energy in petrol is distributed over the number of molecules in the petrol. The energy in a nuclear power plant is distributed over the number of radioactive nuclei. The potential energy of water behind a dam is distributed over only one degree of freedom—the height of the water. As energy is distributed over a larger number of degrees of freedom, it becomes waste heat, incapable of providing any free energy to do work. Thus, the concept of "degrees of freedom" (Fig. 2.6) is central to understanding entropy. For example, 300,000 years after the big bang, the entire universe was filled with hot plasma, at approximately the same temperature. That was a lot of energy. But since the plasma was all at the same temperature, it was in thermal equilibrium, at maximum entropy, which means that the energy was spread over the largest possible number of degrees of freedom. No work or free energy could be extracted from all that maximally spread out, maximum-entropy energy. As the universe expanded, this cosmic background radiation expanded isentropically, so it continued to be at thermal equilibrium and unable to do work. The source of all the free energy and structure in the universe was not this background radiation but the low-entropy gravitational potential energy of unclumped matter, in Fig. 2.4 (Penrose, 2004; Lineweaver and Egan, 2008, 2012). Only low-entropy energy—out of equilibrium—provides an energy gradient that can make winds blow and maintain life (Schroedinger, 1944; Lineweaver and Egan, 2008).

2.3 Plenty of Room at the Bottom

Nanotechnology is a relatively new field, primarily involved with the discovery and exploration of the properties of matter in the size range of roughly 1–100 nm (10^{-9}–10^{-7} m). In 1959, Feynman launched the field of nanotechnology with his paper "There's Plenty of Room at the Bottom." His point was that the size of the smallest parts of technological devices ($\sim10^{-3}$ m) was much bigger than the sizes of the smallest possible parts—atoms and molecules—at the "bottom" ($\sim10^{-10}$ m). Therefore, there was plenty of room ($\sim$7 orders of magnitude in size) between technology and nanotechnology for design miniaturization. His proof of concept was the existence of life. The nanometer-sized biomolecular machines of life continuously manipulate atoms and molecules to perform the useful work we know as metabolism. If molecular evolution could blindly design such machines, why couldn't we design them too?

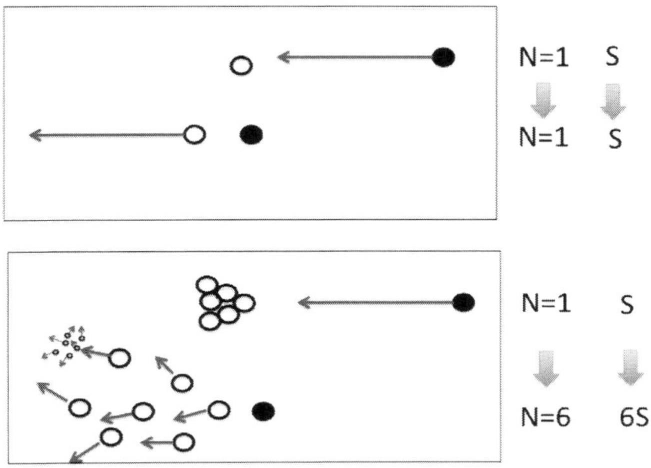

Figure 2.6 Entropy and degrees of freedom. Any increase of entropy is an increase in the number of degrees of freedom over which the given energy is distributed. In the top panel, the kinetic energy K of the black ball of mass m and velocity v is $K = 1/2\ mv^2$. The initial number of degrees of freedom is equal to the number of balls over which this energy is distributed: $N = 1$. The black ball hits the white ball, transferring all its kinetic energy but without increasing the number of degrees of freedom, which remains $N = 1$. Thus, the final entropy is equal to the initial entropy, $S_{\text{final}} = S_{\text{initial}}$. In the lower panel, when the black ball hits six white balls, $N = 1$ becomes $N = 6$ and $S_{\text{final}} = 6S_{\text{initial}}$. Entropy increases by a factor of 6 because the number of degrees of freedom increases by a factor of 6. The six smallest white balls on the left suggest that each larger white ball can begin a cascade by colliding with smaller balls, spreading the initial kinetic energy over an ever-larger number of degrees of freedom, until the energy reaches "the bottom" and becomes waste heat because it is spread over atoms and molecules—the smallest, most numerous balls. Figure modified from Lineweaver and Egan (2012).

Here, we co-opt Feynman's phrase "plenty of room at the bottom" to describe the current state of the energy of the universe. It concisely summarizes the concept that there is still plenty of room (= degrees of freedom) at the bottom for entropy to increase. The entropy of the energy of the universe started out low and has not yet reached its maximum (Fig. 2.3). The available free energy of the universe has not yet been turned into waste heat by being spread over the large number of degrees of freedom at the bottom, among the atoms and molecules or photons, as molecular waste heat or photon waste heat. Since molecular waste heat is contained in matter and matter clumps into black holes, and black holes eventually evaporate into photons, the maximum-entropy state of the energy of the universe—the "bottom"—will be after this last step, when all the energy is distributed over the maximum number of degrees of freedom as photons (Egan and Lineweaver, 2010).

The most efficient conversion of energy into work is a conversion that keeps the energy distributed over the smallest number of degrees of freedom. A heat engine converts chemical energy (with one degree of freedom per molecule of fuel) by first burning the chemical to create heat, thereby distributing its concentrated chemical energy into the more dilute kinetic energy of many molecules. For a heat engine operating between a high temperature T_{in} and a low temperature T_{out}, the maximum efficiency η is the ratio of the work out to the work in, which can be written as (Bejan, 2006):

$$\eta = 1 - T_{out}/T_{in} \tag{2.2}$$

This efficiency depends on the ratio of temperatures. For example, the temperature of an internal combustion engine $T_{in} \approx$ 800 K, while the exhaust $T_{out} \approx 400$ K. Therefore, the efficiency with which the heat inside the cylinder is turned into work cannot be more than ~50% (= 1 – 400/800). This is the maximum efficiency of internal combustion engines because they must have T_{in} low enough to maintain the structural integrity of their cylinders and pistons. Jet engines can get efficiencies as high as ~70% by having much higher values of T_{in}.

Although a solar cell is not a heat engine, we can use Eq. 2.2 to get an idea of the maximum efficiency of a solar cell converting solar photons ($T_{in} \approx 6000$ K) into work at the ambient temperature of the earth ($T_{out} \approx 300$ K). We get $\eta = 1 - 300/6000 \approx 0.95$. Thus, at the earth's surface, 95% of the energy of the solar photons can be converted to useful work.

Life has learned how to take advantage of nanotechnology, not just in its design but in its efficient energy consumption. Animal muscles are an excellent example of energy-efficient biomolecular nanotechnology. They perform work at a single temperature, so they are not heat engines driven by a temperature difference. Muscles convert chemical energy to work without high-temperature combustion. The reason muscles can access so much power so quickly is because the energy in adenosine triphosphate (ATP) can be stored—like water behind a dam—in one degree of freedom. ATP is the petrol that when oxidized supplies the muscle with free energy, and this oxidation occurs without loss of energy to a larger number of degrees of freedom. ATP is not burned. Instead, it is able to drive reactions without spreading its energy into heat first.

Recognizing that muscles must obey the second law, Jaynes (1989) generalized Eq. 2.2 to nonequilibrium situations (specifically animal muscles) by recognizing that the unit of energy E_{in} driving muscles is a single molecule of ATP. We can express E_{in} as an effective temperature with $E_{\text{in}} = 1/2\ N_{\text{in}}\ kT_{\text{in}}$, where $N_{\text{in}} = 1$ is the number of degrees of freedom over which E_{in} is distributed and k is Boltzman's constant. Thus, we have $T_{\text{in}} \approx 2E_{\text{in}}/k$. Plugging this into Eq. 2.2 yields a generalized nonequilibrium equation for maximum efficiency for extracting useful work from energy in one degree of freedom:

$$\eta = 1 - kT_{\text{out}}/(2E_{\text{in}}) \tag{2.3}$$

Just as in Eq. 2.2, the efficiency goes up if the ambient temperature T_{out} can be decreased, and the efficiency also goes up if the amount of energy E_{in}, carried in the energy molecule, goes up. Requiring an efficiency greater than zero also tells us that any biomolecule used as an energy currency must satisfy $E_{\text{in}} > (1/2)\ kT_{\text{out}}$. Inserting values into Eq. 2.3, Jaynes (1989) found that animal muscles have an efficiency of ~70%, much higher than the ~50% maximum efficiency of internal combustion engines.

2.4 Sustainable Maximum Entropy Production?

Life forms are a subset of the organized structures in the universe known as far-from-equilibrium dissipative systems (FarFEDS) (Prigogine, 1978; Schneider and Sagan, 2005; Lineweaver and Egan, 2008). FarFEDS are dissipative structures that, while maintaining their structure, convert low-entropy energy to high-entropy energy. They include galaxies, stars, convection cells, typhoons, fires, humans, and bacteria. All FarFEDS (and thus all life forms) extract free energy from the environment and turn it into waste heat faster than random processes such as diffusion would be able to do. Density, temperature, pressure, and chemical redox gradients in the environment, when steep enough, give rise to FarFEDS, which emerge spontaneously from the gradients to hasten the destruction of the gradients that spawned them. This represents a paradigm shift from "we eat food" to "food has produced us to eat it" (Lineweaver and Egan, 2008).

A growing number of researchers are investigating an extension of the second law, called the maximum entropy production principle (MEPP) (Kleidon and Lorenz, 2005; Dewar, Lineweaver, Niven, and Regenauer-Lieb, 2014). Under this principle, the terrestrial biosphere is a system that was spawned by gradients of free energy. Instead of interpreting the free-energy consumption of life's metabolisms as an imperative of Darwinian evolution, an alternative interpretation based on the MEPP is possible: life originated like a hurricane or a convection cell in order to increase entropy by destroying the gradient that made it. Life has evolved and diversified not only to stay alive but also, quite possibly, to maximize the long-term production of entropy.

If this MEPP hypothesis is correct, one could easily imagine that the natural tendency of all life (and all FarFEDS) is to produce as much entropy as possible, as quickly as possible. Everything should just burn. However, as the fable of the tortoise and the hare shows, there is more than one way to win a race. The total amount of entropy produced during a time t is:

$$S = \int_0^t (dS/dt)\, dt \tag{2.4}$$

Recalling Eq. 2.1, $(dS/dt)_{\text{current}} > (dS/dt)_{\text{sustainable}}$, we can insert our current unsustainable rate of entropy production into Eq. 2.4. However, because $(dS/dt)_{\text{current}}$ is unsustainable, it can only go on for some limited amount of time, t_{collapse}, until the ecosystem collapses and many (or all) of its biological components go extinct. With a sustainable dS/dt, we can integrate much longer, $t_{\text{sustainable}} >> t_{\text{collapse}}$. Thus, we obtain the simple result that a slower, sustainable level of entropy production is consistent with the MEPP:

$$S_{\text{sus}} = \int_0^{t_{\text{sustainable}}} (dS/dt)_{\text{sustainable}}\, dt > S_{\text{collapse}}$$
$$= \int_0^{t_{\text{collapse}}} (dS/dt)_{\text{current}}\, dt \tag{2.5}$$

or, in simpler form, assuming constant dS/dt:

$$t_{\text{sustainable}} (dS/dt)_{\text{sustainable}} > t_{\text{collapse}} (dS/dt)_{\text{current}} \tag{2.6}$$

Life forms, especially diverse ecosystems, unlike hurricanes and fires, allow for slower, more consistent, and continuous exploitation of free-energy gradients over time. For example, the slow and continuous oxidation of aerobic respiration can be more efficient in the long run at producing entropy than the rapid, short-lived oxidation of a forest fire. When life is present, more entropy can be produced over the long run.

Life has the ability to store low-entropy fuel for later use—for example, in the form of sugars or fat or in grain silos—which allows life to persist when low-entropy energy sources are temporarily in short supply. A forest fire that runs out of fuel will go out, but a snake may go without eating for six months at a time. Life forms have the potential to persist and continue to create entropy, when other FarFEDS would fizzle.

Life forms also have the advantage of being able to exploit a wider variety of low-entropy fuels. Life forms store and reproduce information in their DNA, which allows them to evolve to take advantage of changing environmental conditions and to live off the energy gradients created by other life forms. As one life form creates waste from a low-entropy source (e.g., as plants produce oxygen from CO_2 in the process of photosynthesis), another life form (animals) evolves to make use of that waste. Life forms have evolved into intricate systems of interdependence and diversity. Diversity contributes to life's ability to maximize entropy, by evolving catalysts to turn a larger variety of chemical redox potentials into waste heat.

If increasing entropy is the goal, life forms contribute to this goal in ways that other FarFEDS cannot. Life's ability to increase entropy depends, however, on its ability to sustain itself over time. When we life forms quickly use up the stores of low-entropy resources that have been built up over millions of years (freshwater, fossil fuels, mineral deposits, and wild fish and other game), we are behaving like a fire, and like a fire, we will go out and be replaced by less profligate life forms.

2.5 Conclusion

As the nations of the world converge in health and wealth and the human population begins to stabilize, economic development is producing pollution that is beginning to be felt on a global scale. We are burning fossil fuels to produce electricity to make stuff (Leonard, 2010), run desalinization plants to provide freshwater faster than the earth can do it, and make fertilizers to make food faster than is sustainable. Producing this electricity with fossil fuels is giving us a global problem with CO_2. Similarly, burning fossil fuel to produce electricity to run air purifiers to remove the air pollution from burning fossil fuel is a short-term solution causing long-term problems. It is like using an air conditioner to cool an apartment that is too hot because of the heat output of a refrigerator. It is solving the immediate local problem by making the long-term global problem worse. This is unsustainable development and has been described by Hardin (1968) as the tragedy of the commons (see also Buck, 1998). We are borrowing from the future and running up a debt on our children's credit cards.

The earth is not a perpetual motion machine. There are two sources of the earth's capacity to maintain the biosphere and process its pollution. The dominant source is the sun, and the secondary source is the heat of the earth's interior. Saving energy (using energy efficiently) means keeping the energy distributed over a smaller number of degrees of freedom. Animal muscle is an example of nanotechnological design and nanotechnological energy efficiency.

Life can be understood as a product of the MEPP. In this view, life evolved to produce the maximum amount of entropy. Maximum use of low-entropy energy sources over time means that the pace of exploitation of resources does not outstrip the pace of renewal, that biodiversity should be as large as possible, and that populations are kept at sustainable levels. Sustainability and the biosphere's variety of efficient metabolisms are the result of Darwinian evolution, but the origin of life and Darwinian evolution can be understood as a result of a more basic principle of entropy maximization. For the past four billion years, it is possible that the biosphere has sustainably maximized the entropy produced on the earth.

Acknowledgments

We thank Aditya Chopra for help in making Fig. 2.5. We also thank the big bang for making this research possible.

References

Bejan, A. (2006). *Advanced Engineering Thermodynamics.* Hoboken, NJ: Wiley.

Buck, S. J. (1998). *The Global Commons.* Washington, DC: Island Press.

Caldeira, K., and Kasting, J. F. (1992). The life span of the biosphere revisited. *Nature,* **360**, 721.

Daly, H. E. (1996). *Beyond Growth: The Economics of Sustainable Development.* Boston, MA: Beacon Press.

Daly, H. E. (1997a). Georgescu-Roegen versus Solow/Stiglitz. *Ecol. Econ.,* **22**, 261–266.

Daly, H. E. (1997b). Reply to Solow/Stiglitz. *Ecol. Econ.,* **22**, 271–273.

Daly, H. E. (2005). Economics in a full world. *Sci. Am.,* **293**, 3.

Dewar, R., Lineweaver, C. H., Niven, R., and Regenauer-Lieb, K. (2014). *Beyond the Second Law: Entropy Production and Non-Equilibrium Systems.* Heidelberg: Springer.

Diamond, J. (2004). *Collapse: How Societies Choose to Fail or Succeed.* New York, NY: Viking.

Egan, C. A., and Lineweaver, C. H. (2010). A larger estimate of the entropy of the universe. *Astrophys. J.,* **710**, 1825.

Emerson, J., Levy, M.A., Esty, D.C., Mara, V., Kim, C., de Sherbinin, A., Srebotnjak, T., and Jaiteh, M. (2010). *2010 Environmental Performance Index, Yale Center for Environmental Law and Policy,* http://epi.yale.edu/Files.

Feynman, R. P. (1960). There's plenty of room at the bottom: an invitation to enter a new field of physics. *CalTech's Engineering and Science Magazine.* Transcript of a lecture to the American Physical Society, Caltech, 29 December 1959.

Georgescu-Roegen, N. (1971). *The Entropy Law and the Economic Process.* Cambridge, MA: Harvard University Press.

Georgescu-Roegen, N. (1975). Energy and economic myths. *South. Econ. J.,* **41**(93), 347–381.

Hardin, G. (1968). Tragedy of the commons. *Science,* **162**, 1243–1248.

Hardin, G. (1974). Living on a lifeboat. *BioScience,* **24**(10), 561–568.

Jaynes, E. T. (1989). Clearing up mysteries: the original goal. In *Proceedings Volume Maximum Entropy and Bayesian Methods* (Skilling, J., ed.). Dordrecht, Holland: Kluwer Academic, 1–27.

Johnson, D. G. (2000). Population, food, and knowledge. *Am. Econ. Rev.* **90**(1), 1–14.

Kleidon, A. (2010). Life, hierarchy and the thermodynamics machinery of planet earth. *Phys. Life Rev.*, **7**(4), 424–460.

Kleidon, A., and Lorenz, R. D. (2005). *Non-Equillibrium Thermodynamics and the Production of Entropy: Life, Earth and Beyond.* Heidelberg, Germany: Springer.

Kolb, E., and Turner, M. S. (1990). *The Early Universe.* White Plains: NY: Addison-Wesley.

Korenaga, J. (2008). Urey ration and the structure and evolution of earth's mantle. *Rev. Geophys.*, **46**, 241.

Lenton, T. M., and von Bloh, W. (2001). Biotic feedback extends the life span of the biosphere. *Geophys. Res. Lett.*, **28**(9), 1715–1718.

Leonard, A. (2010). *The Story of Stuff,* www.youtube.com/storyofstuffproject#p/u/22/9GorqroigqM.

Lineweaver, C. H. (2010). Spreading the power: comment on "Life, hierarchy, and the thermodynamics machinery of planet earth." *Phys. Life Rev.*, **7**, 463–464.

Lineweaver, C. H., and Egan, C. A. (2008). Life, gravity and the second law of thermodynamics. *Phys. Life Rev.*, **5**, 225–242.

Lineweaver, C. H., and Egan, C. A. (2012). The initial low gravitational entropy of the universe as the origin of design in nature. In *Origins of Design in Nature: A Fresh, Interdisciplinary Look at How Design Emerges in Complex Systems, Especially Life* (Swan, L. S., Gordon, R., and Seckbach, J., ed.). Dordrecht, Netherlands: Springer.

Lovelock, J. E., and Whitfield, M. (1982). Lifespan of the biosphere. *Nature,* **296**, 561.

Lyth, D. H., and Liddle, A. R. (2009). *The Primordial Density Perturbation: Cosmology, Inflation, and the Origin of Structure.* UK: Cambridge University Press.

McMichael, A. J., and Butler, C. D. (2011). Promoting global population health while constraining the environmental footprint. *Annu. Rev. Public Health*, **32**, 179–197.

Moran, D. D., Wachernagel, M., Kitzes, J. A., Goldfinger, S. H., and Boutaud, A. (2008). Measuring sustainable development: nation by nation.

Ecol. Econ., **64**, 470–474, http://www.dcmetrosftp.org/newsletters/NL20061201.html#pop.

Penrose, R. (2004). *The Road to Reality: A Complete Guide to the Laws of the Universe.* London: Jonathan Cape.

Prigogine, I. (1978). Time, structure and fluctuations. *Science*, **201**, 777–785.

Rees, W. E. (1992). Ecological footprints and appropriated carrying capacity: what urban economics leaves out. *Environ. Urban.*, **4**(2), 121–130.

Rockström, J., Steffen, W., Noone, K., et al. (2011). A safe operating space for humanity. *Nature*, **461**, 472.

Rosling, H. (2009). *Hans Rosling: Let My Dataset Change Your Mindset*, www.ted.com/talks/hans_rosling_at_state.html.

Rosling, H. (2010). *Hans Rosling: Global Population Growth, Box by Box*, www.ted.com/talks/hans_rosling_on_global_population_growth.html.

Rosling, H. (2011). *Hans Rosling: The Magic Washing Machine*, www.ted.com/talks/hans_rosling_and_the_magic_washing_machine.html.

Sagan, D. (1990). *Biospheres: Reproducing Planet Earth*. New York: NY: Bantam Books.

Sakharov, A. D. (1967). Violation of CP invariance, C asymmetry, and baryon asymmetry of the universe. *J. Exp. Theor. Phys. Lett.*, **5**, 24–27.

Schneider, E. D., and Sagan, D. (2005). *Into the Cool: Energy Flow, Thermodynamics and Life.* Chicago, IL: University of Chicago Press.

Schroedinger, E. (1944). Cambridge, UK: Cambridge University Press.

Smoot, G. F., Bennett, C. L., Kogut, A., et al. (1992). Structure in the COBE differential microwave radiometer first-year maps. *Astrophys. J.*, **396**, L1–L5.

Solow, R. (1974). The economics of resources or the resources of economics. *Am. Econ. Rev.*, **64**(2), 1–14.

Stiglitz, J. E. (1979). A neoclassical analysis of the economics of natural resources. In *Scarcity and Growth Reconsidered* (Smith, V. K., ed.). Baltimore, MD: RFF and Johns Hopkins Press.

Summers, L. (1991). In an interview cited in Rees, W. E. (6 April 2006). Footprints to sustainability. *UBC Rep.*, **52**(4). University of British Columbia.

Tollefson, J. (2011). Seven billion and counting. *Nature*, **478**, 300.

Trenberth, K. E., Smith, L., Qian, T., Dai, A., and Fasullo, J. (2007). Estimates of the global water budget and its annual cycle using observational and model data. *J. Hydrometeorol.: Spec. Section*, **8**, 758–769.

Togtokh, C. (2011). Time to stop celebrating the polluters. *Nature*, **479**, 269.

United Nations (2013). *Millenium Development Goals Report*, http://mdgs.un.org/unsd/mdg/News.aspx?ArticleId=75.

Wackernagel, M., and Rees, W. (1996). *Our Ecological Footprint: Reducing Human Impact on the Earth.* Gabriola Island, British Columbia, Canada: New Society Publishers.

Wada Y., van Beek, L. P. H., van Kempen, C. M., et al. (2010). Global depletion of groundwater resources. *Geophys. Res. Lett.*, **37**, L20402.

Wardatul, A. (2002). Women's education and fertility rates in developing countries, with special reference to Bangladesh. *Eubios J. Asian Int. Bioeth.*, **12**, 138–143, http://www.eubios.info/EJ124/ej124i.htm.

Wilson, E. O. (2002). The solution. In *The Future of Life*. UK: Little Brown.

Chapter 3

Nanophotonics for Light Trapping

Sudha Mokkapati, Fiona J. Beck, Jonathan Wilson, Er-Chien Wang, and Kylie R. Catchpole
Centre for Sustainable Energy Systems, Australian National University, Building 32, North Road, Canberra, ACT 0200, Australia
sudha.mokkapati@anu.edu.au

Nanophotonics is essential for increasing the absorption in thin-film solar cells. Two particularly promising examples are plasmonics and grating structures. In this chapter, we discuss the physical concepts behind light trapping using these nanophotonic structures and the state of the art in this area.

3.1 Introduction

The worldwide installed capacity of solar photovoltaic (PV) power has escalated from 1.3 GW in 2001 to 15.2 GW in 2008 and 22.9 GW in 2009 and has had an average growth rate of 40% over the last five years. Given that the earth receives more energy from the sun in one day (10^{21} J) (International Energy Agency, 2010) than is used by the world population in one year, PV contribution to the world energy has vast potential. PV can be used to generate electricity as well as

Nanotechnology Toward the Sustainocene
Edited by Thomas A. Faunce

ISBN 978-981-4463-76-8 (Hardcover), 978-981-4463-77-5 (eBook)
www.panstanford.com

fuels like hydrogen. Another advantage of PV, apart from the fact that it is a renewable energy source, is that the world's energy can be produced in a decentralized manner. Setting up small-scale solar power generators in a decentralized manner eliminates the need to transport electricity over long distances, which is inherently lossy. Of the electricity generated from conventional sources, 30% is actually lost in resistive heating while being transported. Decentralized solar electricity generation also eliminates the need for large-capacity transmission lines and transformer stations. Decentralized solar power generators are also more resilient to natural disasters as there will be no single points of failure that can bring down the entire grid.

Currently, 80–90% of the PV market is based on crystalline Si (c-Si) solar cells (Solarbuzz, http://www.solarbuzz.com/). Si is the third-most abundant element on the earth and has near-ideal bandgap energy for maximizing the efficiency of a single-junction solar cell. c-Si solar cells have now exceeded an efficiency of 25% in the laboratory, and silicon modules have reached an efficiency of over 22% (Green, Emery, Hishikawa, and Warta, 2010). High-purity Si used in the fabrication of conventional c-Si solar cells requires expensive and energy-intensive refining of the Si feedstock. Material costs account for ~40% of the total cost of a typical c-Si PV module. Only 25% of the total costs are spent on actual cell fabrication. The rest is the module fabrication cost. An effective approach to reducing the cost per watt of PV-generated electricity to a level comparable with that generated from conventional fossil fuels would be to design and fabricate high-efficiency solar cells based on thin active layers. Conventional c-Si solar cells are fabricated from 180–300 μm thick Si wafers. Fabricating thin-film solar cells with an active layer thickness of hundreds of nanometers to a few microns would reduce the material usage by a factor of 100. In addition to reduced materials usage, thin-film solar cells also have the advantage of reduced carrier collection lengths. The photogenerated carriers in the cell should reach the external contacts before they recombine in order to generate electric current. The distance travelled by the carriers before recombination is called carrier diffusion length. For efficient collection of photogenerated carriers, the carrier diffusion length in the active material should be a few times larger than the thickness of the active layer. Reduced carrier collection lengths facilitate the use of lower-quality active material (material with lower carrier diffusion lengths) for cell fabrication, further reducing the material and deposition costs. This approach also opens up opportunities to use alternative, relatively cheap semiconductors based on earth-abundant materials like Cu, Zn, and Sn for solar cells. These semiconductors are usually of poor quality and can be useful for PVs only in thin-film configuration.

However, a thin active layer compromises the optical absorption in the solar cell. Figure 3.1 shows the spectral irradiance (in $W/m^2/nm$) on the earth's surface for the AM1.5g solar spectrum and the irradiance absorbed by a 2 μm thick Si layer, neglecting reflection losses (i.e., assuming a perfect antireflection coating [ARC] on the front surface). As can be seen from Fig. 3.1, for wavelengths greater than 500 nm, not all of the incident photons are absorbed in the Si layer. Part of the incident energy is lost because of transmission of light through the Si layer. These transmission losses are more significant in the long-wavelength region (700–1180 nm), closer to the band edge of Si. Transmission losses can be reduced by "folding" light multiple times into the absorbing region of the solar cell, thereby increasing the optical path length of light and hence the probability of its absorption inside the solar cell. This process is known as light trapping. By employing light trapping in a solar cell, the "optical thickness" of the active layer is increased several times, keeping its physical thickness unaltered. The ratio of the optical thickness to the physical thickness, that is, the ratio of the path length travelled by photons inside the cell in the presence of light trapping to that in the absence of light trapping is known as path length enhancement. This is an important parameter that enables quantitative comparison of different light-trapping techniques.

Conventionally, light trapping is achieved by modifying the surface of the solar cell to enhance the probability of total internal reflection. By doing so, light gets reflected back into the active volume several times. Theoretically, the most widely studied light-trapping scheme employs the "Lambertian surface." A Lambertian surface is an isotropic scattering surface. For a solar cell with a front Lambertian surface and a perfect rear reflector, the average path length becomes $4n^2w$, or the average path length enhancement with respect to a planar semiconductor of thickness w is $4n^2$, where n is the refractive index of the active material (Patrick and Martin, 1987; Goetzberger, 1981). This estimation is valid only in a weakly absorbing limit, that is, when there is negligible absorption inside the structure or on the surface of the structure. Another important assumption is that the optical mode density in the structure is continuous and is unaffected by wave optical effects. These conditions are satisfied if the optical thickness of the cell is much greater than $\lambda/2n$, where λ is the wavelength of the incident light, n is the refractive index of the material, and the surface texture is random or has a period much larger than λ.

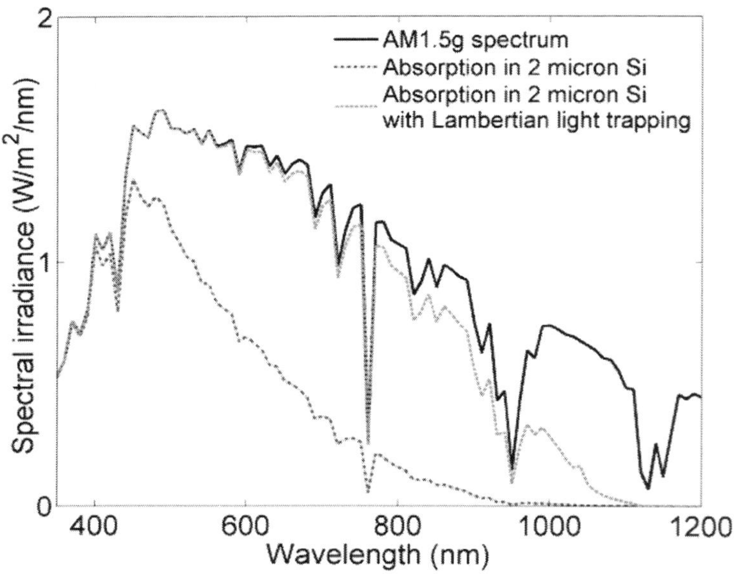

Figure 3.1 The spectral irradiance on the earth's surface. Irradiance absorbed by a 2 μm thick Si layer and a 2 μm thick Si with Lambertian light trapping, neglecting the reflection losses.

It is important to consider two special cases:

1. Thin active layers and a random- or large-period light-trapping structure: For very thin active layers, the optical mode density is not continuous. The active layer now acts as a wave-guiding layer and only supports discrete sets of optical modes. Since the Lambertian (isotropic) surface couples light into all available modes with equal probability, a reduced mode density in the active layer means reduced intensity; hence the maximum path length enhancement that can be achieved in this case is smaller than $4n^2$ (Stuart and Hall, 1997).

2. High-index, thick substrates with a light-trapping structure of a periodicity comparable to the wavelength of light: In this case, the mode density inside the active layer is unaltered, while the number of allowed modes in air is reduced to one. The achievable path length enhancement for such a configuration can exceed the $4n^2$ limit as the light-trapping structure in this case reduces the available modes in air (Yu, Raman, and Tan, 2010b).

Practically, a good approximation to Lambertian light trapping can be achieved by texturing the surface of a Si wafer to form random pyramids. This is the most widely used light-trapping technique employed in the industry. These pyramids have dimensions of the order of 1–10 μm. However, this approach is not suitable for thin solar cells where the active region itself is only a few microns thick.

Thin solar cells need wavelength-scale (or nanophotonic) structures for achieving light trapping. Different approaches to nanophotonic light trapping include plasmonic structures, single-period or dual-period diffraction gratings, and photonic crystals. In this chapter, we will discuss the physics and current status of light trapping using plasmonics and diffraction gratings.

3.2 Plasmonic Solar Cells

Plasmonic structures for light trapping have gained a lot of interest in recent years (Atwater and Polman, 2010) and can be used in three different light-trapping schemes:

(a) Scattering using localized surface plasmons: Discrete metal nanostructures can be used as subwavelength scatterers to couple incident sunlight into thin solar cells. Suitably engineered metal nanoparticles support optically driven localized surface plasmons, which strongly scatter light around resonance. If these particles are in the vicinity of a high-refractive-index substrate (see Fig. 3.2a), the scattered light can be coupled into trapped modes inside the substrate, increasing the amount of light absorbed. This was first demonstrated by Stuart and Hall (1996). In subsequent years, the technique was employed for light trapping in Si solar cells by several researchers (Schaadt, Feng, and Yu, 2005; Derkacs, Lim, Matheu, Mar, and Yu, 2006; Matheu, Lim, Derkacs, Mcpheeters, and Yu, 2008; Pillai, Catchpole, Trupke, and Green, 2007; Keisuke, Katsuaki, and Harry, 2008; Derkacs, Chen, Matheu, Lim, Yu, and Yu, 2008).

(b) Near-field enhancement: This method exploits the highly concentrated near field of nanoparticles embedded within the semiconductor to increase the absorption in the surrounding regions of the cell (see Fig. 3.2b). If the absorption rate within the semiconductor is higher than the rate of Ohmic loss within the metal, the strong local field confinement can result in enhanced photocurrent generation in that region (Rand, Peter, and Stephen, 2004; Seok-Soon, Seok-In, Jang, Dong-Yu, and Yoon-Chae, 2008; Hagglund, Michael, Goran, and Bengt, 2008; Kume, Hayashi, Ohkuma, and Yamamoto, 1995; Konda, Munde, Mustafa et al., 2007; Kirkengen, Bergli, and Galperin, 2007).

(c) Surface plasmon polariton excitations: Nanostructured metal films located on the rear of a thin-film solar cell can couple incident light into surface plasmon polariton modes that propagate along the interface of the semiconductor and the metal or into the guided modes supported by the active volume (see Fig. 3.2c). Such configuration leads to improved cell performance for cells based on strongly absorbing semiconductors (Kristofer, Nils-Krister, Olle, Aliaksandr, and Igor, 2007; Mapel, Singh, Baldo, and Celebi, 2007; Ferry, Marc, Hongbo, Ruud, Harry, and Albert, 2009; Ferry, Munday, and Atwater, 2010).

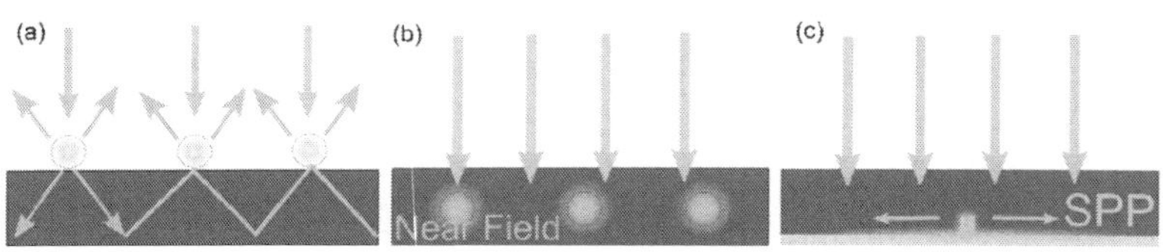

Figure 3.2 Schematic of three different plasmonic light-trapping schemes: (a) subwavelength scattering from discrete metal particles, (b) near-field enhancement from embedded nanoparticles, and (c) surface plasmon polariton excitations on nanostructured metal films.

In the following section, we will discuss light trapping using plasmonic scatterers (configuration shown in Fig. 3.2a).

3.2.1 Localized Surface Plasmons

Metal nanoparticles exhibit interesting optical effects due to strong interaction between the conduction electrons and light incident on them. The free conduction electrons in the metal perform collective oscillations against the positively charged background under the influence of the electric field associated with light. These oscillations are known as localized surface plasmons (Fig. 3.3).

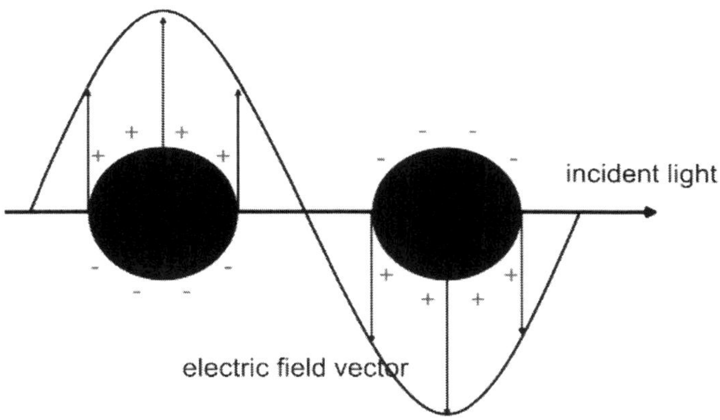

Figure 3.3 Schematic showing the localized surface plasmons set up by the incident light on metal nanoparticles.

Because of the induced oscillations, a metal nanosphere behaves as an oscillating dipole and reradiates (or scatters) energy into the surroundings. The scattering cross section for the nanosphere, C_{scat}, is defined as the power scattered by the sphere divided by the incident power. C_{scat} has units of area. Similarly, the absorption cross section, C_{abs}, is defined as the power absorbed in the nanosphere divided by the incident power. Normalizing the cross sections to the projected area of the nanosphere gives the normalized cross sections (Q_{scat} or Q_{abs}). The normalized cross sections depend on the polarizability of the particle, α (Bohren and Huffman, 1983):

$$Q_{scat} = \frac{k^4}{6\pi^2 a^2}|\alpha|^2 \tag{3.1}$$

$$Q_{\text{abs}} = \frac{k}{\pi a^2}\text{Im}\{\alpha\} \tag{3.2}$$

where k is the wave number of the incident radiation, a is the radius of the particle, and $\alpha = 4\pi a^3 \frac{\varepsilon_s - \varepsilon_m}{\varepsilon_s + 2\varepsilon_m}$.

From the above expressions, when $|\varepsilon_s + 2\varepsilon_m|$ is minimum, the polarizability and hence the scattering and absorption cross sections of the nanosphere exhibit a maximum. This condition is known as the dipolar surface plasmon resonance for the nanospheres, and the wavelength (or frequency) corresponding to the resonance is denoted as λ_{spr} (or ν_{spr}).

The radiation pattern of an ideal dipole in free space is symmetric about its axis, as is illustrated in Fig. 3.4 (Jackson, 1999). However, the presence of a substrate close to the dipole alters its radiation pattern (Mertz, 2000). Figure 3.4 shows the calculated radiation pattern for an ideal dipole in the vicinity of a Si substrate. Most of the dipole radiation is now radiated into the substrate. The optical density of modes in any medium is proportional to n^3, where n is the refractive index of the medium. The available optical modes into which light can couple are much higher in Si than in air because of the higher refractive index of Si, resulting in most of the emitted radiation being coupled into Si.

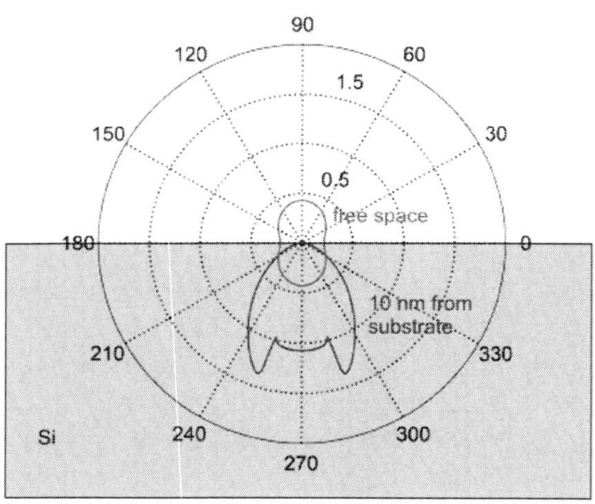

Figure 3.4 Angular distribution of dipole radiation in free space and in the vicinity of a Si substrate.

This brings us to the principle of light trapping using localized surface plasmons in isolated metal nanoparticles. Sunlight, when incident on isolated metal nanoparticles deposited on the surface of a solar cell, excites localized surface plasmons inside the nanoparticles. The nanoparticles now behave as oscillating dipoles and thus reradiate energy into the surroundings. Most of the radiated energy is coupled into the solar cell due to the high density of optical modes in Si. When the scattered radiation reaches the opposite surface of the cell, it is reflected back into the cell by total internal reflection as long as the scattered light lies outside the escape cone for the Si/air interface. The escape cone for the Si/air interface has a half-angle of $\sim 17^{\circ}$. For effective light trapping, light has to be scattered at an angle $>17^{\circ}$ into the substrate. Light trapping can be achieved by depositing the nanoparticles either on the front or on the rear surface of the solar cell, as shown in Fig. 3.5. When nanoparticles are deposited on the rear of the solar cell, light interacts with the nanoparticles after one pass through the cell. So, only long-wavelength light that is weakly absorbed interacts with the nanoparticles, while the short-wavelength light is absorbed before it reaches the nanoparticles.

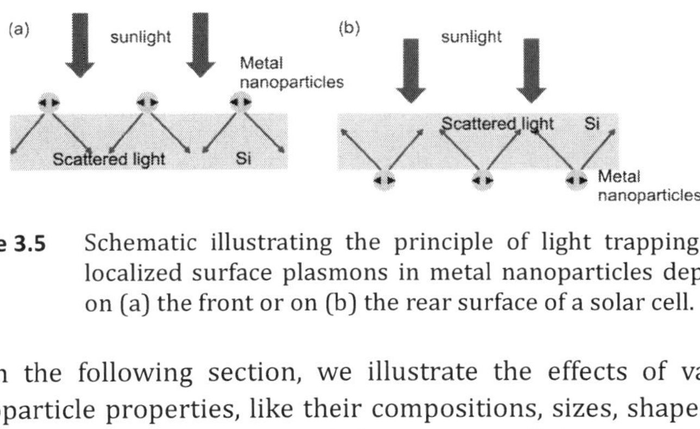

Figure 3.5 Schematic illustrating the principle of light trapping using localized surface plasmons in metal nanoparticles deposited on (a) the front or on (b) the rear surface of a solar cell.

In the following section, we illustrate the effects of various nanoparticle properties, like their compositions, sizes, shapes, and proximity to the substrate, on light-trapping efficiency and design the optimal geometry for maximizing the performance of a given solar cell.

3.2.2 Designing Optimal Geometry for Light Trapping

The simple dipole model described earlier breaks down for larger particles. For larger nanospheres in a homogenous medium, an electrodynamics approach developed by Mie (Bohren and Huffman, 1983; Born and Wolf, 2001; Mie, 1908) is widely used. For solar cell applications we are interested in the scattering behavior of nanoparticles with arbitrary shapes in the vicinity of semiconductor/dielectric stacks. We use the finite-difference time domain (FDTD) technique to numerically solve Maxwell's equations and predict the exact behavior of nanoparticles of varying shapes in the vicinity of solar cells.

We use 3D simulations (Lumerical Solutions, https://www.lumerical.com/) to study the scattering behavior of single-metal nanoparticles as their sizes, shapes, and dielectric environment are altered. Through these simulations we gain a physical insight into the behavior of metal nanoparticles that enables us to design the optimal light-trapping geometry for a given solar cell. Even though we neglect interparticle interactions in the simulations, the trends observed in single-particle simulations correlate very well to experimental results on random arrays of nanoparticles (Beck, Polman, and Catchpole, 2009; Beck, 2010).

Three physical parameters that we use to quantify/evaluate the light-trapping efficiency of metal nanoparticles are the normalized scattering cross sections (Q_{scat} and Q_{abs}) and the fraction of scattered light coupled into the substrate at each scattering event (f_{sub}). For optimal light trapping, we need to maximize Q_{scat} and f_{sub} and minimize Q_{abs}. Q_{scat} can be physically interpreted as the area of the substrate over which the nanoparticle scatters the incident light. Efficient scattering can be obtained with a lower surface coverage of nanoparticles with high Q_{scat}. Maximizing f_{sub} ensures that minimal light is coupled out of the solar cell at each consecutive scattering event, and minimizing Q_{abs} is essential for minimizing parasitic absorption in the nanoparticle itself.

For solar cell applications, the nanoparticle resonance should lie in the 400–1200 nm range. Metals like Au and Ag exhibit scattering resonances in this wavelength range and are thus interesting for light-trapping applications. Ag nanoparticles exhibit high Q_{scat} and very low Q_{abs} compared to Au, Al, or Cu nanoparticles (Beck, Mokkapati, and Catchpole, 2010). A high Q_{scat}:Q_{abs} ratio (>>1) for Ag nanoparticles means that at each scattering event, a large fraction of incident light is scattered and only a small fraction of it is absorbed in the nanoparticle. Due to lower Q_{abs}, Ag is the most promising material for light-trapping applications. For this reason, for the rest of this chapter, we will only discuss the behavior of Ag nanoparticles.

For spherical particles in a homogeneous medium, Mie's theory (Mie, 1908) predicts that the resonance peak will broaden and red-shift as the diameter of the particle increases, due to radiation damping and dynamic depolarization (Meier and Wokaun, 1983). Additionally, higher-order oscillations or multipoles begin to contribute toward the induced dipole moment in the short-wavelength region. From Eq. 3.1 and the definition of α, increasing the refractive index of the surrounding media will also red-shift the particle resonance.

However, for particles fabricated on a high-refractive-index substrate, the resonance behavior can be substantially different, depending on the shape of the particle (Beck, Verhagen, Mokkapati, Polman, and Catchpole, 2011). For spherical particles, the presence of the substrate increases the effective refractive index of the surrounding medium, red-shifting the resonance wavelength of the particles. When the particles have large areas in contact with the substrate, such as disk-shaped or cylindrical particles, the conceptual model of an induced dipole moment is less appropriate. Now two separate resonances occur, one at short wavelengths, concentrated at the top edge of the particle, and one at long wavelengths, concentrated at the bottom (Fig. 3.6). The wavelength at which the resonance at the bottom of the particle occurs depends on the length of the particle in contact with the substrate (for disks and hemispheres, this is equivalent to the diameter). From Fig. 3.6, it is clear that this resonance is significantly red-shifted from the free-space resonance due to the proximity of the high-refractive-index substrate. This shift can be reduced by introducing a low-refractive-index SiO_2 spacer layer between the particle and Si. For a given particle size, the scattering peak of this resonance can be tuned over a wide range of wavelengths by changing the effective dielectric constant of the substrate by changing either the refractive index or the thickness of the spacer layer (Catchpole and Polman, 2008b). Xu, Tazawa, Jin, Nakao, and Yoshimura (2003) have also experimentally demonstrated that the plasmon resonance of nanoparticles can be red-shifted by increasing the effective dielectric constant of the medium surrounding hemispherical particles by introducing a dielectric coating around the particles. Increasing the area of the particle in contact with the substrate will increase the wavelength at which the long-wavelength resonances occur, and large-diameter particles (with $d > 150$ nm) also support higher-order modes at the particle/substrate interface (Beck, Verhagen, Mokkapati, Polman, and Catchpole, 2011).

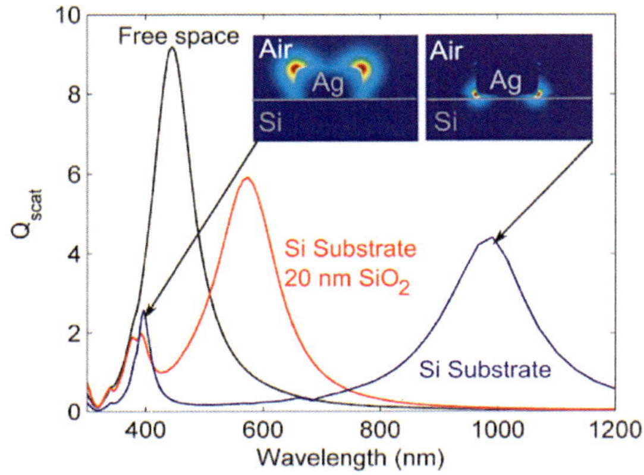

Figure 3.6 Effect of proximity to a Si substrate on the Q_{scat} spectra of a disk-shaped particle with a radius of 50 nm and a height of 50 nm. Data is also shown for the case with a SiO_2 spacer layer between the particle, and the inset shows the |E| profile of a particle directly on a Si substrate at two wavelengths that correspond to the resonances at the top and the bottom of the particle.

For light-trapping applications, it is also necessary to maximize f_{sub}. The coupling efficiency is sensitively dependent on the overlap between the near field of the excited resonance and the substrate (Catchpole and Polman, 2008b). Thus, resonances excited at the bottom of the nanoparticles couple more efficiently to the substrate than the resonance excited at the top of the particle (Beck, Verhagen, Mokkapati, Polman, and Catchpole, 2011). Spherical particles, which have effective dipole moments at roughly the center of mass of the particle, have been shown to have substantially lower f_{sub} values compared to the long-wavelength resonances of hemispheres and cylinders (Catchpole and Polman, 2008a). Introducing a spacer layer between the particle and the substrate will reduce the overlap of the near field and hence f_{sub}; the extent of this reduction depends sensitively on the spacer layer thickness, as shown in Fig. 3.7.

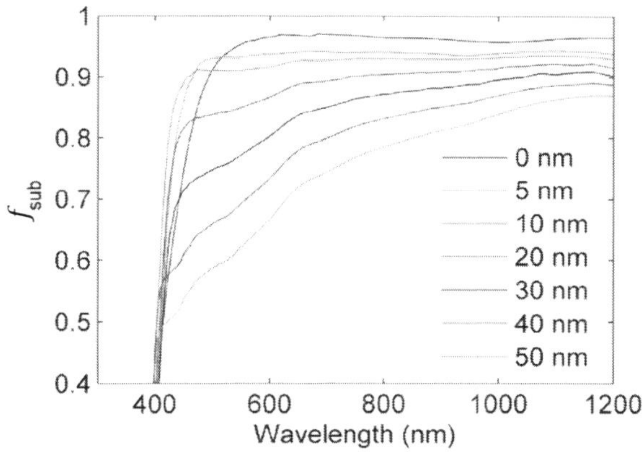

Figure 3.7 The effect of the spacer (SiO_2) thickness on f_{sub}. The particles considered are hemispheres and have a diameter of 200 nm.

Nanoparticles should be chosen to have large areas in contact with the substrate, and as such, disk-shaped or hemispherical particles are ideal. This ensures that the main scattering resonance occurs at the particle/substrate interface and is coupled to the substrate. The scattering peak can then be tuned to the desired wavelength region by choosing the diameter of the particle or introducing thin spacer layers. However, the spacer layer thickness should be minimized to ensure high f_{sub} values.

As mentioned earlier, light trapping can be achieved by depositing metal nanoparticles either on the front or on the rear surface of the solar cell. Experimental external quantum efficiency (EQE) enhancement measurements compared with a solar cell without metal nanoparticles show that for the front illumination configuration, there is a decrease in the EQE of the plasmonic solar cell relative to that of the reference cell (Beck, Polman, and Catchpole, 2009; Lim, Mar, Matheu, Derkacs, and Yu, 2007). For wavelengths shorter than the particle plasmon resonance wavelength, the scattered light is out of phase with respect to the incident light. Inside the solar cell the scattered light and the light that has been transmitted into the solar cell without interacting with the nanoparticles interfere destructively because of the phase difference. This destructive interference leads to lower net intensity inside the cell for short wavelengths, leading to a decrease in the EQE of the cell with respect to the reference cell. Above the plasmon resonance wavelength of the particles, the scattered light is in phase with respect to the incident light, leading to greater net intensities and an enhanced EQE with respect to the reference cell. These undesirable interference effects at short wavelengths can be avoided by depositing nanoparticles on the rear of the solar cell. In this configuration, the short-wavelength light is absorbed inside the cell before it reaches the nanoparticles, preventing suppression in the EQE at short wavelengths (Beck, Polman, and Catchpole, 2009; Lim, Mar, Matheu, Derkacs, and Yu, 2007).

3.2.3 Experimental Results

On the basis of the studies presented so far, we designed an optimal light-trapping geometry for 22 μm thick bifacial c-Si solar cells (Beck, Mokkapati, and Catchpole, 2010). Figure 3.8 shows a schematic of the cell geometry. The cells were fabricated with a double dielectric layer structure of 10 nm of SiO_2 and 8 nm of Si_3N_4 on both their front and rear surfaces. On the basis of the theoretical studies presented earlier, the total thickness of the dielectric stack separating the nanoparticles from the solar cell was kept below 20 nm to ensure efficient coupling of the light scattered by the nanoparticles into the cell.

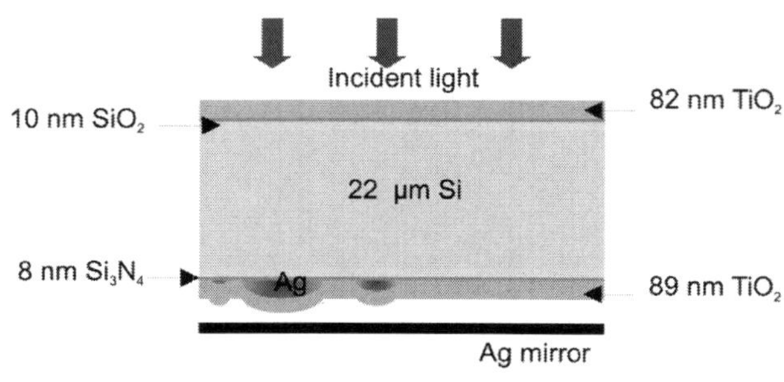

Figure 3.8 Schematic of the experimental geometry showing the dielectric layer structure of the bifacial Si solar cell, with particles on the rear surface.

Random Ag nanoparticles were fabricated on one half of the rear surface of the finished solar cells by depositing a thin layer of Ag followed by thermal annealing. The nanoparticles were fabricated on the rear surface of the cell to avoid the undesirable interference effects occurring at wavelengths smaller than the particle plasmon resonance wavelength for the front illumination configuration, as discussed earlier. The thickness of the Ag layer was chosen to get large nanoparticles, without forming large, irregular islands. For smaller Ag thickness, smaller nanoparticles are formed that have a scattering resonance in the short-wavelength region and also higher Q_{abs}. For thicker Ag layers, very irregular islands of Ag were obtained, which would support higher-order plasmonic oscillations that are not optimal for light-trapping applications. To red-shift the particle plasmon resonance into the long-wavelength region, the cells were then coated with TiO_2 by atmospheric pressure chemical vapor deposition. TiO_2 on the front surface of the cell acts as an ARC.

Figure 3.9 shows experimental EQE measurements for the cells in different configurations. At a wavelength of 750 nm transmission losses become significant for the cell and the data for cells with nanoparticles and/or mirrors exhibits EQE enhancements compared to the cell with ARC due to the reduction in transmission at the rear of the cell. The cell with only a mirror on the rear performs as well

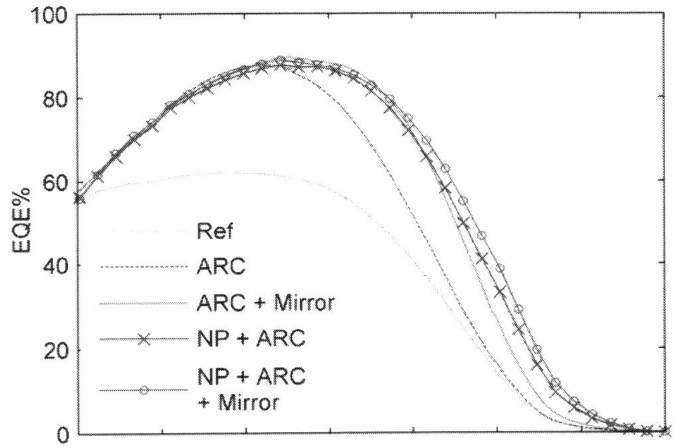

Figure 3.9 Experimental EQE measurements on a 22 μm c-Si solar cell: with a TiO_2 ARC; with a TiO_2 ARC and a mirror; with rear-located nanoparticles and a TiO_2 ARC; and with rear-located nanoparticles, a TiO_2 ARC, and a mirror. The EQE spectrum of the same cell before TiO_2 and Ag deposition is shown for reference. Figure reproduced with permission from Beck, Mokkapati, and Catchpole (2010).

as the cells with nanoparticles on the rear, up to a wavelength of ~870 nm. This is due to the fact that specular reflection from the mirror increases the path length of light in Si to approximately 44 μm, which absorbs 90% of the incident light up to a wavelength of 870 nm. Beyond this wavelength the nanoparticle arrays perform better than cells with only a rear mirror. By scattering light at high angles the nanoparticles can trap a significant portion of the light inside the Si by total internal reflection, leading to an increase in the path length of light in the cell, compared to cells with only a rear mirror. When a mirror is included behind the nanoparticles, light that is not initially scattered by the particles into the Si substrate is reflected back to provide multiple scattering opportunities, increasing the EQE further. The cell with a detached Ag mirror and the particles exhibits a 13.0% increase in the maximum short circuit current density, J_{sc}, compared to the reference cell.

3.3 Periodic Dielectric Structures

1D, 2D, or 3D periodic dielectric structures (gratings/photonic crystals) have the potential to enhance the optical absorption of solar cells in several ways. Figure 3.10 shows a schematic representation of dielectric structures used in different forms to enhance solar cell performance. Optimized 1D dielectric gratings, or Bragg stacks, can be used as back reflectors that double the path length of light in the active volume of a solar cell (Fig. 3.10a). Single-period or biperiodic dielectric structures can be used for reducing the reflection from the illuminated surface of the solar cell or for trapping light inside the active volume of the cell (Fig, 3.10b). Light trapping can be achieved either by coupling light into the guided modes of the thin, waveguide-like active region or by coupling light into diffraction modes that propagate outside the loss cone in the active volume. 3D periodic dielectric structures or photonic crystals can be used as highly efficient omnidirectional reflectors (Bermel, Luo, Zeng, Kimerling, and Joannopoulos, 2007) (Fig. 3.10c). Alternatively, the active volume itself can be patterned in 3D to confine light and increase the absorption of long-wavelength light (Chutinan, Kherani, and Zukotynski, 2009; El Daif, Drouard, Gomard et al., 2010) (Fig. 3.10d). Several of these structures can be used in conjunction with each other to achieve one or more of the above-mentioned effects simultaneously.

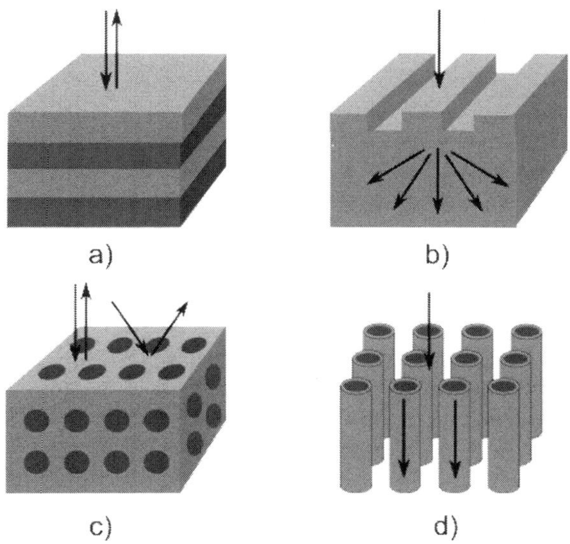

Figure 3.10 Schematic of periodic dielectric structures used for enhancing solar cell performance: (a) 1D (Bragg) stacks used as back reflectors; (b) 2D gratings, used for antireflection and/or light trapping; and (c and d) 3D photonic crystals, which are highly effective as omnidirectional reflectors or trap light by a wave-guiding effect.

3.3.1 Gratings for Back Reflectors and Antireflection

A Bragg stack, or a distributed Bragg reflector (DBR), is the simplest (1D) form of grating. Light reflected from different layers of the stack interferes constructively or destructively, leading to high reflectance or transmittance. Bragg stacks are efficient only for a narrow wavelength range and are also very sensitive to the incidence angle. Their wavelength-selective nature makes them unsuitable for use as broadband antireflective layers for solar cells. However, they can be used as low-loss, efficient rear reflectors in thin solar cells to efficiently reflect a narrow band of light transmitted through the solar cell. Enhanced solar cell performance has been predicted and demonstrated by using optimized Bragg stacks as rear reflectors in conjunction with 2D/3D dielectric gratings (Ning-Ning, Michel, Lirong et al., 2007; Zeng, Bermel, Yi et al., 2008; Zeng, Yi, Hong et al., 2006; Bermel, Luo, Zeng, Kimerling, and Joannopoulus, 2007; Mutitu, Shi, Chen et al., 2008; Zhao, Zuo, Zhou, Li, Diao, and Wang, 2010; Dayu and Rana, 2008).

2D dielectric gratings with characteristic features much smaller than that of the wavelength of incident light act as efficient antireflection layers (Dewan, Marinkovic, Noriega, Phadke, Salleo, and Knipp, 2009; Sai, Kanamori, Arafune, Ohshita, and Yamaguchi, 2007; Kanamori, Sasaki, and Hane, 1999; Gaylord, Baird, and Moraram, 1986; Chih-Hung, Wei-Lun, Nicholas, Peng, and Bin, 2007; Chih-Hung, Peng, and Bin, 2008; Eisele, Nebel, and Stutzmann, 2001; Francisco and Ignacio, 2006; Song, Yu, and Lee, 2010; Gombert, Glaubitt, Rose et al., 1999; Gombert, Glaubitt, Rose et al., 2000; Forberich, Dennler, Scharber, Hingerl, Fromherz, and Brabec, 2008; Auslender, Levy, and Hava, 1998; Escoubas, Simon, Torchio et al., 2011; Zhu, Yu, Burkhard et al., 2008; Gombert, Rose, Heinzel et al., 1998; Bräuer and Bryngdahl, 1994). Only the principal diffraction order, or the 0th diffraction order, propagates in the medium when subwavelength-scale periodic gratings are used. Hence, these structures do not provide any light trapping.

An example of a subwavelength-scale periodic structure that acts as an efficient antireflection layer is a moth eye structure (Fig. 3.11). Such a structure can be treated as a homogeneous medium that has a gradually changing refractive index between that of air and the semiconductor active layer and hence provides better impedance matching and couples light efficiently into the semiconductor, with minimal reflection losses. Effective medium theory has been developed and is widely used for the analysis of single-period and biperiodic subwavelength-scale periodic dielectric structures (Raguin and Morris, 1993; Ono, Kimura, Ohta, and Nishida, 1987; Gaylord, Baird, and Moharam, 1986; Glytsis and Gaylord, 1992; Bräuer and Bryngdahl, 1994). Rigorous coupled wave analysis (RCWA) simulations have been widely used to optimize such structures for antireflection layers (Boden and Bagnall, 2010).

Subwavelength-scale periodic dielectric structures for antireflection layers have been fabricated using techniques like electron beam lithography followed by reactive ion etching (Kanamori, Sasaki, and Hane, 1999) and fast atom beam etching through alumina templates (Sai and Kondo, 2011). Cheap and large-area fabrication techniques, like embossing in a sol-gel (Gombert, Glaubitt, Rose et al., 2000; Gombert, Rose, Heinzel et al., 1998) or using spin-coated or self-assembled dielectric spheres as etch patterns/masks (Chih-Hung, Wei-Lun, Nicholas, Peng, and Bin, 2007; Chih-Hung, Peng, and Bin, 2008; Zhu, Yu, and Burkhard, 2008), have also been developed for fabrication of subwavelength-scale gratings that could be employed as antireflection layers on solar cells.

Figure 3.11 A moth eye structure fabricated using electron beam lithography and reactive ion etching for application as an antireflection layer. Figure reproduced with permission from Kanamori, Sasaki, and Hane (1999).

Nanocone arrays on amorphous Si (a-Si):H thin-film cells have demonstrated enhanced absorption because of the antireflection property of the arrays (Zhu, Yu, and Burkhard, 2008). An enhancement of 25% in the efficiency of a 2 μm thick c-Si solar cell has been achieved by incorporating antireflection gratings fabricated by interference lithography. An antireflection grating reduced the reflectance of the cell to less than 10% in the 300–1200 nm range (Song, Yu, and Lee, 2010). An EQE enhancement of 3.5% was achieved in organic solar cells by employing a moth eye antireflection structure (Forberich, Dennler, Scharber, Hingerl, Fromherz, and Brabec, 2008).

3.3.2 Gratings for Light Trapping

2D gratings with periods larger than that of the incident wavelength (in the medium) support higher-order diffraction modes that propagate at an angle with respect to the surface normal. The propagation angle for a diffraction order can be calculated using the grating equation

$$n_1 \sin\theta_{1\mathrm{m}} = n_2 \sin\theta_{2\mathrm{m}} = \frac{m\lambda}{d} \tag{3.3}$$

where θ is the angle of propagation in each medium, m is the grating order (an integer), λ is the wavelength of light, and d is the period of the grating. Light trapping can be achieved by designing the gratings such that most of the light incident on the gratings is coupled to diffraction orders propagating outside the escape cone in the active volume of the solar cell. A key advantage of relatively large-period diffraction gratings is that they are less wavelength sensitive, which is important in achieving light trapping for thin films that are weakly absorbing over a broad wavelength range. Relatively larger periods would also be expected to provide better coupling over a range of incident angles over the day, since small periods can only couple to normally incident light (Yu, Raman, and Fan, 2010a).

Diffraction gratings can be numerically modelled by methods such as RCWA (Moharam and Gaylord, 1982; Moharam, Grann, Pommet, and Gaylong, 1995) or FDTD simulations. However, there is a wide parameter space over which the simulations need to be run, and the numerical results do not offer any physical insight into why a particular set of grating parameters should result in optimal performance. It has been shown that simplified modal analysis (Clausnitzer, Kämpfe, Kley et al., 2005; Sheng, Stepleman, and Sanda, 1982; Tishchenko, 2005) can be used to predict the optimal parameters for rectangular gratings for coupling most of the incident light into diffraction orders propagating outside the loss cone (Catchpole, 2007). The modal analysis also provides insight into the physical processes that dictate the optimal grating parameters and can be used to predict the parameters for the best-performing rectangular gratings on a thin-film solar cell without the need for extensive simulations.

Fabry–Pérot resonances and interference between different grating modes determine the efficiency of coupling light to different diffraction orders. For efficiently coupling incident light into higher-order diffraction modes, the period of the grating should be chosen such that at least two grating modes are excited in the wavelength region of interest. Once the period of the grating is determined the height of the grating should be chosen such that a net phase shift of π is introduced between the two grating modes. The optimal grating parameters determined on the basis of the above principles match very well with the parameters determined using RCWA (Catchpole, 2007).

Modal analysis is very effective for predicting the behavior of rectangular gratings or pillar gratings (Catchpole and Green, 2007). Another simple analytical model has been developed to understand and predict/optimize the behavior of diffraction gratings of arbitrary shape. The technique is applicable for gratings with periods larger than the wavelength of incident light, as it is based on the scalar diffraction theory (Born and Wolf, 2001; Goodman, 1996). The approach outlined here provides us with a conceptual understanding of the phenomena determining the light coupling into various diffraction orders from a grating of an arbitrary shape. In this approach, the transmission function for the periodic arrangement of structures is determined and the corresponding far-field diffraction pattern is computed by applying a Fourier transform (Sinzinger and Testorf, 1995). The results of the analytical modeling agree well with the numerical results obtained using FDTD modeling (Mokkapati, Beck, and Catchpole, 2011).

The above-mentioned approaches are a simple means of optimizing a large-period ($d > \lambda$) grating performance. However, the fundamental question of whether diffraction gratings can or cannot match or exceed the path length enhancement of $4n^2$ obtained using an ideal Lambertian surface on a weakly absorbing bulk substrate still needs to be addressed. The ray optics approach used for analyzing the light trapping in a bulk substrate using a Lambertian surface cannot be used to analyze the case of dielectric gratings with wavelength-scale periodicity or when only a few modes are allowed in the substrate (i.e., for thin substrates with thicknesses comparable to the wavelength of incident light). Yu, Raman, and Fan (2010a) studied the case of periodic gratings on semiconductors using a statistical temporal coupled mode theory. Their analysis shows that a maximum absorption enhancement of πn can be obtained using large, single-period ($d > \lambda$) gratings. This upper bound on the absorption enhancement is lower than the Lambertian limit of $4n^2$.

For thin active regions, light trapping is achieved by resonantly coupling light into the guided modes in the active volume. Using a biperiodic grating structure increases the number of resonances that can be accessed, thus increasing the maximum achievable absorption enhancement. Bermel, Luo, Zeng, Kimerling, and Joannopoulus (2007) have experimentally demonstrated that biperiodic gratings result in better performance than single-period gratings. Numerical studies by Mallick, Agrawal, and Peumans (2010) show that the Lambertian limit for light trapping can be exceeded for normal incidence using biperiodic gratings. Tobias, Luque, and Marti (2008) also showed that biperiodic gratings are better than single-period gratings and can exceed the Lambertian limit for path length enhancement in a narrow band using the theory of periodic radiation arrays. For a biperiodic grating with a period larger than the incident wavelength ($d > \lambda$), the upper bound on absorption enhancement approaches that of the Lambertian ($4n^2$). The maximum absorption enhancement can be further increased to $4\pi n^2$ by reducing the period of the grating such that $d = \lambda$ (Yu, Raman, and Fan, 2010a).

Having determined the limit of light trapping with grating structures, the question arises as to what types of structures can approach that limit. An important parameter to consider here is the symmetry of the grating. A symmetric grating can efficiently couple out the light from the active volume of the cell to an outgoing mode just as it couples incoming light into the active volume (Fig. 3.12a). According to the reciprocity theorem, the coupling efficiency of a diffraction mode to the outgoing wave is the same as the coupling efficiency of the incident wave to the diffraction modes. It has been proposed that the out-coupling of light from the solar cell active volume can be reduced by introducing asymmetry into the grating structure (Llopis and Tobías, 2005; Heine and Morf, 1995), as illustrated in Fig. 3.12b.

However, until recently, it has been unclear whether asymmetry is essential or not for achieving better light-trapping efficiencies. From the simple analytical model based on scalar diffraction theory presented earlier it can be seen that the diffraction orders into which most light is coupled are determined by the phase shift introduced into the incident light by the grating element. This phase shift is the same for light incident from the air side on to the grating or from the active volume on to the grating. For achieving light trapping and reducing the out-coupling of light from the active volume to air, the following two conditions have to be satisfied: (a) the phase shift introduced by the grating element should be such that light is coupled into higher-order diffraction modes inside the cell, and (b) higher-order diffraction modes do not exist in air. For small-period gratings, higher-order diffraction modes do not exist in air and introducing asymmetry into the grating structure directs incident light into higher-order diffraction modes inside the cell for small aspect ratios, compared to symmetric structures. For large-period gratings ($d > \lambda$), higher-order diffraction modes exist in air and introducing asymmetry into the grating structure does not reduce out-coupling of light.

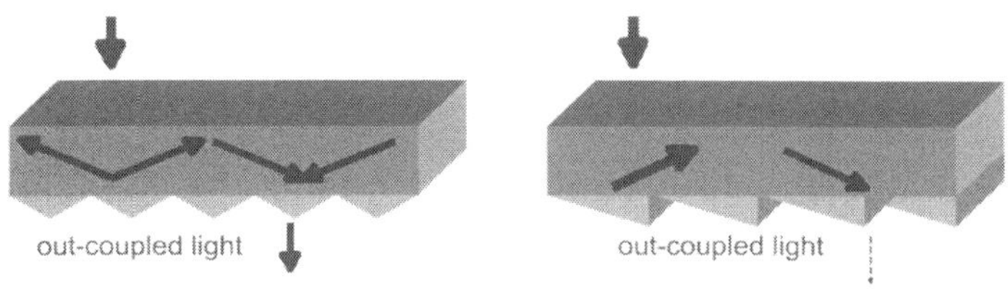

Figure 3.12 Schematic of (a) out-coupling of light from a solar cell with a symmetric grating structure and (b) reduced out-coupling by the introduction of asymmetry into the grating structure.

These results are consistent with the extensive analysis presented by Yu, Raman, and Fan (2010a) for estimating the upper bound of absorption enhancement using gratings. Their analysis shows that for large, single-period ($d > \lambda$) gratings, the upper bound on the absorption enhancement does not increase with the introduction of asymmetry into the grating. However, the maximum achievable absorption enhancement can be doubled to $2\pi n$ for single-period gratings with a period equal to the wavelength of the incident light by the introduction of asymmetry into the grating structure. Han and Chen (2010) have demonstrated using group theory that the Lambertian light-trapping limit can be exceeded using asymmetric subwavelength-scale biperiodic gratings.

From the above analysis, wavelength-scale, asymmetric biperiodic gratings should result in the best wavelength-scale light trapping. We have optimized the grating parameters for maximizing the short circuit current density, J_{sc}, that can be obtained from a 3 μm thick c-Si solar cell. Figure 3.13 shows the absorptance in a 3 μm thick c-Si layer with and without TiO_2 gratings on the rear. The gratings considered here are biperiodic pyramid or skewed pyramid structures on the rear of the Si layer. The absorptance data shown is for the optimized grating parameters. The increase in absorptance in structures with the gratings is entirely due to light trapping. The symmetric grating structure results in increased absorptance for wavelengths larger than 680 nm. The skewed grating structure increases the absorptance of the Si layer for wavelengths larger than 500 nm. In the 680–1100 nm range, the skewed gratings result in higher absorptance than the symmetric gratings. The maximum expected short circuit current density, J_{sc}, of the reference (planar) cell is enhanced by 39% and 52%, respectively, with the symmetric and asymmetric gratings on the rear of the Si cell.

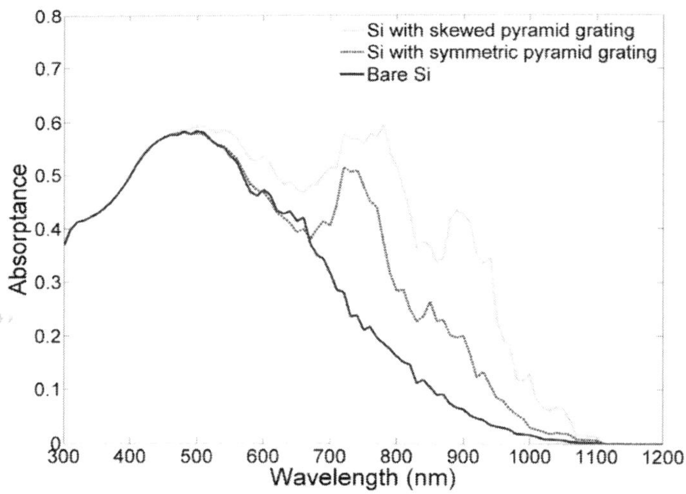

Figure 3.13 Absorptance as a function of wavelength in a 3 μm thick c-Si layer with or without a biperiodic TiO_2 grating on the rear surface. The data is shown for optimal symmetric and skewed pyramid grating parameters.

For practical applications, the electrical performance of the solar cell in the presence of a grating structure or a photonic crystal structure also needs be considered. If a grating is fabricated in the active layer of the solar cells, it tends to lead to increased surface recombination. Additionally, if the active layer is deposited on a substrate with a grating structure, steep features can make subsequent deposition of high-quality material difficult. Currently, methods for evaluating the overall cell performance by integrating electrical and optical characteristics of the cell are being developed (Peters, Rüdiger, Bläsi, and Platzer, 2010; Stiebig, Haase, Zahren, Rech, and Senoussaoui, 2006).

3.4 Summary

Potentially, all of the world's energy needs can be satisfied using solar power. Radiation from the sun can be used to generate electricity, heating, and also fuels such as hydrogen. Apart from being a renewable source of energy, solar power also has the advantage of being able to supply energy in a decentralized manner.

Solar electricity is still a factor of 2 more expensive than electricity generated from conventional sources, like fossil fuels. Reducing the materials costs associated with solar cells while maintaining high efficiencies is essential to bring down the cost of solar electricity. To achieve this goal, it is necessary to fabricate high-efficiency solar cells with thin absorber layers.

Nanophotonics is essential for increasing the absorption in thin-film solar cells. Two particularly promising examples are plasmonics and grating structures. In this chapter, we have discussed the physical concepts behind light trapping using these nanophotonic structures and the state of the art this area.

References

Atwater, H. A., and Polman, A. (2010). Plasmonics for improved photovoltaic devices. *Nat. Mater.*, **9**, 205–213.

Auslender, M., Levy, D., and Hava, S. (1998). One-dimensional antireflection gratings in (100) silicon: a numerical study. *Appl. Opt.*, **37**, 369–373.

Beck, F. J. (2010). *Designing Plasmonic Nanoparticles for Light Trapping Applications in Solar Cells*. Canberra, Australia: Australian National University.

Beck, F. J., Mokkapati, S., and Catchpole, K. R. (2010). Plasmonic light-trapping for Si solar cells using self-assembled, Ag nanoparticles. *Prog. Photovolt.: Res. Appl.*, **18**, 500–504.

Beck, F. J., Polman, A., and Catchpole, K. R. (2009). Tunable light trapping for solar cells using localized surface plasmons. *J. Appl. Phys.,* **105**, 114310.

Beck, F. J., Verhagen, E., Mokkapati, S., Polman, A., and Catchpole, K. R. (2011). Resonant SPP modes supported by discrete metal nanoparticles on high-index substrates. *Opt. Express,* **19**, A146–A156.

Bermel, P., Luo, C., Zeng, L., Kimerling, L. C., and Joannopoulos, J. D. (2007). Improving thin-film crystalline silicon solar cell efficiencies with photonic crystals. *Opt. Express,* **15**, 16986–17000.

Boden, S. A., and Bagnall, D. M. (2010). Optimization of moth-eye antireflection schemes for silicon solar cells. *Prog. Photovol: Res. Appl.,* **18**, 195–203.

Bohren, C. F., and Huffman, D. R. (1983). *Absorption and Scattering of Light by Small Particles.* Weinheim, Baden-Württemberg, Germany: Wiley-VCH Verlag & Co. KGaA.

Born, M., and Wolf, E. (2001). *Principles of Optics.* Cambridge, UK: Cambridge University Press.

Bräuer, R., and Bryngdahl, O. (1994). Design of antireflection gratings with approximate and rigorous methods. *Appl. Opt.,* **33**, 7875–7882.

Catchpole, K. R. (2007). A conceptual model of the diffuse transmittance of lamellar diffraction gratings on solar cells. *J. Appl. Phys.,* **102**, 013102.

Catchpole, K. R., and Green, M. A. (2007). A conceptual model of light coupling by pillar diffraction gratings. *J. Appl. Phys.,* **101**, 063105.

Catchpole, K. R., and Polman, A. (2008a). Design principles for particle plasmon enhanced solar cells. *Appl. Phys. Lett.,* **93**, 191113.

Catchpole, K. R., and Polman, A. (2008b). Plasmonic solar cells. *Opt. Express,* **16**, 21793–21800.

Chih-Hung, S., Peng, J., and Bin, J. (2008). Broadband moth-eye antireflection coatings on silicon. *Appl. Phys. Lett.,* **92**, 061112.

Chih-Hung, S., Wei-Lun, M., Nicholas, C. L., Peng, J., and Bin, J. (2007). Templated fabrication of large area subwavelength antireflection gratings on silicon. *Appl. Phys. Lett.,* **91**, 231105.

Chutinan, A., Kherani, N. P., and Zukotynski, S. (2009). High-efficiency photonic crystal solar cellarchitecture. *Opt. Express,* **17**, 8871–8878.

Clausnitzer, T., Kämpfe, T., Kley, E. B., et al. (2005). An intelligible explanation of highly-efficient diffraction in deep dielectric rectangular transmission gratings. *Opt. Express,* **13**, 10448–10456.

Dayu, Z., and Rana, B. (2008). Photonic crystal enhanced light-trapping in thin film solar cells. *J. Appl. Phys.,* **103**, 093102.

Derkacs, D., Chen, W. V., Matheu, P. M., Lim, S. H., Yu, P. K. L., and Yu, E. T. (2008). Nanoparticle-induced light scattering for improved performance of quantum-well solar cells. *Appl. Phys. Lett.,* **93**, 091107.

Derkacs, D., Lim, S. H., Matheu, P., Mar, W., and Yu, E. T. (2006). Improved performance of amorphous silicon solar cells via scattering from surface plasmon polaritons in nearby metallic nanoparticles. *Appl. Phys. Lett.,* **89**, 093103.

Dewan, R., Marinkovic, M., Noriega, R., Phadke, S., Salleo, A., and Knipp, D. (2009). Light trapping in thin-film silicon solar cells with submicron surface texture. *Opt. Express,* **17**, 23058–23065.

Eisele, C., Nebel, C. E., and Stutzmann, M. (2001). Periodic light coupler gratings in amorphous thin film solar cells. *J. Appl. Phys.,* **89**, 7722–7726.

El Daif, O., Drouard, E., Gomard, G., et al. (2010). Absorbing one-dimensional planar photonic crystal for amorphous silicon solar cell. *Opt. Express,* **18**(S3), A293–A299.

Escoubas, L., Simon, J.-J., Torchio, P., et al. (2011). Bringing some photonic structures for solar cells to the fore. *Appl. Opt.,* **50**(9), C329–C339.

Ferry, E. V., Marc, A. V., Hongbo, B. T. L., Ruud, E. I. S., Harry, A. A., and Albert, P. (2009). Improved red-response in thin film a-Si:H solar cells with soft-imprinted plasmonic back reflectors. *Appl. Phys. Lett.,* **95**, 183503.

Ferry, V. E., Munday, J. N., and Atwater, H. A. (2010). Design considerations for plasmonic photovoltaics. *Adv. Mater.,* **22**, 4794–4808.

Forberich, K., Dennler, G., Scharber, M. C., Hingerl, K., Fromherz, T., and Brabec, C. J. (2008). Performance improvement of organic solar cells with moth eye anti-reflection coating. *Thin Solid Films,* **516**, 7167–7170.

Francisco, L., and Ignacio, T. (2006). Texture profile and aspect ratio influence on the front reflectance of solar cells. *J. Appl. Phys.,* **100**, 124504.

Gaylord, T. K., Baird, W. E., and Moharam, M. G. (1986). Zero-reflectivity high spatial-frequency rectangular-groove dielectric surface-relief gratings. *Appl. Opt.,* **25**, 4562–4567.

Glytsis, E. N., and Gaylord, T. K. (1992). High-spatial-frequency binary and multilevel stairstep gratings: polarization-selective mirrors and broadband antireflection surfaces. *Appl. Opt.,* **31**, 4459–4470.

Goetzberger, A. (12–15 May 1981). *Optical Confinement in Thin Si-Solar Cells by Diffuse Back Reflectors,* paper presented at the 15th Photovoltaic Specialists Conference, Kissimmee, Florida.

Gombert, A., Glaubitt, W., Rose, K., et al. (1999). Subwavelength-structured antireflective surfaces on glass. *Thin Solid Films,* **351**, 73–78.

Gombert, A., Glaubitt, W., Rose, K., et al. (2000). Antireflective transparent covers for solar devices. *Sol. Energy,* **68**, 357–360.

Gombert, A., Rose, K., Heinzel, A., et al. (1998). Antireflective submicrometer surface-relief gratings for solar applications. *Sol. Energy Mater. Sol. Cells,* **54**, 333–342.

Goodman, J. W. (1996). *Introduction to Fourier Optics*, 2nd ed. New York, NY: McGraw-Hill.

Green, M. A., Emery, K., Hishikawa, Y., and Warta, W. (2010). Solar cell efficiency tables (version 36). *Prog. Photovolt: Res. Appl.,* **18**, 346–352.

Hagglund, C., Michael, Z., Goran, P., and Bengt, K. (2008). Electromagnetic coupling of light into a silicon solar cell by nanodisk plasmons. *Appl. Phys. Lett.,* **92**, 053110.

Han, S. E., and Chen, G. (2010). Toward the Lambertian limit of light trapping in thin nanostructured silicon solar cells. *Nano Lett.,* **10**(11), 4692–4696.

Heine, C., and Morf, R. H. (1995). Submicrometer gratings for solar energy applications. *Appl. Opt.,* **34**, 2476–2482.

International Energy Agency. World Energy Outlook 2010 Executive Summary. http://www.worldenergyoutlook.org/media/weowebsite/2010/WEO2010_es_english.pdf

Jackson, J. D. (1999). *Classical Electrodynamics.* New York, NY: John Wiley & Sons.

Kanamori, Y., Sasaki, M., and Hane, K. (1999). Broadband antireflection gratings fabricated upon silicon substrates. *Opt. Lett.,* **24**, 1422–1424.

Keisuke, N., Katsuaki, T., and Harry, A. A. (2008). Plasmonic nanoparticle enhanced light absorption in GaAs solar cells. *Appl. Phys. Lett.,* **93**, 121904.

Kirkengen, M., Bergli, J., and Galperin, Y. M. (2007). Direct generation of charge carriers in c-Si solar cells due to embedded nanoparticles. *J. Appl. Phys.,* **102**, 093713.

Konda, R. B., Mundle, R., Mustafa, H., et al. (2007). Surface plasmon excitation via Au nanoparticles in n-CdSe/p-Si heterojunction diodes. *Appl. Phys. Lett.,* **91**, 191111.

Kristofer, T., Nils-Krister, P., Olle, I., Aliaksandr, R., and Igor, V. Z. (2007). Surface plasmon increase absorption in polymer photovoltaic cells. *Appl. Phys. Lett.,* **91**, 113514.

Kume, T., Hayashi, S., Ohkuma, H., and Yamamoto, K. (1995). Enhancement of photoelectric conversion efficiency in copper phthalocyanine solar cell: white light excitation of surface plasmon polaritons. *Jpn. J. Appl. Phys.,* **34**(12A), 6448.

Lim, S. H., Mar, W., Matheu, P., Derkacs, D., and Yu, E. T. (2007). Photocurrent spectroscopy of optical absorption enhancement in silicon photodiodes via scattering from surface plasmon polaritons in gold nanoparticles. *J. Appl. Phys.,* **101**, 104309.

Llopis, F., and Tobías, I. (2005). The role of rear surface in thin silicon solar cells. *Sol. Energy Mater. Sol. Cells,* **87**, 481–492.

Lumerical Solutions Inc. https://www.lumerical.com/

Mallick, S. B., Agrawal, M., and Peumans, P. (2010). Optimal light trapping in ultra-thin photonic crystal crystalline silicon solar cells. *Opt. Express,* **18**, 5691–5706.

Mapel, J. K., Singh, M., Baldo, M. A., and Celebi, K. (2007). Plasmonic excitation of organic double heterostructure solar cells. *Appl. Phys. Lett.,* **90**, 121102.

Matheu, P., Lim, S. H., Derkacs, D., Mcpheeters, C., and Yu, E. T. (2008). Metal and dielectric nanoparticle scattering for improved optical absorption in photovoltaic devices. *Appl. Phys. Lett.,* **93**, 113108.

Meier, M., and Wokaun, A. (1983). Enhanced fields on large metal particles: dynamic depolarization. *Opt. Lett.,* **8**, 581–583.

Mertz, J. (2000). Radiative absorption, fluorescence, and scattering of a classical dipole near a lossless interface: a unified description. *J. Opt. Soc. Am. B,* **17**, 1906–1913.

Mie, G. (1908). Beiträge zur optik trüber medien, speziell kolloidaler metallösungen. *Ann. Phys.,* **25**, 25.

Moharam, M. G., and Gaylord, T. K. (1982). Diffraction analysis of dielectric surface-relief gratings. *J. Opt. Soc. Am.,* **72**, 1385–1392.

Moharam, M. G., Grann, E. B., Pommet, D. A., and Gaylord, T. K. (1995). Formulation for stable and efficient implementation of the rigorous coupled-wave analysis of binary gratings. *J. Opt. Soc. Am. A,* **12**, 1068–1076.

Mokkapati, S., Beck, F. J., and Catchpole, K. R. (2011). Analytical approach for design of blazed dielectric gratings for light trapping in solar cells. *J. Phys. D: Appl. Phys.,* **44**, 055103.

Mutitu, J. G., Shi, S., Chen, C., et al. (2008). Thin film solar cell design based on photonic crystal and diffractive grating structures. *Opt. Express,* **16**, 15238–15248.

Ning-Ning, F., Michel, J., Lirong, Z., et al. (2007). Design of highly efficient light-trapping structures for thin-film crystalline silicon solar cells. *IEEE Trans. Electron. Devices,* **54**(8), 1926–1933.

Ono, Y., Kimura, Y., Ohta, Y., and Nishida, N. (1987). Antireflection effect in ultrahigh spatial-frequency holographic relief gratings. *Appl. Opt.,* **26**, 1142–1146.

Patrick, C., and Martin, A. G. (1987). Light trapping properties of pyramidally textured surfaces. *J. Appl. Phys.,* **62**, 243–249.

Peters, M., Rüdiger, M., Bläsi, B., and Platzer, W. (2010). Electro optical simulation of diffraction in solar cells. *Opt. Express,* **18**, A584–A593.

Pillai, S., Catchpole, K. R., Trupke, T., and Green, M. A. (2007). Surface plasmon enhanced silicon solar cells. *J. Appl. Phys.,* **101**, 093105.

Raguin, D. H., and Morris, G. M. (1993). Analysis of antireflection-structured surfaces with continuous one-dimensional surface profiles. *Appl. Opt.,* **32**, 2582–2598.

Rand, P. B., Peter, P., and Stephen, R. F. (2004). Long-range absorption enhancement in organic tandem thin-film solar cells containing silver nanoclusters. *J. Appl. Phys.,* **96**, 7519–7526.

Sai, H., Kanamori, Y., Arafune, K., Ohshita, Y., and Yamaguchi, M. (2007). Light trapping effect of submicron surface textures in crystalline Si solar cells. *Prog. Photovolt.: Res. Appl.,* **15**, 415–423.

Sai, H., and Kondo, M. (2011). Light trapping effect of patterned back surface reflectors in substrate-type single and tandem junction thin-film silicon solar cells. *Sol. Energy Mater. Sol. Cells,* **95**, 131–133.

Schaadt, D. M., Feng, B., and Yu, E. T. (2005). Enhanced semiconductor optical absorption via surface plasmon excitation in metal nanoparticles. *Appl. Phys. Lett.,* **86**, 063106.

Seok-Soon, K., Seok-In, N., Jang, J., Dong-Yu, K., and Yoon-Chae, N. (2008). Plasmon enhanced performance of organic solar cells using electrodeposited Ag nanoparticles. *Appl. Phys. Lett.,* **93**, 073307.

Sheng, P., Stepleman, R. S., and Sanda, P. N. (1982). Exact eigenfunctions for square-wave gratings: application to diffraction and surface-plasmon calculations. *Phys. Rev. B,* **26**, 2907.

Sinzinger, S., and Testorf, M. (1995). Transition between diffractive and refractive micro-optical components. *Appl. Opt.,* **34**, 5970–5976.

Solarbuzz: Solar Market Research and Analysis. http://www.solarbuzz.com/

Song, Y. M., Yu, J. S., and Lee, Y. T. (2010). Antireflective submicrometer gratings on thin-film silicon solar cells for light-absorption enhancement. *Opt. Lett.,* **35**, 276–278.

Stiebig, H., Haase, C., Zahren, C., Rech, B., and Senoussaoui, N. (2006). Thin-film silicon solar cells with grating couplers: an experimental and numerical study. *J. Non-Cryst. Solids,* **352**, 1949–1952.

Stuart, H. R., and Hall, D. G. (1996). Absorption enhancement in silicon-on-insulator waveguides using metal island films. *Appl. Phys. Lett.,* **69**, 2327–2329.

Stuart, H. R., and Hall, D. G. (1997). Thermodynamic limit to light trapping in thin planar structures. *J. Opt. Soc. Am. A,* **14**, 3001–3008.

Tishchenko, A. V. (2005). Phenomenological representation of deep and high contrast lamellar gratings by means of the modal method. *Opt. Quant. Electron.,* **37**(1–3), 309–330.

Tobias, I., Luque, A., and Marti, A. (2008). Light intensity enhancement by diffracting structures in solar cells. *J. Appl. Phys.,* **104**, 034502.

Xu, G., Tazawa, M., Jin, P., Nakao, S., and Yoshimura, K. (2003). Wavelength tuning of surface plasmon resonance using dielectric layers on silver island films. *Appl. Phys. Lett.,* **82**, 3811–3813.

Yu, Z., Raman, A., and Fan, S. (2010a). Fundamental limit of light trapping in grating structures. *Opt. Express,* **18**, A366–A380.

Yu, Z., Raman, A., and Fan, S. (2010b). Fundamental limit of nanophotonic light trapping in solar cells. *Proc. Natl. Acad. Sci. U. S. A.,* **107**, 17491–17496.

Zeng, L., Bermel, P., Yi, Y., et al. (2008). Demonstration of enhanced absorption in thin film Si solar cells with textured photonic crystal back reflector. *Appl. Phys. Lett.,* **93**, 221105.

Zeng, L., Yi, Y., Hong, C., et al. (2006). Efficiency enhancement in Si solar cells by textured photonic crystal back reflector. *Appl. Phys. Lett.,* **89**, 111111.

Zhao, L., Zuo, Y. H., Zhou, C. L., Li, H. L., Diao, H. W., and Wang, W. J. (2010). A highly efficient light-trapping structure for thin-film silicon solar cells. *Sol. Energy,* **84**(1), 110–115.

Zhu, J., Yu, Z., Burkhard, G. F., et al. (2008). Optical absorption enhancement in amorphous silicon nanowire and nanocone arrays. *Nano Lett.,* **9**, 279–282.

Chapter 4

Growth and Characterization of GaAs Nanowires

Qiang Gao, Hannah J. Joyce, Hark Hoe Tan, and Chennupati Jagadish
Electronic Materials Engineering Department, Research School of Physics and Engineering, Australian National University, Building 58, John Carver Building, Mills Road, Canberra, ACT 0200, Australia
Qiang.Gao@anu.edu.au

Semiconductor nanowires have exhibited significant potential to be building blocks for the next generation of electronic and optoelectronic devices. However, the great challenge we are facing now is to fabricate nanowires with well-controlled dimensions, orientation, structure, phase purity, and chemical composition. Compared to the conventional epitaxy of two-dimensional layers, nanowires have exhibited a very different growth mechanism. An understanding of these growth behaviors is critical if high-quality nanowires are to be realized for device applications. We review here the fundamental growth behavior of GaAs nanowires by metal-organic chemical vapor deposition via the vapor-liquid-solid (VLS) mechanism. Transmission and scanning electron microscopy have been used to characterize the nanowire morphology and the crystal structure.

Nanotechnology Toward the Sustainocene
Edited by Thomas A. Faunce

ISBN 978-981-4463-76-8 (Hardcover), 978-981-4463-77-5 (eBook)
www.panstanford.com

4.1 Introduction

With the significant advances in nanoscience and nanotechnology, semiconductor nanowires have attracted enormous attention in recent years due to their unique physical properties and great potential as building blocks for the next generation of electronic and optoelectronic devices. Group III–V compound semiconductor nanowires are of particular interest for optoelectronic and photonic devices due to their high optical emission efficiency compared to their indirect-bandgap group IV counterparts such as Ge and Si. They are also of great interest in cost-effective solar energy use due to the enhanced light absorption and improved carrier collection efficiency (Czaban, Thompson, and Lapierre 2009; Tian, Kempa, and Lieber, 2009; Hu and Chen, 2007; Lewis, 2007). Indeed, a broad range of group III–V semiconductor nanowire-based devices have already been demonstrated, including solar cells (Czaban, Thompson, and Lapierre, 2009; Tian, Kempa, and Lieber, 2009; Hu and Chen, 2007), photodetectors (Pettersson, Tragardh, Persson, Landin, Hessman, and Samuelson, 2006; Wang, Gudiksen, Duan, Cui, and Lieber, 2001), light-emitting diodes (Minot, Kelkensberg, and Kouwen, 2007; Qian, Gradecak, Li, Wen, and Lieber, 2005), lasers (Chin, Vaddiraju, Maslov, Ning, Sunkara, and Meyyappan, 2006; Hua, Motohisa, Kobayashi, Hara, and Fukui, 2009; Minot, Kelkensberg, and Kouwen, 2007; Qian, Gradecak, Li, Wen, and Lieber, 2005), single-photon emitters (Borgstrom, Zwiller, Muller, and Imamoglu, 2005; Heinrich, Huggenberger, Heinde et al., 2010), resonant tunneling diodes (Bjork, Ohlsson, Thelander et al., 2002), single-electron transistors and memory devices (Thelander, Martensson, Bjork et al., 2003; Thelander, Nilsson, Jensen, and Samuelson, 2005; Bjork, Thelander, Hansen et al., 2004), field emission sources (De Jonge, Heeres, Bakkers, Roest, Kaiser, and Oosterkamp, 2007), and integrated photonic circuits (Barrelet, Bao, Loncar, Park, Capasso, and Lieber, 2006; Park, Barrelet, Wu, Tian, Qian, and Lieber, 2008).

Unlike conventional bulk materials or 2D epitaxial layers, semiconductor nanowires grow as freestanding quasi-1D structures on a wafer, offering great flexibility to construct a true 3D architecture via a bottom-up approach, which is not possible otherwise. Our group at the Australian National University, together with our national and international collaborators, has carried out extensive research work on the growth and characterization of various group III–V semiconductor nanowires, including GaAs (Zou, Paladugu, Wang et al., 2007; Joyce, Gao, Tan et al., 2007; Kang, Gao, Joyce et al., 2010; Joyce, Gao, Tan et al., 2009; Joyce, Gao, Tan et al., 2008; Parkinson, Lloyd-Hughes, Gao et al., 2007; Wang, Joyce, Gao et al., 2011; Wang, Wang, Joyce et al., 2011; Joyce, Gao, Wong-Leung et al., 2011; Joyce, Docherty, Gao, et al., 2013), InAs (Joyce, Gao, Wong-Leung et al., 2011; Joyce, Wong-Leung, Gao, Tan, and Jagadish, 2010; Zhang, Zou, Paladugu et al., 2009), InP (Maharjan, Pemasiri, Kumar et al., 2009; Mishra, Titova, Hoang et al., 2007; Paiman, Gao, Joyce et al., 2010; Paiman, Gao, Tan et al., 2009; Pemasiri, Kontazeri, Gass et al., 2009; Perera, Pemasiri, Fickenscher et al., 2010; Reece, Paiman, Abdul-Nabi et al., 2009 Joyce, Docherty, Gao, et al., 2013; Reece, Toe, Wang et al., 2011), and their related heterostructures (Guo, Zou, Paladugu et al., 2006; Paladugu, Zou, Guo et al., 2007; Titova, Hoang, Jackson et al., 2006; Smith, Jackson, Yarrison-Rice, and Jagadish, 2010; Perera, Fickenscher, Jackson et al., 2008a; Paladugu, Zou, Guo et al., 2008b; Paladugu, Zou, Guo et al., 2008a; Paladugu, Zou, Auchterlonie et al., 2007; Paladugu, Zou, Guo et al., 2009a; Paladugu, Zou, Guo et al., 2009b; Montazeri, Fickenscher, Smith et al., 2010; Parkinson, Joyce, Gao et al., 2009; Perera, Fickenscher, Jackson et al., 2008b). Due to the scope limit of this chapter, we only focus on our recent work on high-quality GaAs nanowires for use in optoelectronic devices. This work has produced significant advances in understanding the growth mechanism of GaAs nanowires and related heterostructures by metal-organic chemical vapor deposition (MOCVD) and in nanowire characterization by extensive electron microscopy.

We start with a general introduction to nanowire growth in Section 4.2. It introduces the concept of the VLS mechanism using Au nanoparticles as seeds to initiate an anisotropic 1D growth via MOCVD. The fundamental principles underlying the VLS mechanism in MOCVD and the growth procedure used in our work are also discussed in this section. Sections 4.3 and 4.4 present the effects of two crucial growth parameters in MOCVD—temperature and growth rate—on the morphology and crystallographic structure of GaAs nanowires. We show the unexpected growth behavior of GaAs nanowires compared to conventional 2D epitaxy and exemplify how growth parameters can affect the structural properties in these sections. A summary will be given in Section 4.5.

4.2 Nanowire Growth

The biggest challenge to growing nanowires is the precise control of nanowire dimensions, orientation, structure, phase purity, and chemical composition. Size and shape are crucial in the nanoscale regime; even small variations in size can have a large effect on overall device performance. For instance, nanowire diameter and its uniformity determine the degree of confinement and consequently affect the behavior of charge carriers in quantum electronic devices. For freestanding nanowires, 1D growth relies on the anisotropic growth from a catalyst. It then elongates in a particular direction with the highest growth rate, which is normally the [111]B direction for group III–V semiconductors, while the other directions are constrained by much slower growth rates.

4.2.1 VLS Mechanism

Numerous mechanisms have been proved to be responsible for freestanding nanowires, for example, self-catalyzed growth (Plissard, Larriue, Wallart, and Caroff, 2011), selective-area epitaxy (Hua, Motohisa, Kobayashi, Hara, and Fukui, 2009; Ooike, Motohisa, and Fukui, 2004; Cantoro, Brammertz, Richard et al., 2009), and Au-assisted VLS (Wagner and Ellis, 1964) or vapor-solid-solid (VSS) growth (Dick, Deppert, Martensson, Mandl, Samuelson, and Seifert, 2005). The VLS mechanism is the most widely cited growth mechanism and also used in this work. The VLS mechanism was originally proposed by Wagner and Ellis in 1964 to explain the anisotropic growth of Si whiskers catalyzed by Au particles. VLS was named due to the copresence of vapor, liquid, and solid phases, as schematically illustrated in Fig. 4.1. The Si precursor species were supplied in the vapor phase. At growth temperature the metallic Au particle formed a liquid eutectic alloy with Si and became a sink for Si atoms supplied from the surrounding vapor. With further supply of Si this Au–Si alloy particle became supersaturated and the Si then precipitated at the liquid particle–solid semiconductor interface to form a solid crystalline Si wire. Two possibilities were proposed to explain this mechanism. First, due to the high sticking coefficient of the liquid phase, the liquid droplet is the favorable collection site for the vapor-phase reaction species and thus assists wire growth. The second possibility is that the Au/Au–Ga particle is a chemical catalyst that lowers the activation energy barriers to enhance the decomposition of the gas-phase precursors, resulting in preferential growth under the Au particle (Verheijen, Immink, De Smet, Borgstrom, and Bakkers, 2006).

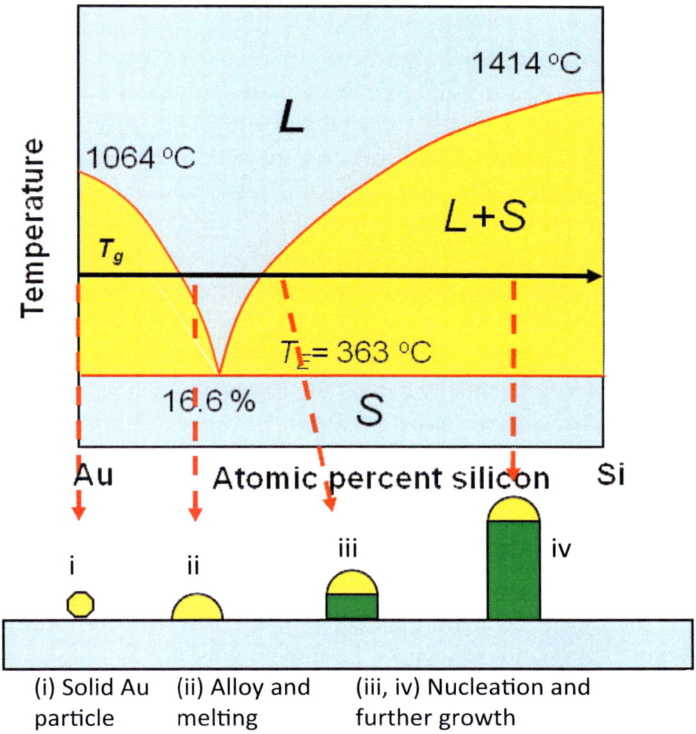

Figure 4.1 Schematic of the VLS growth of Si whiskers using an Au particle. The Au–Si binary phase diagram illustrated in the upper portion indicates the various phases the Au–Si particle can adopt under different temperatures and in different composition ranges. The red arrows and labels (i), (ii), and (iii) illustrate the different stages of wire growth and the corresponding states of the Au–Si particle.

4.2.2 MOCVD Growth and Principles

The same principle of the VLS mechanism was applied to grow group III–V nanowires, where group III species are believed to alloy with Au to form liquid eutectic droplets at growth temperature and group V species were mainly supplied from the vapor phase. This Au-assisted nanowire growth mechanism has been utilized in a number of growth systems, including laser-assisted catalytic growth (Duan and Lieber, 2000), molecular beam epitaxy (MBE) (Plante and LaPierre, 2006), chemical beam epitaxy (CBE) (Persson, Ohlsson, Jeppesen, Samuelson, 2004), and MOCVD (Hiruma, Murakoshi, Yazawa, and Katsuyama, 1996). Of all these techniques, the most common technique for group III–V nanowire growth is MOCVD, which is also used for many applications in electronics and optoelectronics (Behet, Hovel, Kohl, Kusters, Opitz, and Heime, 1996). This technique offers great flexibility and high accuracy in nanowire design and is readily scalable for industrial mass fabrication. Indeed, Au-assisted MOCVD growth has produced nanowires with abrupt heterostructure interfaces, high purity, controlled doping, and perfect phase purity. For these reasons, Au-assisted MOCVD is our choice for group III–V nanowire growth.

In MOCVD growth, a number of parameters can be tightly controlled to achieve the desired material properties. These parameters include growth temperature, the V/III ratio, and the growth rate. This chapter will focus on growth temperature and the growth rate, which give unexpected effects compared to conventional 2D epitaxy. Growth temperature is widely acknowledged as a crucial factor in group III–V semiconductor growth (Stringfellow, 1999). The growth rate is mainly controlled by the flow rate of the group III precursors in conventional 2D epitaxy. However, the flow rate of group V elements also plays an important role in the nanowire growth rate, both axially and radially (Joyce, Gao, Tan et al., 2008). We investigate how these growth parameters affect nanowire growth and how they can be tailored to achieve desirable nanowire properties.

There are two major growth modes taking place during Au-assisted group III–V nanowire growth by MOCVD—axial growth and radial growth. Axial growth represents the elongating growth of the nanowire along the [111]B direction in this work. There are three major contributions to axial growth, as illustrated in Fig. 4.2: (i) reaction species directly impinging on the nanoparticle, which is the main contribution for vertical nanowire growth; (ii) species adsorbed on nanowire sidewalls; and (iii) species adsorbed on the substrate.

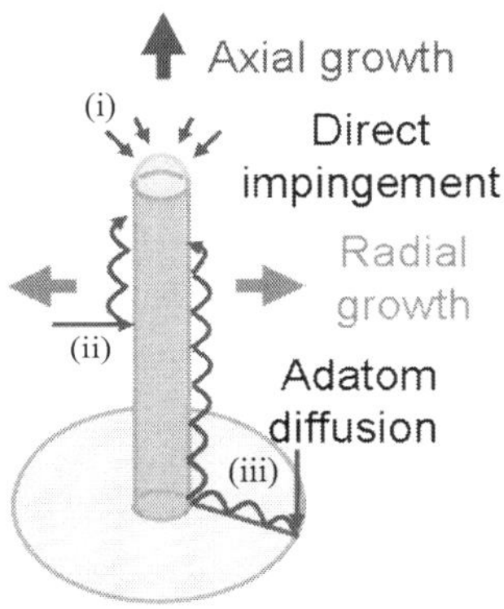

Figure 4.2 Schematic of axial and radial nanowire growth. It shows (i) direct impingement of precursor species on the nanoparticle at the nanowire tip, (ii) species adsorbed on nanowire sidewalls, which diffuse toward the tip, and (iii) species adsorbed on substrate, which diffuse toward the nanowire tip along the substrate and nanowire sidewalls.

These adatoms diffuse along the concentration gradient toward the Au nanoparticle, where they are incorporated into axial growth. These two contributions greatly depend on the diffusion length of adatoms and the growth condition. If these adatoms cannot reach the Au nanowire growth front, they will deposit on the nanowire sidewalls or substrate surface, resulting in radial growth, also known as lateral growth or conventional epitaxy. These growths follow a simple vapor-solid growth mechanism and do not involve the Au nanoparticle. Due to the radial growth, nanowires often exhibit wider bases and narrower tips. The tapering parameter is used here to quantify this tapering effect and is defined as the increment in nanowire radius Δr per increment in nanowire length Δl, that is:

$$\text{Tapering parameter} = \Delta r/\Delta l \tag{4.1}$$

Measurements of Δr and Δl were taken at consecutive intervals of approximately 1 μm along each nanowire, from the Au nanoparticle–

nanowire interface to the base. That is, the change in the nanowire radius Δr was measured over each length interval of $\Delta l \approx 1$ μm. For each interval the tapering parameter was determined. This was then averaged for each nanowire. At least 10 nanowires were examined for each data point.

4.2.3 Growth Procedure and Characterization Technique

All nanowires in this work were grown by horizontal-flow MOCVD at a pressure of 100 mbar using Au nanoparticles to initiate nanowire growth. Trimethylgallium (TMG) was used as the precursor for the group III element Ga. The group V element As was provided by arsine (AsH_3). GaAs nanowires were grown on semi-insulating GaAs (111)B substrates. Before growth, Au nanoparticles 50 nm in diameter were applied on the substrate surface, as described in previous reports (Kim, Joyce, Gao et al., 2006; Mishra, Titova, Hoang et al., 2007). Prior to growth initiation, the substrate was annealed at 600°C under an AsH_3 ambient to desorb surface organics and contaminants. After cooling to the desired temperature, TMG was introduced to initiate nanowire growth. The standard flow rates of TMG and AsH_3 were $III_0 = 1.2 \times 10^{-5}$ mol/min and $V_0 = 5.4 \times 10^{-4}$ mol/min, respectively, unless mentioned elsewhere, which gives a standard V/III ratio of 46.

GaAs nanowires were grown by either a single-temperature or a two-temperature procedure. The single-temperature procedure involved 30 minutes of growth at a constant temperature T_g between 350°C and 500°C. For the two-temperature procedure, growth initiated first at a nucleation temperature T_n of 450°C. The temperature was then ramped down to the subsequent "growth" temperature T_g, between 350°C and 400°C. The total growth time was 30 minutes, including the cooling time typically of about 3 minutes.

Field emission scanning electron microscopy (FESEM) (Hitachi S4500) is used to identify general nanowire morphology, including diameter and height. In this work, FESEM is operated at low acceleration voltages of 3 kV to explicate the nanowire surface morphology. Transmission electron microscopy (TEM) (FEI Tecnai F30) was used to identify crystal structures and crystal defects, such as twins and stacking faults.

4.3 Effects of Growth Temperature

A range of growth temperatures from 330°C to 500°C have been tested in single-temperature and two-temperature growth procedures and showed a significant effect on nanowire morphology. Figure 4.3 shows FESEM images of nanowires grown by the single-temperature procedure at various T_g (from 375°C to 500°C). When grown at a T_g of 450°C and above, nanowires were straight and epitaxially aligned in the vertical [111]B direction. However, nanowires become severely tapered at higher temperatures (e.g., 500°C). In contrast, nanowires grown at a T_g of 400°C or lower exhibit irregular kinked morphologies and cannot grow in a vertical direction. The lowest T_g to produce straight-aligned nanowires was determined to be 410°C in our previous study (Joyce, Gao, Tan et al., 2007).

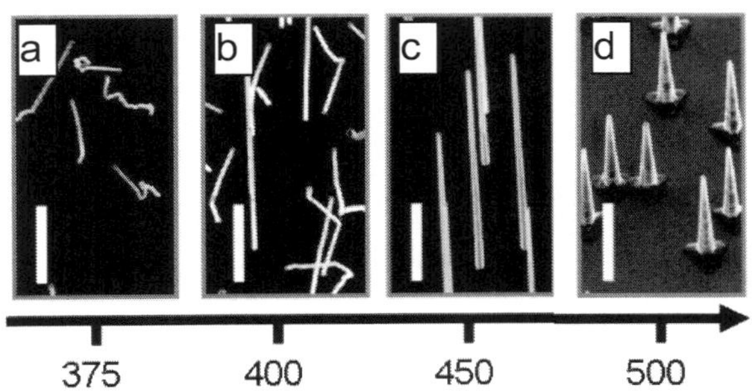

Figure 4.3 FESEM images of GaAs nanowires grown by the single-temperature procedure at different growth temperatures T_g, as indicated. Scale bars are 1 μm. Samples are tilted at 40°. Adapted from Joyce, Gao, Wong-Leung et al. (2011) with permission from IEEE INC.

As shown in Fig. 4.4, nanowires grown by the two-temperature procedure differ significantly from those grown by the single-temperature procedure. At a T_g of 375°C–400°C, all nanowires are straight and vertically aligned at the [111]B direction, where they kink via the single-temperature procedure. The lowest T_g to produce such straight nanowires was 350°C in our previous study (Mishra, Titova, Hoang et al., 2007). In addition, the tapering effect in these nanowires is very little, which will be discussed later.

It is proposed that a liquid eutectic Au–Ga particle is essential for the initiation of vertically aligned [111]B nanowires and the continuation of epitaxial straight nanowire growth. This Ga-rich eutectic alloy forms only when TMG is supplied at growth initiation. The kinked nanowire growth is the consequence of the solid or partially solid state of the alloyed Au nanoparticles at a very low T_g (<410°C) (Kodambaka, Tersoff, Reuter, and Ross, 2007; Tchernycheva, Harmand, Patriarche, Travers, and Cirlin, 2006). This is reasonable because these T_g are close to the eutectic point of the Au–Ga system (339.3°C). Note that the cited T_g are not the actual temperatures of the substrate surface. The nominal temperature is measured by a thermocouple in the center of the susceptor, which is generally a few tens of degrees higher than the actual substrate temperature. However, in the two-temperature procedure, the melted eutectic alloy experiences a temperature hysteresis between melting and solidification (Tchernycheva, Harmand, Patriarche, Travers, and Cirlin, 2006); therefore, the liquid state of the eutectic alloy can be maintained at lower T_g. But a minimum T_n = 410°C is required, which is the same as the minimum T_g in the one-temperature procedure.

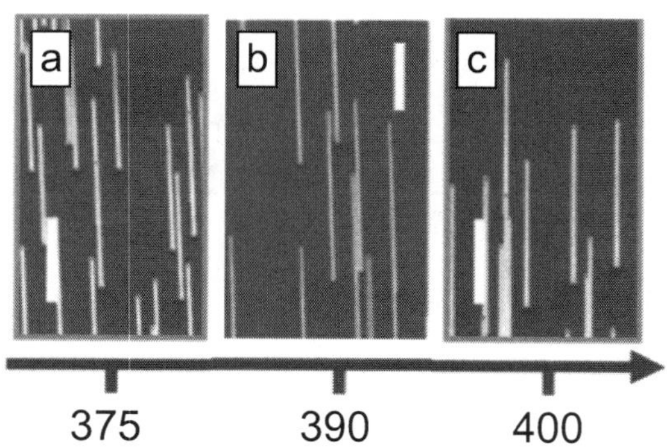

Figure 4.4 FESEM images of GaAs nanowires grown by the two-temperature procedure at different growth temperatures T_g, as indicated, and with the nucleation temperature T_n, at 450°C. Scale bars are 1 μm. Samples are tilted at 40°.

Figure 4.5 plots the nanowire diameters measured at various distances from the nanowire tip of selected straight nanowires grown by both one-temperature and two-temperature procedures. These measurements were obtained from FESEM images of nanowires transferred to Si substrates. Only the upper section of the nanowires was examined because nanowire base growth is influenced by the brief high-temperature nucleation step. The nanowire-tapering parameter was calculated from the slopes of straight lines fitted to the data in Fig. 4.5. It was 17 nm/µm for nanowires grown by the one-temperature procedure at a T_g of 450°C but only less than 3 nm/µm for nanowires grown by the two-temperature procedure at a T_g less than 400°C and a T_n of 450°C. This temperature-dependent reduction in tapering can be explained by the different growth mechanisms of axial growth and radial growth. As illustrated in Fig. 4.2, axial growth is mainly induced by species that impinge directly upon the nanoparticle. This process is temperature independent and relies on a mass-flow controlled mechanism. However, radial growth is attributed to the diffusion of adsorbed species on the substrate and nanowire sidewalls. These diffusions are kinetically limited, and the species are less likely to be incorporated into radial growth at lower growth temperatures.

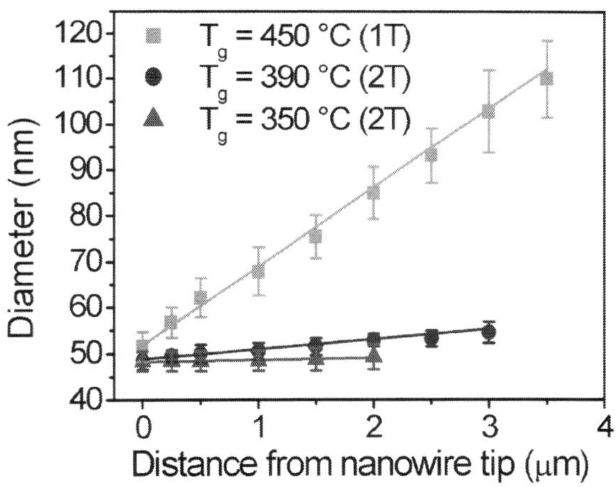

Figure 4.5 A plot of nanowire diameter measured at distances from the nanowire tip for three T_g. Nanowires were grown by either a single-temperature or a two-temperature procedure. Error bars represent standard deviations of samples of 10 nanowires at each point. Adapted from Joyce, Gao, Tan et al. (2007) with permission from American Chemical Society.

In addition to improved morphology, the most surprising effect resulting from low-temperature growth using the two-temperature procedure is the high crystallographic quality. In conventional 2D epitaxy, higher temperature (>650°C) is normally required to obtain good crystal quality. However, in the GaAs nanowire growth, a high T_g of 450°C by the single-temperature procedure results in high density of twin defects, as indicated by arrows in Fig. 4.6a,b. In marked contrast, we cannot find any twins or other planar defects in nanowires grown at the lower T_g of 390°C by the two-temperature procedure (Fig. 4.6c–f). This result is consistent with recent TEM studies of zinc blende GaP nanowires, which showed a distinct reduction in twin density with decreasing growth temperature (Johansson, Karlsson, Svensson et al., 2006).

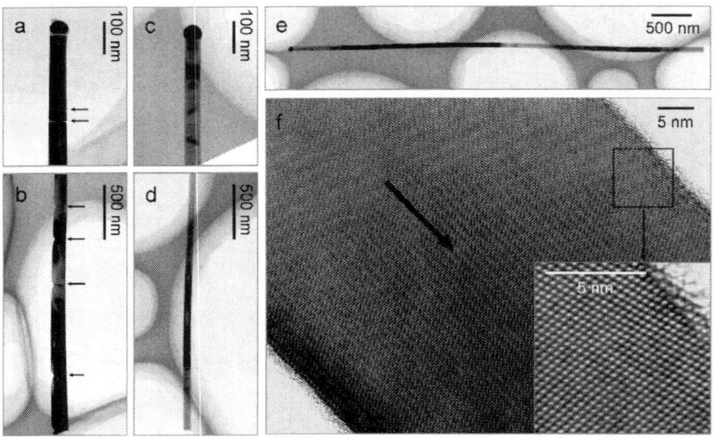

Figure 4.6 (a–e) Bright-field TEM and (f) HRTEM images of GaAs nanowires. (a, b) Nanowires grown at a T_g of 450°C by the single-temperature procedure, with arrows showing twin defects. (c–f) Nanowires grown at a low T_g of 390°C by the two-temperature procedure (T_n of 450°C), showing a nanowire tip (c), a nanowire middle (d), an entire nanowire (e), and a section of a nanowire (f). The thick arrow in (f) indicates the nanowire growth direction. Adapted from Joyce, Gao, Tan et al. (2007) with permission from American Chemical Society.

4.4 Effects of Growth Rate

The growth rate had a profound effect on nanowire morphology (Joyce, Gao, Tan et al., 2009). In conventional epitaxy of 2D group III–V materials, a rapid growth rate is always associated with lower-quality material. Unexpectedly, we observed exactly the opposite effect for GaAs nanowires, as will be discussed later.

Higher growth rates were achieved by increasing both group III and group V flows while keeping the V/III ratio constant. Four different V/III ratios, of 12, 23, 46, and 93, were studied here. The growth time was 15 minutes for a standard group III flow rate (III_0) and scaled inversely with group III flow for others to achieve comparable lengths for all nanowires. Figure 4.7 shows FESEM images of nanowires grown at different growth rates (i.e., group III precursor flow rates) with a V/III ratio of 46. All nanowires grow epitaxially in the [111]B direction, normal to the substrate surface. Obviously, the nanowires grown at rapid growth rates (higher III flow) are markedly less tapered.

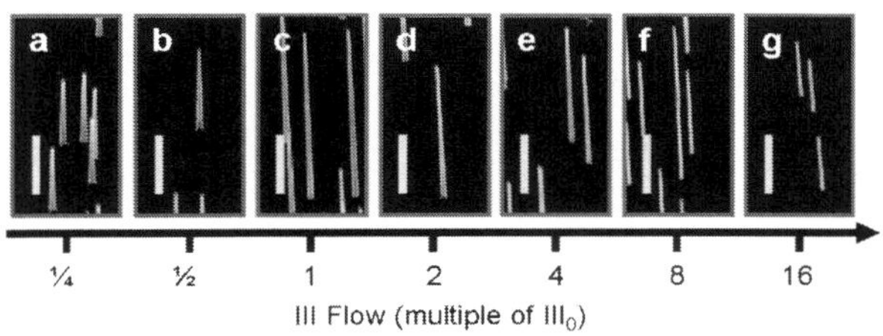

Figure 4.7 FESEM images of nanowires grown at a constant V/III ratio of 46 and the indicated group III flows. Samples are tilted at 40°. Scale bars are 1 μm. Figure reproduced with permission from Joyce, Gao, and Tan et al. (2009).

To quantify the tapering effect, the tapering parameter is plotted versus group III precursor flow rates, as shown in Fig. 4.8.

Similar to the effects of low growth temperature, this reduction in tapering with a higher growth rate can be explained by the different growth mechanisms of axial growth and radial growth.

Consistent with the kinetically limited growth mechanism for radial growth, the radial growth rate increases only marginally with a high group III flow rate. However, the mass-flow-controlled axial growth rate increased almost linearly with the group III flow rate, resulting in the reduction of nanowire tapering.

Furthermore, the crystal quality was examined by TEM. Surprisingly, the nanowire grown at a high growth rate ($III = 4\ III_0$; $V = 4\ V_0$) is completely twin free, as shown in Fig. 4.9, which is a great advantage compared to nanowires grown at slower rates, as shown in Fig. 4.6a,b. This is an unexpected result because increasing precursor flows increases the supersaturation of precursor species in liquid alloys, which may cause twin defects according to theoretical and experimental studies of both planar and nanowire growth (Hurle, 1995; Hurle and Rudolph, 2004; Johansson, Karlsson, Dick et al., 2008). This unexpected result is possibly due to the reduced surface and interfacial tensions resulting from high flow rates of As and Ga precursor species. It was reported that in the melt growth of bulk GaAs material, the addition of excess As is known to lower the surface tension of the melt, which suppresses twinning (Hurle, 1995). We expect that the high flow rate of the As species significantly lowers the surface tension at the Au liquid–vapor interface. On the other hand, a higher Ga concentration in Au is expected to reduce the nanoparticle surface and interfacial tensions since Ga has lower surface and interfacial energies than pure Au (Peng, Meister, Chan, Zhang, and Cui, 2007; Song, Jung, Kim et al., 2008). Therefore, the high flow rates of As and Ga species can reduce surface and interface tensions to prevent formation of twin defects.

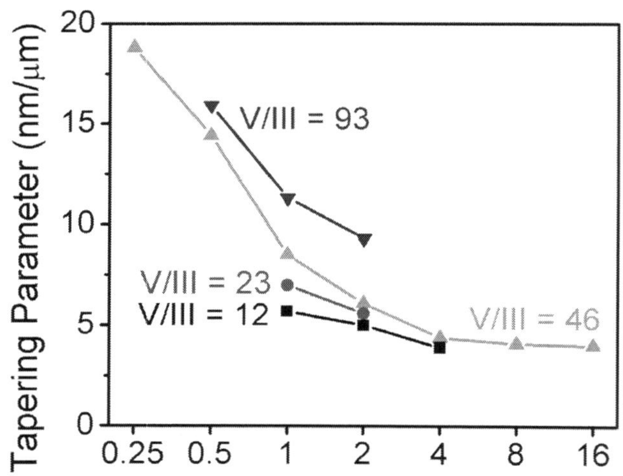

Figure 4.8 The relationship between the tapering parameter and the group III precursor flow rate. Straight lines group data sets for a constant V/III ratio. The abscissa scale is logarithmic. Standard deviations for each data point were less than 3 nm/μm. Figure adapted with permission from Joyce, Gao, and Tan et al. (2009).

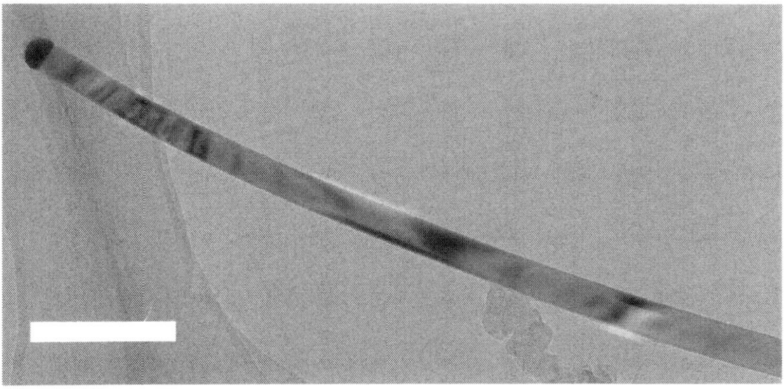

Figure 4.9 Bright-field TEM of a GaAs nanowire grown at a rapid growth rate (III = 4 III_0; V = 4 V_0). There are no lattice defects through this whole nanowire, where the contrast is due to bend contours. The scale bar is 250 nm.

4.5 Summary

The above discussion demonstrates how the growth behavior changed in the GaAs nanowire growth. By using unfavorable growth conditions for conventional 2D growth, such as low growth temperatures and high growth rates, we can significantly improve nanowire properties. This knowledge greatly enhances our understanding of GaAs nanowire growth and is expected to be applied in the synthesis of other group III–V nanowire materials.

Acknowledgments

We wish to thank our collaborators at the Australian National University (Jennifer Wong-Leung, Jung Hyun Kang, Suriati Paiman, and Timothy Burgess) and the University of Queensland (Jin Zou, Mohanchand Paladugu, Xin Zhang, and Yanan Guo). We thank the Australian Research Council for financial support. The Australian National Fabrication Facility, established under the Australian Government's National Collaborative Research Infrastructure Strategy, is gratefully acknowledged for providing access to the facilities used in this research.

References

Barrelet, C. J., Bao, J. M., Loncar, M., Park, H. G., Capasso, F., and Lieber, C. M. (2006). Hybrid single-nanowire photonic crystal and microresonator structures. *Nano Lett.,* **6**, 11–15.

Behet, M., Hovel, R., Kohl, A., Kusters, A. M., Opitz, B., and Heime, K. (1996). MOVPE growth of III-V compounds for optoelectronic and electronic applications. *Microelectron. J.,* **27**, 297–334.

Bjork, M. T., Ohlsson, B. J., Thelander, C., et al. (2002). Nanowire resonant tunneling diodes. *Appl. Phys. Lett.,* **81**, 4458–4460.

Bjork, M. T., Thelander, C., Hansen, A. E., et al. (2004). Few-electron quantum dots in nanowires. *Nano Lett.,* **4**, 1621–1625.

Borgstrom, M. T., Zwiller, V., Muller, E., and Imamoglu, A. (2005). Optically bright quantum dots in single nanowires. *Nano Lett.,* **5**, 1439–1443.

Cantoro, M., Brammertz, G., Richard, O., et al. (2009). Controlled III/V nanowire growth by selective-area vapor-phase epitaxy. *J. Electrochem. Soc.,* **156**, H860–H868.

Chin, A. H., Vaddiraju, S., Maslov, A. V., Ning, C. Z., Sunkara, M. K., and Meyyappan, M. (2006). Near-infrared semiconductor subwavelength-wire lasers. *Appl. Phys. Lett.,* **88**, 63115–63115.

Czaban, J. A., Thompson, D. A., and Lapierre, R. R. (2009). GaAs core-shell nanowires for photovoltaic applications. *Nano Lett.,* **9**, 148–154.

De Jonge, N., Heeres, E. C., Bakkers, E., Roest, A. L., Kaiser, M., and Oosterkamp, T. H. (2007). Electron emission from individual indium arsenide semiconductor nanowires. *Nano Lett.,* **7**, 536–540.

Dick, K. A., Deppert, K., Martensson, T., Mandl, B., Samuelson, L., and Seifert, W. (2005). Failure of the vapor-liquid-solid mechanism in Au-assisted MOVPE growth of InAs nanowires. *Nano Lett.,* **5**, 761–764.

Duan, X. F., and Lieber, C. M. (2000). Laser-assisted catalytic growth of single crystal GaN nanowires. *J. Am. Chem. Soc.,* **122**, 188–189.

Guo, Y. N., Zou, J., Paladugu, M., et al. (2006). Structural characteristics of GaSb/GaAs nanowire heterostructures grown by metal-organic chemical vapor deposition. *Appl. Phys. Lett.,* **89**, 231917.

Heinrich, J., Huggenberger, A., Heindel, T., et al. (2010). Single photon emission from positioned GaAs/AlGaAs photonic nanowires. *Appl. Phys. Lett.,* **96**, 21117.

Hiruma, K., Murakoshi, H., Yazawa, M., and Katsuyama, T. (1996). Self-organized growth of GaAs/InAs heterostructure nanocylinders by organometallic vapor phase epitaxy. *J. Cryst. Growth,* **163**, 226–231.

Hu, L., and Chen, G. (2007). Analysis of optical absorption in silicon nanowire arrays for photovoltaic applications. *Nano Lett.,* **7**, 3249–3252.

Hua, B., Motohisa, J., Kobayashi, Y., Hara, S., and Fukui, T. (2009). Single GaAs/GaAsP coaxial core-shell nanowire lasers. *Nano Lett.,* **9,** 112–116.

Hurle, D. T. J. (1995). A mechanism for twin formation during Czochralski and encapsulated vertical Bridgman growth of III-V compound semiconductors. *J. Cryst. Growth,* **147**, 239–250.

Hurle, D. T. J., and Rudolph, P. (2004). A brief history of defect formation, segregation, faceting, and twinning in melt-grown semiconductors. *J. Cryst. Growth,* **264**, 550–564.

Johansson, J., Karlsson, L. S., Dick, K. A., et al. (2008). Effects of growth conditions on the crystal structure of gold-seeded GaP nanowires. *J. Cryst. Growth,* **310**, 5102–5105.

Johansson, J., Karlsson, L. S., Svensson, C. P. T., et al. (2006). Structural properties of (111)B-oriented III-V nanowires. *Nat. Mater.,* **5**, 574–580.

Joyce, H. J., Docherty, C. J., Gao, Q., et al. (2013). Electronic properties of GaAs, InAs and InP nanowires studied by terahertz spectroscopy. *Nanotechnology*, **24**. 214006.

Joyce, H. J., Gao, Q., Tan, H. H., et al. (2007). Twin-free uniform epitaxial GaAs nanowires grown by a two-temperature process. *Nano Lett.,* **7**, 921–926.

Joyce, H. J., Gao, Q., Tan, H. H., et al. (2008). High purity GaAs nanowires free of planar defects: growth and characterization. *Adv. Funct. Mater.,* **18**, 3794–3800.

Joyce, H. J., Gao, Q., Tan, H. H., et al. (2009). Unexpected benefits of rapid growth rate for III-V nanowires. *Nano Lett.,* **9**, 695–701.

Joyce, H. J., Gao, Q., Wong-Leung, J., et al. (2011). Tailoring GaAs, InAs, and InGaAs nanowires for optoelectronic device applications. *IEEE Journal of Selected Topics in Quantum Electronics*, **17**, 766–778.

Joyce, H. J., Wong-Leung, J., Gao, Q., Tan, H. H., and Jagadish, C. (2010). Phase perfection in zinc blende and wurtzite III-V nanowires using basic growth parameters. *Nano Lett.,* **10**, 908–915.

Kang, J. H., Gao, Q., Joyce, H. J., et al. (2010). Novel growth and properties of GaAs nanowires on Si substrates. *Nanotechnology,* **21**, 35604.

Kim, Y., Joyce, H. J., Gao, Q., et al. (2006). Influence of nanowire density on the shape and optical properties of ternary InGaAs nanowires. *Nano Lett.,* **6**, 599–604.

Kodambaka, S., Tersoff, J., Reuter, M. C., and Ross, F. M. (2007). Germanium nanowire growth below the eutectic temperature. *Science,* **316**, 729–732.

Lewis, N. S. (2007). Toward cost-effective solar energy use. *Science,* **315**, 798–801.

Maharjan, A., Pemasiri, K., Kumar, P., et al. (2009). Room temperature photocurrent spectroscopy of single zincblende and wurtzite InP nanowires: art. no. 193115. *Appl. Phys. Lett.,* **94**, 93115.

Minot, E. D., Kelkensberg, F., Van Kouwen, M., et al. (2007). Single quantum dot nanowire LEDs. *Nano Lett.,* **7**, 367–371.

Mishra, A., Titova, L. V., Hoang, T. B., et al. (2007). Polarization and temperature dependence of photoluminescence from zincblende and wurtzite InP nanowires. *Appl. Phys. Lett.,* **91**(26), 263104-1-3.

Montazeri, M., Fickenscher, M., Smith, L. M., et al. (2010). Direct measure of strain and electronic structure in GaAs/GaP core-shell nanowires. *Nano Lett.,* **10**, 880–886.

Ooike, N., Motohisa, J., and Fukui, T. (2004). Fabrication of GaAs nanowire devices with self-aligning W-gate electrodes using selective-area MOVPE. *Thin Solid Films,* **464**, 220–224.

Paiman, S., Gao, Q., Joyce, H. J., et al. (2010). Growth temperature and V/III ratio effects on the morphology and crystal structure of InP nanowires: art. no. 445402. *J. Phys. D: Appl. Phys.,* **43**, 45402–45402.

Paiman, S., Gao, Q., Tan, H. H., et al. (2009). The effect of V/III ratio and catalyst particle size on the crystal structure and optical properties of InP nanowires. *Nanotechnology,* **20**, 225606.

Paladugu, M., Zou, J., Auchterlonie, G. J., et al. (2007). Evolution of InAs branches in InAs/GaAs nanowire heterostructures. *Appl. Phys. Lett.,* **91**, 33115.

Paladugu, M., Zou, J., Guo, Y. N., et al. (2007). Novel growth phenomena observed in axial InAs/GaAs nanowire heterostructures. *Small,* **3**, 1873–1877.

Paladugu, M., Zou, J., Guo, Y. N., et al. (2008a). Polarity driven formation of InAs/GaAs hierarchical nanowire heterostructures. *Appl. Phys. Lett.,* **93**, 1908.

Paladugu, M., Zou, J., Guo, Y. N., et al. (2008b). Nature of heterointerfaces in GaAs/InAs and InAs/GaAs axial nanowire heterostructures. *Appl. Phys. Lett.,* **93**, 1911.

Paladugu, M., Zou, J., Guo, Y. N., et al. (2009a). Crystallographically driven Au catalyst movement during growth of InAs/GaAs axial nanowire heterostructures. *J. Appl. Phys.,* **105**, 73503.

Paladugu, M., Zou, J., Guo, Y. N., et al. (2009b). Formation of hierarchical InAs nanoring/GaAs nanowire heterostructures. *Angew. Chem., Int. Ed.,* **48**, 780–783.

Park, H. G., Barrelet, C. J., Wu, Y. N., Tian, B. Z., Qian, F., and Lieber, C. M. (2008). A wavelength-selective photonic-crystal waveguide coupled to a nanowire light source. *Nat. Photonics,* **2**, 622–626.

Parkinson, P., Joyce, H. J., Gao, Q., et al. (2009). Carrier lifetime and mobility enhancement in nearly defect-free core-shell nanowires measured using time-resolved terahertz spectroscopy. *Nano Lett.,* **9**, 3349–3353.

Parkinson, P., Lloyd-Hughes, J., Gao, Q., et al. (2007). Transient terahertz conductivity of GaAs nanowires. *Nano Lett.,* **7**, 2162–2165.

Pemasiri, K., Montazeri, M., Gass, R., et al. (2009). Carrier dynamics and quantum confinement in type II ZB-WZ InP nanowire homostructures. *Nano Lett.,* **9**, 648–654.

Peng, H. L., Meister, S., Chan, C. K., Zhang, X. F., and Cui, Y. (2007). Morphology control of layer-structured gallium selenide nanowires. *Nano Lett.,* **7**, 199–203.

Perera, S., Fickenscher, M. A., Jackson, H. E., et al. (2008a). Nearly intrinsic exciton lifetimes in single twin-free GaAs/AlGaAs core-shell nanowire heterostructures. *Appl. Phys. Lett.,* **93**, 53110.

Perera, S., Fickenscher, M. A., Jackson, H. E., et al. (2008b). Nearly intrinsic exciton lifetimes in single twin-free GaAs/AlGaAs core-shell nanowire heterostructures. *Appl. Phys. Lett.,* **93**, 053110-1-3.

Perera, S., Pemasiri, K., Fickenscher, M. A., et al. (2010). Probing valence band structure in wurtzite InP nanowires using excitation spectroscopy. *Appl. Phys. Lett.,* **97**, 23106–23106.

Persson, A. I., Ohlsson, B. J., Jeppesen, S., and Samuelson, L. (2004). Growth mechanisms for GaAs nanowires grown in CBE. *J. Cryst. Growth,* **272**, 167–174.

Pettersson, H., Tragardh, J., Persson, A. I., Landin, L., Hessman, D., and Samuelson, L. (2006). Infrared photodetectors in heterostructure nanowires. *Nano Lett.,* **6**, 229–232.

Plante, M. C., and LaPierre, R. R. (2006). Growth mechanisms of GaAs nanowires by gas source molecular beam epitaxy. *J. Cryst. Growth,* **286**, 394–399.

Plissard, S., Larrieu, G., Wallart, X., and Caroff, P. (2011). High yield of self-catalyzed GaAs nanowire arrays grown on silicon via gallium droplet positioning. *Nanotechnology,* **22**, 75602–75602.

Qian, F., Gradecak, S., Li, Y., Wen, C. Y., and Lieber, C. M. (2005). Core/multishell nanowire heterostructures as multicolor, high-efficiency light-emitting diodes. *Nano Lett.,* **5**, 2287–2291.

Reece, P. J., Paiman, S., Abdul-Nabi, O., et al. (2009). Combined optical trapping and microphotoluminescence of single InP nanowires. *Appl. Phys. Lett.,* **95**, 1109.

Reece, P. J., Toe, W. J., Wang, F., et al. (2011). Characterization of semiconductor nanowires using optical tweezers. *Nano Lett.,* **11**, 2375–2381.

Smith, L. M., Jackson, H. E., Yarrison-Rice, J. M., and Jagadish, C. (2010). Insights into single semiconductor nanowire heterostructures using time-resolved photoluminescence. *Semicond. Sci. Technol.,* **25**, 024010.

Song, M. S., Jung, J. H., Kim, Y., et al. (2008). Vertically standing Ge nanowires on GaAs(110) substrates. *Nanotechnology,* **19**, 25602.

Stringfellow, G. B. (1999). *Organometallic Vapor-Phase Epitaxy: Theory and Practice.* San Diego, CA: Academic Press.

Tchernycheva, M., Harmand, J. C., Patriarche, G., Travers, L., and Cirlin, G. E. (2006). Temperature conditions for GaAs nanowire formation by Au-assisted molecular beam epitaxy. *Nanotechnology,* **17**, 4025–4030.

Thelander, C., Martensson, T., Bjork, M. T., et al. (2003). Single-electron transistors in heterostructure nanowires. *Appl. Phys. Lett.,* **83**, 2052–2054.

Thelander, C., Nilsson, H. A., Jensen, L. E., and Samuelson, L. (2005). Nanowire single-electron memory. *Nano Lett.,* **5**, 635–638.

Tian, B., Kempa, T. J., and Lieber, C. M. (2009). Single nanowire photovoltaics. *Chem. Soc. Rev.,* **38**, 16–24.

Titova, L. V., Hoang, T. B., Jackson, H. E., et al. (2006). Temperature dependence of photoluminescence from single core-shell GaAs-AlGaAs nanowires. *Appl. Phys. Lett.,* **89**, 173126.

Verheijen, M. A., Immink, G., De Smet, T., Borgstrom, M. T., and Bakkers, E. (2006). Growth kinetics of heterostructured GaP-GaAs nanowires. *J. Am. Chem. Soc.,* **128**, 1353–1359.

Wagner, R. S., and Ellis, W. C. (1964). Vapor-liquid-solid mechanism of single crystal growth. *Appl. Phys. Lett.,* **4**, 89–90.

Wang, J. F., Gudiksen, M. S., Duan, X. F., Cui, Y., and Lieber, C. M. (2001). Highly polarized photoluminescence and photodetection from single indium phosphide nanowires. *Science,* **293**, 1455–1457.

Wang, Y. B., Joyce, H. J., Gao, Q., et al. (2011). Self-healing of fractured GaAs nanowires. *Nano Lett.,* **11**, 1546–1549.

Wang, Y. B., Wang, L. F., Joyce, H. J., et al. (2011). Super deformability and Young's modulus of GaAs nanowires. *Adv. Mater.,* **23**, 1356–1360.

Zhang, X., Zou, J., Paladugu, M., et al. (2009). Evolution of epitaxial InAs nanowires on GaAs (111)B. *Small,* **5**, 366–369.

Zou, J., Paladugu, M., Wang, H., et al. (2007). Growth mechanism of truncated triangular III-V nanowires. *Small,* **3**, 389–393.

Chapter 5

The Synthesis, Structure, and Properties of Titania-Coated Silica Nanowires

Avi Shalav and Robert G. Elliman

Department of Electronic Materials Engineering, Research School of Physics and Engineering, Australian National University, Building 58, John Carver Building, Mills Road, Canberra, ACT 0200, Australia

avi.shalav@anu.edu.au

The development of nanostructured materials offers unique opportunities to address many technological problems. Of particular interest are functional surfaces that can be utilized using renewable energy sources, such as sunlight, for a range of sustainable applications. Anatase titania has very good electrical and photocatalytic properties, but while nanostructures of this phase offer many advantages, they are particularly difficult to fabricate. This chapter presents an entirely novel approach to this problem, namely, the coating of substoichiometric silica glass nanowires with thin anatase titania films. We compare films using different coating techniques and demonstrate that these methods can uniformly coat silica nanowires with anatase titania to produce large-surface-area nanostructures.

Nanotechnology Toward the Sustainocene

Edited by Thomas A. Faunce

ISBN 978-981-4463-76-8 (Hardcover), 978-981-4463-77-5 (eBook)

www.panstanford.com

5.1 Introduction

5.1.1 Applications of TiO_2

Titania (TiO_2) is one of the most investigated single-crystalline systems in surface science and metal oxide literature (Diebold, 2003). TiO_2 is a nontoxic semiconductor that has a bandgap ideal for energy transfer mechanisms with other materials and can also be directly excited with ultraviolet radiation, making it an ideal material for photogenerated catalysis. TiO_2 has therefore found many applications, including artificial photosynthesis (Yang, Yu, van der Linden, Wu, and Mul, 2010), electrochemical dye-sensitized solar cells (DSSCs) (Oregan and Gratzel, 1991), self-cleaning surfaces (Linsebigler, Lu, and Yates, 1995), conversion of CO_2 to gaseous hydrocarbons (Tan, Zou, and Hu, 2006), H_2 production via water splitting (Kudo and Miseki, 2009), disinfection (McCullagh, Robertson, Bahnemann, and Robertson, 2007) via decomposition of bacteria and viruses and enhanced drug delivery (Li, Wang, Lu et al., 2009), degradation and elimination of offensive odors (Anpo, Dohshi, Kitano, Hu, Takeuchi, and Matsuoka, 2005), and oxidation of contaminants (Kostedt, Drwiega, Mazyck et al., 2005). As a result, TiO_2-based materials have the potential to directly address a number of United Nations' millennium development goals, particularly in applications involving environmental sustainability. For example, TiO_2 is a key constituent of modern organic cells and DSSCs, providing an efficient n-type material for photoexcited electrons. TiO_2 is also a well-known photocatalyst, able to split water and produce reactive oxygen species in the presence of sunlight and water, providing unique surfaces for water remediation and self-cleaning products.

5.1.1.1 Solar cells

Crystalline Si photovoltaic (PV) cells exhibit some of the highest power conversion efficiencies but still have relatively high production costs. Solar cells based on semiconducting hybrid inorganic/organic architectures have become promising alternatives since substituted conjugated polymer materials generally have higher absorption coefficients, are easily processed, and are flexible. Electrochemical DSSCs sensitized by ruthenium dyes still have the highest efficiencies among organic molecular devices yet suffer from encapsulation issues since liquid electrolytes are often used. Current research studies on DSSCs and hybrid devices are merging, with the replacement of the liquid electrolytes by solid-state polymer hole conductors in DSSCs and the use of inorganic nanocrystals in polymer-hybrid devices. Unfortunately, many nanocrystals investigated over recent years use toxic precursors. TiO_2 nanoparticles, although considered to be nontoxic and offering physical and chemical stability, are polar and therefore incompatible with organic solvents. The current DSSC technology is based on the use of TiO_2 nanoparticles that are used to separate the charge carriers from the photoexcited dye. Studies have shown that nanowires ensure the rapid collection of carriers for DSSC applications (Law, Greene, Johnson, Saykally, and Yang, 2005) and offer much better performance than TiO_2 nanoparticles (Suzuki, Ngamsinlapasathian, Yoshida, and Yoshikawa, 2006).

5.1.1.2 Photolysis (water splitting)

Since the early 1970s, hydrogen (H_2) has been promoted as a potential clean fuel for motive transport and is now utilized within commercially available fuel cells to generate electricity. Unfortunately, H_2 does not occur terrestrially in vast quantities and is often generated using energy-intensive processes. TiO_2 has been shown to decompose water into its elements—H_2 and O_2 gas—catalyzing this process in the presence of ultraviolet light (Fujishima and Honda, 1972). Intense research over the last decade has shown that chemically modifying TiO_2 (Khan, Al-Shahry, and Ingler, 2002) and/or adding additional dopants (Li, Peng, Jiang, Lu, and Li, 2007; Moshfegh, 2009), one can obtain an improved photoresponse and electrode performance, even upon visible illumination. Films of sufficient thickness are typically required so that a large proportion of the incident solar radiation can be absorbed. Unfortunately, the charge carriers generated have a very short diffusion length and recombine before they can partake in photolysis at the TiO_2/H_2O interface. TiO_2-based nanowire materials have a high optical density and a high surface-to-volume ratio, potentially minimizing the efficiency-limiting effects due to large optical path lengths required by thin-film-type devices (Liu, Snapp, and Park, 2011).

5.1.1.3 Production of reactive oxygen species

One of the most important photocatalytic reactions using TiO_2 involves the creation of reactive oxygen species, such as hydroxyl radicals ($OH^{\bullet}$) and superoxide anions ($O_2^{-\bullet}$), in the presence of O_2 and water. Ultraviolet light incident on a TiO_2 nanoparticle can excite an electron (e^-) from the valence band (VB) to the conduction band (CB), leaving behind a hole (h^+) (Brunet, Lyon, Hotze, Alvarez, and Wiesner, 2009), as illustrated in Fig. 5.1. These photoexcited electrons can reduce molecular O_2 to $O_2^{-\bullet}$, while the holes can oxidize hydroxyl ions (OH^-) or H_2O to $OH^{\bullet}$. Furthermore, additional reactive oxygen species, including singlet oxygen ($^1O_2$) and hydrogen peroxide (H_2O_2), can be produced via the superoxide. Reactive oxygen species are well known to damage molecules and cell structures, processes that enable applications such as self-cleaning surfaces (Linsebigler, Lu, and Yates, 1995), disinfection (McCullagh, Robertson, Bahnemann, and Robertson, 2007), degradation and elimination of offensive odors (Anpo, Dohshi, Kitano, Hu, Takeuchi, and Matsuoka, 2005), and oxidation of contaminants (Kostedt, Drwiega, Mazyck et al., 2005).

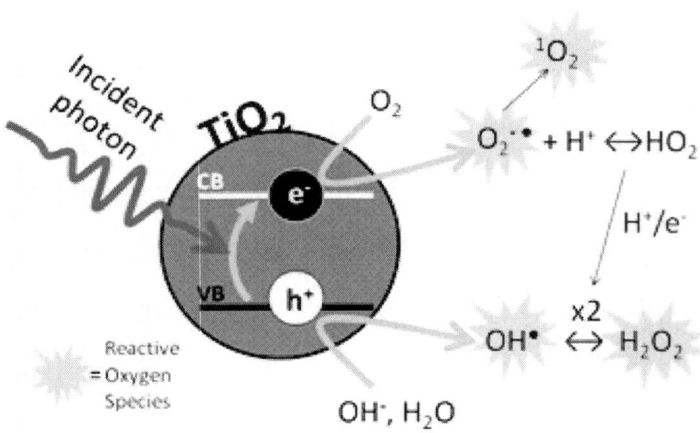

Figure 5.1 Mechanisms for reactive oxygen species production from nanosized TiO_2. Figure adapted from Brunet, Lyon, Hotze, Alvarez, and Wiesner (2009).

5.1.2 Advantages and Limitations of Nanostructured TiO_2

Nanostructured TiO_2 surfaces provide much larger photoreactive surface areas compared to simple thin films, enhancing photocatalytic energy transfer mechanisms, conversion efficiencies, and charge collection. The most efficient photocatalytic devices generally incorporate porous, nanostructured matrices of anatase TiO_2 nanoparticles. The lower-temperature (typically <600°C) anatase phase of TiO_2 has higher electron mobilities and photocatalytic activities, while also providing a more open structure than the higher-temperature, more thermodynamically stable, rutile crystal configuration. Nanocrystalline anatase TiO_2 also offers preferred gas-sensing characteristics, including faster response times, lower resistance, and higher sensitivity to surface absorbents when compared to traditional ZrO_2- and rutile TiO_2–phase gas-sensing-based devices (Galatsis, Li, Wlodarski et al., 2002). Such devices include humidity, oxygen, and organic vapor sensors, which can be further enhanced using other metallic dopants (Comini, Faglia, Sberveglieri, Li, Wlodarski, and Ghantasala, 2000).

Titania (TiO_2) nanowires offer many advantages over TiO_2 nanoparticles, including the elimination of filtration processes for nanoparticle recovery from sensing and water remediation applications (Wilhelm and Stephan, 2007), and also provide electrical pathway enhancement for DSSCs (Law, Greene, Johnson, Saykally, and Yang, 2005; Suzuki, Ngamsinlapasathian, Yoshida, and Yoshikawa, 2006). TiO_2-based nanowires could also be used as an alternative to the carbon-based anodes of Li-ion batteries since TiO_2 nanostructures are well known to have increased reaction areas, shortened Li^+ diffusion, and enhanced solubility/capacity (Centi and Perathoner, 2009; Ortiz, Hanzu, Djenizian, Lavela, Tirado, and Knauth, 2009; Yang, Choi, Kerisit et al., 2009). Unfortunately, TiO_2 nanowires are particularly difficult to manufacture, and common methods, for example, the vapor-liquid-solid (VLS) process, typically require high temperatures for the vapor phase, which yields the less reactive rutile phase. Although anatase TiO_2 nanowires have been prepared using electrochemical or hydrothermal approaches (Suzuki, Pavasupree, Yoshikawa, and Kawahata, 2007), layer-by-layer assembly, template synthesis, sol-gel methods, and hydrolysis/thermolysis methods, to name a few, these methods require suitable templates or special equipment and technologies that suffer from low yields and/or high costs (Wang, Yu, Zhou, Wang, and Liu, 2009). The resulting nanostructures are also mechanically fragile.

5.1.3 The Growth of SiO_x Nanowires via Active Oxidation

Substoichiometric silica (SiO_x) nanowires are typically grown via the VLS growth mechanism using a variety of often time-consuming and expensive techniques (Rao, Deepak, Gundiah, and Govindaraj, 2003). A simple alternative method involves the synergy between Au-mediated Si substrate decomposition via active oxidation and nanowire nucleation. Under active oxidation conditions, typically at higher temperatures and low partial pressures of oxygen (Gelain, Cassuto, and Legoff, 1971; Lander and Morrison, 1962), volatile SiO is produced via the reaction $2Si + O_2 \rightarrow 2SiO$, directly from the substrate surface where the nanowires are to be grown. Typically, thin (~10 nm) Au films are thermally deposited on a Si wafer, with their native oxide intact, and annealed at higher temperatures in an Ar or N_2 ambient, with a residue O_2 partial pressure estimated to be 1–10 parts per million (ppm). Some of the Au decomposes the native oxide and alloys with the underlying Si substrate, enhancing the active oxidation processes. The remaining Au on the native oxide forms small islands, providing nucleation sites for nanowire growth. This synergetic process is illustrated in Fig. 5.2. Dense nanowire films, or "turfs," can be readily grown without the need for an expensive or dangerous external precursor, such as silane. Nanowires grown using this method typically have diameters from 10 nm to 500 nm, with typical lengths of tens of micrometers (refer to Fig. 5.2e). These insulating nanowire substrates may provide an ideal support surface, or "backbone," for a range of semiconducting materials.

5.1.4 TiO_2–SiO_x Hybrid Materials

Solution-based TiO_2 can be easily deposited on large-surface-area glass substrates. Bulk silica is thermally and chemically stable and does not decompose due to the strong photocatalytic action of TiO_2 or from the reactive oxygen species created. The surface properties of the silica are particularly important for the adhesion of dispersed titanium alkoxide coatings since surface nonisolated hydroxyl groups are required and vary depending on the silica substrate (Srinivasan, Datye, Smith, and Peden, 1994). High-silica surface areas (e.g., fumed silica) have been proven to be favorable (Srinivasan, Datye, Smith, and Peden, 1994), further emphasizing the potential of SiO_x nanowires as a support for photocatalytic TiO_2.

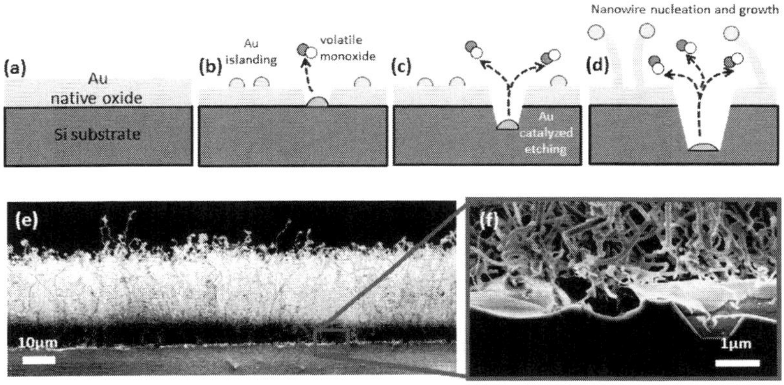

Figure 5.2 SiO_x nanowires grown via Si substrate active oxidation. (a) An Au film is evaporated on to the native oxide of a Si substrate. (b) At a higher *T*, Au particles make islands on the oxide surface and also decompose it in places, resulting in Au-alloyed particles in contact with the substrate. These alloyed particles enhance the active oxidation process. (c) Active oxidation of the wafer produces volatile monoxide species, forming etch pits in the substrate. (d) Au particles on the native oxide surface nucleate nanowires via the VLS mechanism. (e) A typical nanowire turf cross section. (f) A close-up of the nanowire substrate interface showing nanowire growth and etching.

Si–O–Ti chemical bonding at the interface is likely to reduce particle (Erjavec, Dominko, Umek et al., 2008) and crystalline growth, retarding the transformation to the rutile phase (Lee, Othman, Eom, Lee, Kim, and Kim, 2008), and also result in higher concentrations of TiO_2-active centers (Murashkevich, Lavitskaya, Alisienok, and Zharskii, 2009). These properties imply that silica is an ideal supporting substrate or binder/buffer layer (e.g., to protect cellulose fibers for textile applications) (Veronovski, Sfiligoj-Smole, and Viota, 2010) for catalytic TiO_2-based materials and devices. Although many studies investigating photocatalytic TiO_2–SiO_2 mesoporous films (Wang, Wang, Li, and Lu, 2009) and nanoparticles (Murashkevich, Lavitskaya, Alisienok, and Zharskii, 2009) have appeared over the last decade, only a few recent papers exist investigating rather complex TiO_2–SiO_2 nanowire-type structures (Erjavec, Dominko, Umek et al., 2008).

5.1.5 Synthesis and Structural Properties of Silica–Titania Core-Shell Nanowires

Primary SiO_x nanowires were grown via the VLS mechanism where the vapor precursor, SiO, is obtained directly from the Si substrate via active oxidation in a N_2 ambient (1200 sccm) at 1100°C. Polished p-type (10–20 Ω·cm) Si (100) samples, approximately 1×1 cm, were coated with 10 nm of Au via thermal evaporation or radio-frequency (RF) sputtering. These samples were then annealed at 1100°C in a high-purity N_2 ambient (1200 sccm) for different times, resulting in dense SiO_x nanowire networks with different average nanowire diameters, ranging from 40 nm to 400 nm. Extended annealing times resulted in nanowires with larger average diameters due to lateral growth caused by the ongoing adsorption of the SiO vapor. Rapid growth of nanowires occurs within the first few minutes of growth, with diameters typically in the 20–50 nm range, similar to the diameters of the catalyzing Au particles on their tips. Prolonged annealing results in lateral growth as SiO vapor is adsorbed onto the already grown nanowires, and nanowire samples with average diameters from 40 nm to 400 nm were readily obtained for shorter and longer anneal times, respectively. Further details of the nanowire growth process can be found elsewhere (Kim, Shalav, and Elliman, 2010; Shalav, Kim, and Elliman, 2010). These nanowire networks with average nanowire diameters of 40 nm and 400 nm were then coated with titania using either a simple droplet-coating method using a Ti-alkoxide solution or atomic layer deposition (ALD) using a gaseous Ti-isopropoxide precursor. Postdeposition annealing in O_2 or N_2 at low (500°C–600°C) and high (800°C–1100°C) temperatures was then performed to test the thermal stability and structural integrity of the resulting coatings. The crystal structure of the deposited and annealed TiO_2 coatings was investigated via X-ray diffraction (XRD). The surface morphology of the as-deposited and annealed nanowire networks was compared using a Zeiss UltraPlus analytical field-emission scanning electron microscope (SEM).

5.1.6 Atomic Layer Deposition

Primary SiO_x nanowires were coated with 20 nm of TiO_2 using a Cambridge NanoTech Savannah ALD system. During deposition, the substrate temperature was kept at 200°C and the 20 nm TiO_2 layer was built up by consecutive pulse cycles of a gaseous Ti-isopropoxide precursor and water vapor. Stable, conformal coatings were readily applied to nanowires with diameters greater than about 140 nm in diameter, but SiO_x nanowires with diameters less than about 90 nm were found to decompose during deposition (Shalav, Venkatachalam, Reichardt, Fischer, and Elliman, 2009). Such etching of the primary SiO_x nanowires is unexpected as sputtered TiO_2 layers on thin films of SiO_x (80 nm) are particularly stable without noticeable mixing or decomposition of the SiO_x film (Seong, Kim, Kim, Lee, and Hahn, 2009). Although decomposition of solid SiO and SiO_2 can occur in the presence of elemental metals (e.g., via the reaction $SiO_2 + Ti \rightarrow TiO_2 + Si$), these reactions only occur at higher temperatures (Biehl, Schubert, and Kubel, 2001; Vitanov, Stefanov, Harizanova, and Ivanova, 2008), much greater than the ALD temperature of 200°C. It is known, however, that organic halides can react with solid SiO in the presence of metal catalysts to produce silane at temperatures in the range of 200°C–1200°C (Bokerman, Cannady, and Kuivila, 1991). The presence of halide contaminants (Cl is a known contaminant present at concentrations <200 ppm) in the ALD source gas may therefore facilitate such a reaction to effectively etch oxygen-rich nanowires and would preferentially etch thinner nanowires if their composition was more silicon rich (e.g., closer to amorphous SiO) than the thicker wires.

The behavior of the thinner nanowires (<90 nm) discussed above is in contrast to that of thicker (>140 nm) nanowires in which conformal TiO_2 coatings are readily produced via ALD, as shown in the inset of Fig. 5.3d. In this case, XRD measurements show that the TiO_2 coatings are deposited as a combination of amorphous and anatase phases and that the amorphous phase crystallizes to the anatase phase during postdeposition annealing at 600°C in oxygen, as shown in Fig. 5.3.

To address the decomposition of SiO_x nanowires during ALD an alternative strategy was investigated in which a 25–50 nm buffer layer of HfO_2 was first deposited on the as-grown SiO_x nanowires by ALD prior to the deposition of TiO_2. The gaseous Hf precursor, Hf dimethylamide, has a trace Cl concentration less than 10 ppm, up to 20 times less than that present for the Ti precursor, and nanowire decomposition is therefore less likely. The resulting ALD HfO_2 films also have a higher average atomic number than SiO_x or TiO_2, allowing for easy film constituent identification using electron backscattering (EBS) analysis. Such coatings proved to be effective in eliminating nanowire decomposition, with no apparent decomposition observed for HfO_2-coated nanowires of various diameters during subsequent ALD coating with a 10–20 nm layer of TiO_2. Figure 5.4 shows typical cross-sectional SEM and corresponding EBS micrographs of the resulting SiO_x–HfO_2–TiO_2 core-shell-shell structure for nanowire diameters greater than 150 nm, approximately 100 nm and less than 50 nm, respectively.

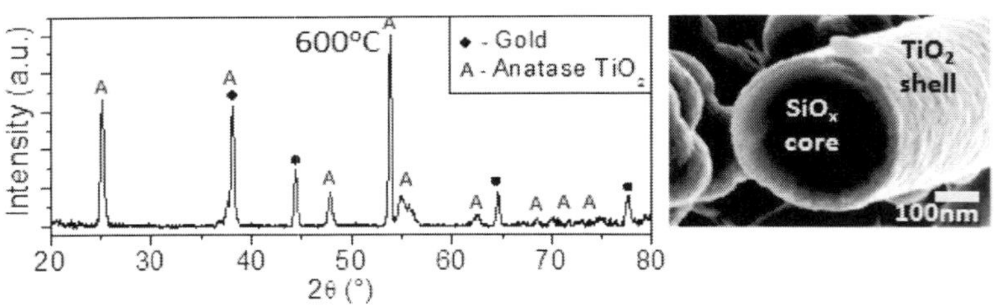

Figure 5.3 XRD spectra of ALD-coated SiO_x nanowires (diameters > 140 nm) and a typical nanowire SEM cross section (on the right) after annealing at 600°C in O_2.

5.1.7 Droplet Coatings

Droplet coating was achieved by placing a droplet of a Ti-alkoxide precursor dissolved in isopropanol (mixing ratio 1:12) directly onto the SiO_x nanowire films. The nanowire films have very good wicking properties, and a single droplet quickly disperses throughout the nanowire network. The effect of additional droplets of the Ti-alkoxide on the resulting coating thickness was also investigated. However, preliminary studies showed that the application of multiple droplets of Ti-alkoxide caused the nanowire networks to crack, or "tear," due to shrinkage during solvent evaporation (Shalav, Venkatachalam, Reichardt, Fischer, and Elliman, 2009). To minimize these effects, a short, 15-minute, low-temperature anneal at 500°C in O_2 was performed after each droplet deposition to evaporate the organic solvents present and to crystallize the TiO_2 on the SiO_x nanowires. The TiO_2 coating obtained from a single Ti-alkoxide droplet could not be resolved via cross-sectional SEM analysis, suggesting that the deposited TiO_2 shell thickness was less than 10 nm. However, measurements suggest that additional droplets effectively add 8 ± 2 nm to the radius of the shell thickness, with five consecutive droplets resulting in a coating thickness of 40–50 nm. For primary SiO_x nanowires with diameters close to or less than 100 nm, such coatings cause the nanowires to deform and stick together. Such effects were not observed for SiO_x nanowires with diameters to the order of a few hundred nanometers. This suggests that the thinner nanowires are deformed by capillary forces produced by the Ti-alkoxide solution and fused together by the deposited TiO_2.

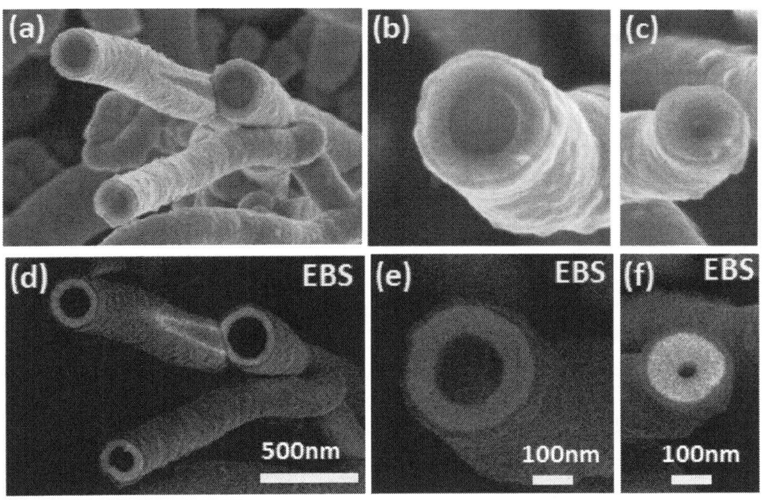

Figure 5.4 Typical cross-sectional SEM and corresponding EBS micrographs of deliberately broken TiO_2-HfO_2- SiO_x core-shell-shell structures for nanowire diameters (a, d) greater than 150 nm, (b, e) approximately 100 nm, and (c, f) less than 50 nm. The brighter rings observed in the EBS images are the HfO_2 layers deposited via ALD.

XRD analysis of the films shows that they contain mixed phases of TiO_2, depending on the postdeposition annealing temperature, with anatase being the dominant phase for samples annealed at 500°C and rutile being the dominant phase for samples annealed at 900°C. A cross-sectional SEM comparison of thicker (>100nm) nanowires confirm that the nanowires are conformally coated with uniform films of TiO_2. Futher comparisons also suggests that the anatase-coated samples are smoother than the rutile-coated samples, an effect that is attributed to faceting of larger rutile crystallites. This process is further investigated in the next section.

5.1.8 Effect of High-Temperature Annealing

The thermal stability of the SiO_x–HfO_2–TiO_2 heterostructure nanowires was investigated by additional annealing at temperatures greater than 1000°C in N_2 (again at 1200 sccm), and their associated XRD and SEM micrographs were compared with those obtained at lower temperatures. The rutile phase of TiO_2 is expected to dominate at these higher temperatures, and the average grain size of rutile crystallites is known to increase with an increasing annealing temperature (Kim, Kim, Jung, and Hong, 2005; Rath, Anand, Mohapatra et al., 2009). Figure 5.5 shows typical SEM micrographs and corresponding XRD spectra after an initial 500°C and a post-1050°C anneal. The XRD spectra are dominated by monoclinic HfO_2 for both samples, consistent with previous studies of HfO_2 (Ohtaka, Fukui, Funakoshi, Utsumi, Irifune, and Kikegawa, 2002). The high-temperature anneal also resulted in a highly facetted rutile TiO_2 outer layer surrounding the monoclinic HfO_2 buffer layer, which, in turn, coats an amorphous SiO_x core. No hafnium titanium oxide or silicide XRD peaks are observed, suggesting that the crystallized coatings are stable even at high processing temperatures.

Interestingly, annealing the samples close to or above 1100°C reproduces the Si active oxidation conditions used to initially grow the SiO_x nanowires. In this case the SiO vapor produced from the substrate uniformly coats the already coated nanowires with an additional SiO_x sheath demonstrating that SiO_x can be uniformly deposited on precoated nanowire samples by simple thermal annealing, allowing the development of more complex multishelled nanowire configurations.

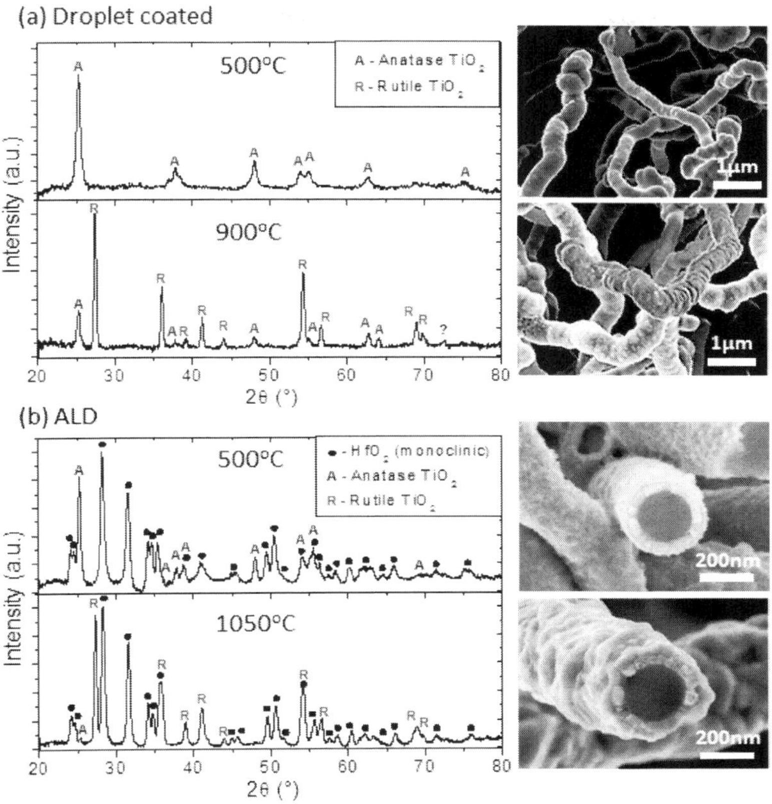

Figure 5.5 XRD spectra of (a) droplet- and (b) ALD-coated SiO_x nanowires and their corresponding SEM images (on the right) after annealing at low (500°C) and high (900°C–1050°C) temperatures.

5.2 Mechanical Properties of Silica–Titania Core-Shell Nanowires

Nanoindentation has been recently employed to ascertain the mechanical properties of the TiO_2-coated SiO_x nanowires (Anderson, Shalav, Ruffell, and Elliman, 2011). Firstly, as-grown nanowire layers with thicknesses approximately 30 μm in height were grown, as previously described, with average diameters between 400 nm and 500 nm. These samples were then coated with thin TiO_2

(approximately 10 nm) via simple Ti-alkoxide droplet deposition. Anatase and rutile TiO_2 coatings were subsequently obtained after annealing for one hour in O_2 at 500°C and 900°C, respectively. An ultra-micro-indentation system (UMIS) nanoindentor with a large, spherical ruby indentation tip (with a radius of 1 mm) was used to obtain characteristic load displacement curves loads of 3, 5, 7, 10, 20, and 50 mN. From these curves and utilization of the Oliver and Pharr method, the "global" effective elastic modulus and the hardness of the as-grown and coated nanowire layers were calculated. Unfortunately, a number of inherent difficulties exist when applying nanoindentation techniques to nanowire layers since the nanowire layers do not behave like traditional bulk materials or thin films for which the method had been developed. Under higher loads, it is likely that both brittle fracture and subsequent plastic deformation of the nanowire layer occur, as shown in Fig. 5.6.

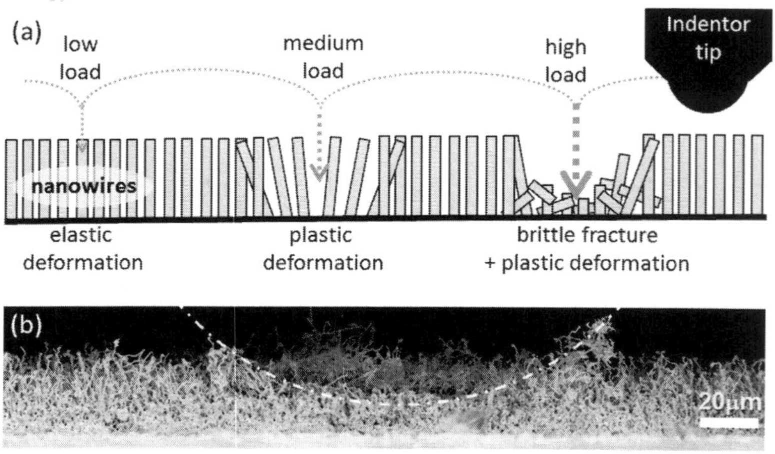

Figure 5.6 (a) The effect of increasing loads on the nanowire layers. High loads result in brittle fracture and plastic deformation of the nanowire surface. (b) An SEM micrograph showing brittle fracture and plastic deformation after a high-load nanoindentation. The dashed line shows the arc of a circle with a radius of 1 mm, corresponding to the tip dimensions.

A comparison of SEM micrograph images of the indent sites after nanoindentation at different loads allows for the direct observation of the onset of brittle fracture of the nanowires. At lower loads, typically <10 mN for the anatase-coated samples and <7 mN for the rutile-coated samples, no brittle fracture is observed, suggesting that only elastic and plastic deformation occurs at these values. The anatase-coated nanowire layer, therefore, appears more robust than the rutile-coated samples since brittle fracture does not occur until slightly higher load values. Figure 5.7 shows the calculated hardness of the nanowire layers as a function of indentation depth. The hardness increases for higher loads and corresponding depths, most likely due to compaction and densification of the nanowire layer. At a given lower load, the rutile measurements show a slightly higher hardness value at lower depths, Anatase thin films are known to have lower hardness and are therefore more ductile, properties that the coated silica nanowires also seem to exhibit.

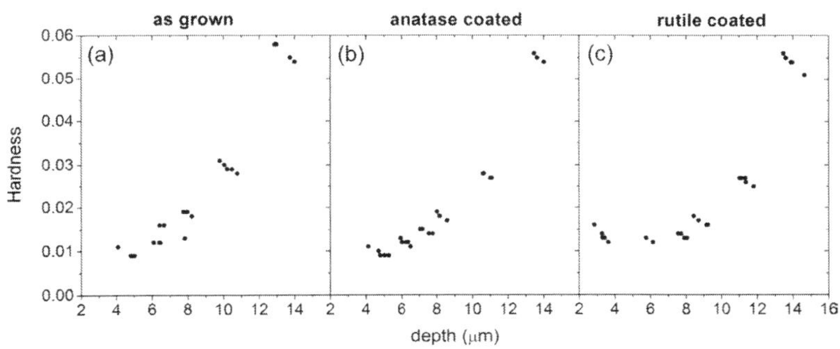

Figure 5.7 Calculated hardness values as a function of the indentation depth for (a) as-grown, (b) anatase-coated, and (c) rutile-coated SiO_x nanowire layers. At higher loads (and higher depths), the hardness value increases, most likely due to compaction and densification of the nanowire layer.

5.3 Conclusion

It has been shown that SiO_x nanowire layers, grown via the active oxidation of a Si substrate, can be conformally coated with TiO_2 thin films. Both solution coating with a Ti-alkoxide or ALD methods have been shown to produce large-surface-area, mechanically stable structures that could potentially be utilized for important sunlight-mediated applications, including solar cells and water remediation. As-deposited films consist of amorphous and anatase phases and were shown to be selectively converted to either the anatase or the rutile phase via low-temperature (500°C–600°C) or high-temperature (800°C–900°C) annealing, respectively, with higher-temperature (1050°C) annealing producing a facetted film morphology due to an increase in the average grain size of the rutile film.

Acknowledgments

We would like to thank Dr. Dinesh K. Venkatachalam for ALD sample preparation and Mr. Peter Anderson and Dr. Simon Ruffell for the nanoindentation analysis. The Australian Research Council is gratefully acknowledged for financial support. The Australian National University nodes of the ANFF and the AMMRF, both established under the Australian National Cooperative Research Infrastructure Strategy, are acknowledged for access to the facilities used in this work.

References

Anpo, M., Dohshi, S., Kitano, M., Hu, Y., Takeuchi, M., and Matsuoka, M. (2005). The preparation and characterization of highly efficient titanium oxide-based photofunctional materials. *Annu. Rev. Mater. Res.*, **35**, 1–27.

Biehl, E., Schubert, U., and Kubel, F. (2001). Reduction of solid silicon monoxide by elemental metals. *New J. Chem.*, **25**(8), 994–998.

Bokerman, G. N., Cannady, J. P., and Kuivila, C. S. (inventors) (1990). Silane products from reaction of solid silicon monoxide with organic halides. US Patent 5,051,247, filed Sep. 4, 1990, and issued Sep. 24, 1991.

Brunet, L., Lyon, D. Y., Hotze, E. M., Alvarez, P. J. J., and Wiesner, M. R. (2009). Comparative photoactivity and antibacterial properties of C_{60} fullerenes and titanium dioxide nanoparticles. *Environ. Sci. Technol.*, **43**(12), 4355–4360.

Centi, G., and Perathoner, S. (2009). The role of nanostructure in improving the performance of electrodes for energy storage and conversion. *Eur. J. Inorg. Chem.*, **26**, 3851–3878.

Comini, E., Faglia, G., Sberveglieri, G., Li, Y. X., Wlodarski, W., and Ghantasala, M. K. (2000). Sensitivity enhancement towards ethanol and methanol of TiO_2 films doped with Pt and Nb. *Sens. Actuators, B,* **64**(1–3), 169–174.

Diebold, U. (2003). The surface science of titanium dioxide. *Surf. Sci. Rep.,* **48**(5–8), 53–229.

Erjavec, B., Dominko, R., Umek, P., et al. (2008). RuO_2-wired high-rate nanoparticulate TiO_2 (anatase): suppression of particle growth using silica. *Electrochem. Commun.,* **10**(6), 926–929.

Fujishima, A., and Honda, K. (1972). Electrochemical photolysis of water at a semiconductor electrode. *Nature,* **238**(5358), 37–38.

Galatsis, K., Li, Y. X., Wlodarski, W., et al. (2002). Comparison of single and binary oxide MoO_3, TiO_2 and WO_3 sol-gel gas sensors. *Sens. Actuators, B,* **83**(1–3), 276–280.

Gelain, C., Cassuto, A., and Legoff, P. (1971). Kinetics and mechanism of low-pressure, high-temperature oxidation of silicon II. *Oxid. Met.,* **3**(2), 139–151.

Khan, S. U. M., Al-Shahry, M., and Ingler, W. B. (2002). Efficient photochemical water splitting by a chemically modified n-TiO_2. *Science,* **297**(5590), 2243–2245.

Kim, J. Y., Kim, D. W., Jung, H. S., and Hong, K. S. (2005). Influence of anatase-rutile phase transformation on dielectric properties of sol-gel derived TiO_2 thin films. *Jpn. J. Appl. Phys.,* **44**(8), 6148–6151.

Kim, T.-H., Shalav, A., and Elliman, R. G. (2010). Active-oxidation of Si as the source of vapor-phase reactants in the growth of SiO_x nanowires on Si. *J. Appl. Phys.,* **108**(7), 076102.

Kostedt, W. L., Drwiega, J., Mazyck, D. W., et al. (2005). Magnetically agitated photocatalytic reactor for photocatalytic oxidation of aqueous phase organic pollutants. *Environ. Sci. Technol.,* **39**(20), 8052–8056.

Kudo, A., and Miseki, Y. (2009). Heterogeneous photocatalyst materials for water splitting. *Chem. Soc. Rev.,* **38**(1), 253–278.

Lander, J. J., and Morrison, J. (1962). Low voltage electron diffraction study of oxidation and reduction of silicon. *J. Appl. Phys.,* **33**(6), 2089–2093.

Law, M., Greene, L. E., Johnson, J. C., Saykally, R., and Yang, P. D. (2005). Nanowire dye-sensitized solar cells. *Nat. Mater.,* **4**(6), 455–459.

Lee, J. W., Othman, M. R., Eom, Y., Lee, T. G., Kim, W. S., and Kim, J. (2008). The effects of sonification and TiO_2 deposition on the micro-characteristics of the thermally treated SiO_2/TiO_2 spherical core-shell particles for photo-catalysis of methyl orange. *Microporous Mesoporous Mater.,* **116**(1–3), 561–568.

Li, Q. N., Wang, X. M., Lu, X. H., et al. (2009). The incorporation of daunorubicin in cancer cells through the use of titanium dioxide whiskers. *Biomaterials,* **30**(27), 4708–4715.

Li, Y. X., Peng, S. Q., Jiang, F. Y., Lu, G. X., and Li, S. B. (2007). Effect of doping TiO_2 with alkaline-earth metal ions on its photocatalytic activity. *J. Serb. Chem. Soc.,* **72**(4), 393–402.

Linsebigler, A. L., Lu, G. Q., and Yates, J. T. (1995). Photocatalysis on TiO_2 surfaces: principles, mechanisms, and selected results. *Chem. Rev.,* **95**(3), 735–758.

Liu, M. Z., Snapp, N. D., and Park, H. (2011). Water photolysis with a cross-linked titanium dioxide nanowire anode. *Chem. Sci.,* **2**(1), 80–87.

McCullagh, C., Robertson, J. M. C., Bahnemann, D. W., and Robertson, P. K. J. (2007). The application of TiO_2 photocatalysis for disinfection of water contaminated with pathogenic micro-organisms: a review. *Res. Chem. Intermed.,* **33**(3–5), 359–375.

Moshfegh, A. Z. (2009). Nanoparticle catalysts. *J. Phys. D,* **42**(23), 1–30.

Murashkevich, A. N., Lavitskaya, A. S., Alisienok, O. A., and Zharskii, I. M. (2009). Fabrication and properties of SiO_2/TiO_2 composites. *Inorg. Mater.,* **45**(10), 1146–1152.

Ohtaka, O., Fukui, H., Funakoshi, K., Utsumi, W., Irifune, T., and Kikegawa, T. (2002). Phase relations and EOS of ZrO_2 and HfO_2 under high-temperature and high-pressure. *High Pressure Res.,* **22**(1), 221–226.

Oregan, B., and Gratzel, M. (1991). A low-cost, high-efficiency solar-cell based on dye-sensitized colloidal TiO_2 films. *Nature,* **353**(6346), 737–740.

Ortiz, G. F., Hanzu, I., Djenizian, T., Lavela, P., Tirado, J. L., and Knauth, P. (2009). Alternative Li-ion battery electrode based on self-organized titania nanotubes. *Chem. Mater.,* **21**(1), 63–67.

Rao, C. N. R., Deepak, F. L., Gundiah, G., and Govindaraj, A. (2003). Inorganic nanowires. *Prog. Solid State Chem.,* **31**(1–2), 5–147.

Rath, H., Anand, S., Mohapatra, M., et al. (2009). Effect of thermal annealing on the structure and microstructure of TiO2 thin films. *Indian J. Phys.,* **83**(4), 559–565.

Seong, S. G., Kim, E. J., Kim, Y. S., Lee, K. E., and Hahn, S. H. (2009). Influence of deposition atmosphere on photocatalytic activity of TiO_2/SiO_x double-layers prepared by RF magnetron sputtering. *Appl. Surf. Sci.,* **256**(1), 1–5.

Shalav, A., Bullock, J., Anderson, P., Ruffell, S., White, J., and Elliman, R.G. (2012). The mechanical and photochemical properties of titania coated silica nanowire layers. *ECS J. Solid State Sci. Technol.*, **1**(5), Q114–Q118.

Shalav, A., Kim, T., and Elliman, R. G. (2010). SiO_x nanowires grown via the active oxidation of silicon. *IEEE J. Sel. Top. Quantum Electron.*, **17**(4), 785–793.

Shalav, A., Venkatachalam, D. K., Reichardt, F., Fischer, F., and Elliman, R. G. (2009). *Titania Coated Silica Nanowires,* paper presented at MRS Fall Meeting Conference Proceedings, Boston, Massachussetts, 1206-M16-05.

Srinivasan, S., Datye, A. K., Smith, M. H., and Peden, C. H. F. (1994). Interaction of titanium isopropoxide with surface hydroxyls on silica. *J. Catal.*, **145**(2), 565–573.

Suzuki, Y., Ngamsinlapasathian, S., Yoshida, R., and Yoshikawa, S. (2006). Partially nanowire-structured TiO_2 electrode for dye-sensitized solar cells. *Cent. Eur. J. Chem.*, **4**(3), 476–488.

Suzuki, Y., Pavasupree, S., Yoshikawa, S., and Kawahata, R. (2007). Direct synthesis of an anatase-TiO_2 nanofiber/nanoparticle composite powder from natural rutile. *Phys. Status Solidi A,* **204**(6), 1757–1761.

Tan, S. S., Zou, L., and Hu, E. (2006). Photocatalytic reduction of carbon dioxide into gaseous hydrocarbon using TiO_2 pellets. *Catal. Today,* **115**(1–4), 269–273.

Veronovski, N., Sfiligoj-Smole, M., and Viota, J. L. (2010). Characterization of TiO_2/TiO_2-SiO_2 coated cellulose textiles. *Text. Res. J.*, **80**(1), 55–62.

Vitanov, P., Stefanov, P., Harizanova, A., and Ivanova, T. (2008). XPS characterization of thin $(Al_2O_3)_x(TiO_2)_{1-x}$ films deposited on silicon. *J. Phys. Conf. Series,* **113**, 012036.

Wang, Z. Y., Wang, P. Y., Li, J. G., and Lu, Q. S. (2009). Structure and optical properties of ordered mesoporous titania-silica composite thin films. *J. Nanosci. Nanotechnol.*, **9**(4), 2574–2581.

Wang, D., Yu, B., Zhou, F., Wang, C., and Liu, W. (2009). Synthesis and characterization of anatase TiO_2 nanotubes and their use in dye-sensitized solar cells. *Mater. Chem. Phys.*, **113**(2–3), 602–606.

Wilhelm, P., and Stephan, D. (2007). Photodegradation of rhodamine B in aqueous solution via SiO_2@TiO_2 nano-spheres. *J. Photochem. Photobiol., A,* **185**(1), 19–25.

Yang, C. C., Yu, Y. H., van der Linden, B., Wu, J. C. S., and Mul, G. (2010). Artificial photosynthesis over crystalline TiO_2-based catalysts: fact or fiction? *J. Am. Chem. Soc.*, **132**(24), 8398–8406.

Yang, Z. G., Choi, D., Kerisit, S., et al. (2009). Nanostructures and lithium electrochemical reactivity of lithium titanites and titanium oxides: a review. *J. Power Sources,* **192**(2), 588–598.

Chapter 6

Global Health and Environmental Implications of Mimicking Biological Ion Channels Using Nanotubes

Tamsyn A. Hilder
Computational Biophysics Group, Research School of Biology,
Building 46, R. N. Robertson Building, Biology Place, Australian National University,
Canberra, ACT 0200, Australia
tamsyn.hilder@anu.edu.au

This chapter discusses both positive and negative impacts of synthetic nanotube devices designed to be selective to water molecules, cations (positively charged ions), or anions (negatively charged ions) for applications in medicine and the environment.

6.1 Introduction

Biological ion channels exist in all living cells, for example, plants, bacteria, and humans. Their function is to regulate the movement of charged particles, such as sodium, chloride, and potassium ions, across the cell membrane by providing a water-filled pore through

Nanotechnology Toward the Sustainocene
Edited by Thomas A. Faunce

ISBN 978-981-4463-76-8 (Hardcover), 978-981-4463-77-5 (eBook)
www.panstanford.com

which ions can pass. Ion channels regulate all electrical activities in the nervous system by allowing specific ions to move across the cell membrane (Hille, 2001). By disrupting the balance between ions in the extracellular (outside) and intracellular (inside) compartments of a cell, it is possible to damage the cellular equilibrium and even cause cell death.

Engineered channels, such as nanotubes, are far less complex than biological systems; yet, they have the ability to perform some of the functions biological ion channels carry out. Single-walled nanotubes may be manufactured from various materials, for example, carbon, boron nitride, and silicon carbide. They are ideal candidates to mimic biological ion channels because of the following reasons:

i. Their nanoscale diameter, which resembles that of biological ion channels. Ion channels have diameters of approximately 1 nm (10^{-9} m). The nanotube diameter can vary depending on the fabrication method used. For example, carbon nanotubes fabricated using electric arc discharge produce diameters ranging from 1.1 nm to 1.5 nm (Farhat, de La Chapelle, Loiseau et al., 2001) and using gas-phase catalytic growth can be 0.7 nm in diameter (Nikolaev, Bronikowski, Bradley et al., 1999).
ii. The ability to precisely control their length (Wang, Liang, Wang, Zhang, and Rahman, 2007) to target cells with a specific membrane thickness. For example, gram-negative bacteria have thinner cell membranes than human cells, allowing these bacteria cells to be targeted without affecting normal cells.
iii. The fact that they can embed in the cell membrane and form an open channel spanning the entire membrane (Lopez, Nielsen, Moore, and Klein, 2004; Hilder, Gordon, and Chung, 2010).
iv. Their drastically enhanced flow of water and ions compared to biological ion channels, which will be highlighted in Section 6.2.
v. The ability to modify their surface chemistry to target specific cells, such as specific bacteria or cancer cells. For example, folate receptors are often expressed in large numbers on cancer cells compared to healthy cells. Nanoparticles with

folate receptor targeting have been successfully delivered to folate receptor–expressing cells (Hilgenbrink and Low, 2005).

In Section 6.2, we highlight recent work that illustrates the ability of various nanotubes to mimic the function of biological ion channels and thus their enormous potential in the development of novel nanoscale devices and products. Our computational work has shown that it is possible to design a nanotube with the ability to either reject all charged particles or selectively reject positive or negative ions.

In mimicking the function of biological ion channels, we may be able to fabricate nanoscale devices for various applications, including desalination, water purification, and ultrasensitive detection. Water shortage and water contamination are important global challenges. Over a third of the world's population lives in a region where water is scarce (Elimelech and Phillip, 2011), and about one-sixth of the population doesn't have access to safe drinking water (Botes and Cloete, 2010). Nanotube-based applications may provide significant benefits to both industry and local communities, such as through more efficient desalination membranes and improved access to safe drinking water. Nanotube-based devices may also enable the detection of very small concentrations of contaminants in water, such as heavy metals and bacteria. In addition, we may be able to design new pharmaceutical products for the treatment of bacterial infections, cancer, cystic fibrosis, and many other diseases. Section 6.3.1 examines the positive potential impacts of nanotube-based devices for a secure and sustainable future.

Unfortunately, since nanotubes can affect cells, they could also have negative health and environmental risks, and issues such as safe handling and disposal will need to be carefully considered. To facilitate the development of these nanotube-based devices, it is vital that we have an in-depth understanding of their impact on global health and the environment. Section 6.3.2 examines the negative implications of developing a nanotube-based device that mimics biological ion channels. The dissemination of their impact will enable policy makers and the public to make informed decisions and have a better understanding of the technology that may result from such research.

6.2 Mimicking Biological Ion Channels

Interest in mimicking the function of biological ion channels is rapidly gaining momentum. Several attempts have been made to design (Kalra, Garde, and Hummer, 2003; Won and Aluru, 2007; Corry, 2008; Suk, Raghunathan, and Aluru, 2008; Hilder, Gordon, and Chung, 2009a) and fabricate (Fornasiero, Park, Holt et al., 2008) nanotubes that can selectively allow water molecules to pass through, while rejecting charged particles, for application in water purification. These nanotube-based water channels have shown flow rates substantially higher than both biological water channels and currently used desalination membranes.

Recently, the possibility of engineering ion-selective nanotubes has also been explored. Ion-selective nanotube-based devices, if successfully engineered, could be used as antibiotics or could be targeted to specific cells to modify their functions or render them unviable. The placement of a rim of partially charged atoms at the open-ended entrance makes it possible for ions to surmount the energy barrier at the entrance and enter the nanotube interior. This effect has been confirmed in experiments by Majumder, Chopra, and Hinds (2005), whereby the nanotube tips were functionalized with various molecules, including negatively charged dye molecules. Moreover, various computational studies have been conducted illustrating the ability of nanotubes to be selective to both positive and negative ions (Joseph, Mashl, Jakobsson, and Aluru, 2003; Park, Sinnott, and Aluru, 2006; Hilder, Gordon, and Chung, 2009b; Won and Aluru, 2009).

The following subsections highlight recent results where we have demonstrated the possibility of mimicking various biological ion channels using nanotubes, namely, biological water channels (aquaporin), and channels selective to either positively charged or negatively charged ions. To investigate the conductance of water molecules and ions through synthetic nanotube channels we use two computational tools, (i) molecular dynamics and (ii) distributional molecular dynamics. Distributional molecular dynamics was devised by our group to study the conduction of ions across nanoscale channels at longer time scales than possible using molecular dynamics (Gordon, Krishnamurthy, and Chung, 2009; Hilder, Gordon, and Chung, 2010). Employing this computational tool enables our theoretical results to be directly compared to future experimental measurements of ion conduction.

6.2.1 Water Channels (Aquaporin)

We show (Hilder, Gordon, and Chung, 2009a) computationally that a boron nitride nanotube with a diameter of 0.69 nm embedded in a silicon nitride membrane is able to reject salt ions at seawater concentrations while conducting water molecules. The water flow rate is as high as 10.7 water molecules per nanosecond, or 0.93 L m^{-2} hr^{-1}. Moreover, at the same operating pressure, the water conduction is four to five times faster than currently used premium-grade seawater reverse osmosis membranes. The boron-nitride-nanotube-embedded silicon nitride membrane we designed can, in principle, achieve 100% salt rejection. Furthermore, a unique water structure forms within the nanotube, developing from a single-file chain, as shown in Fig. 6.1, to an ordered spiral and 4-gonal tube as the nanotube diameter increases.

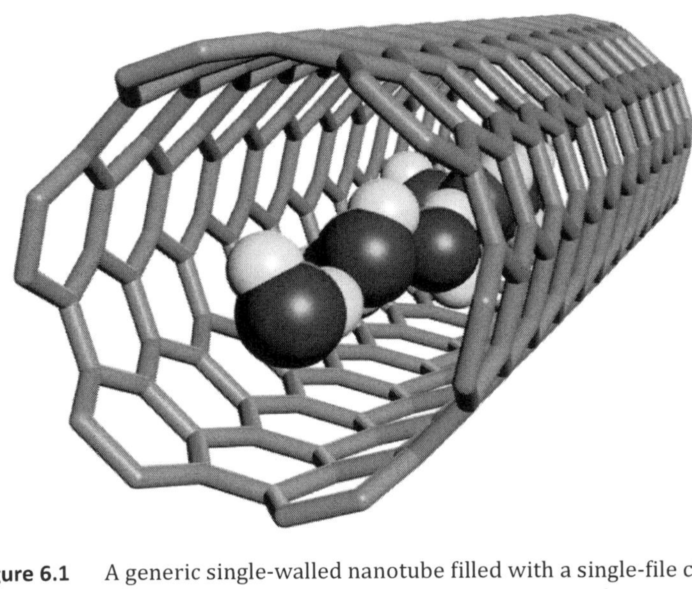

Figure 6.1 A generic single-walled nanotube filled with a single-file chain of water molecules. Figure reprinted in part with permission from Hilder, Yang, Gordon, Rendell, and Chung (2012). Copyright 2012, American Chemical Society.

Recently, we examined the conduction of water and ions through silicon carbide nanotubes 3.6 nm in length and of various diameters, ranging from 0.9 nm to 1.2 nm (Hilder, Yang, Gordon, Rendell, and Chung, 2012). All investigated nanotubes have water conduction rates an order of magnitude higher than the biological water channel aquaporin and current reverse osmosis membranes. Moreover, Khademi and Sahimi (2011) find that silicon carbide nanotubes exhibit enhanced water flow when compared to carbon nanotubes. As expected, the water conduction increases as the nanotube radius increases, as shown in Fig. 6.2, but ions are no longer rejected. In particular, we find that a silicon carbide nanotube approximately 0.9 nm in diameter rejects all ions and has an osmotic permeability of approximately 7.34 × 10^{-13} cm^3/s. These silicon carbide nanotubes present advantages over carbon nanotubes for ion-selective membranes since chemical functionalization of their surfaces is not required to achieve selectivity.

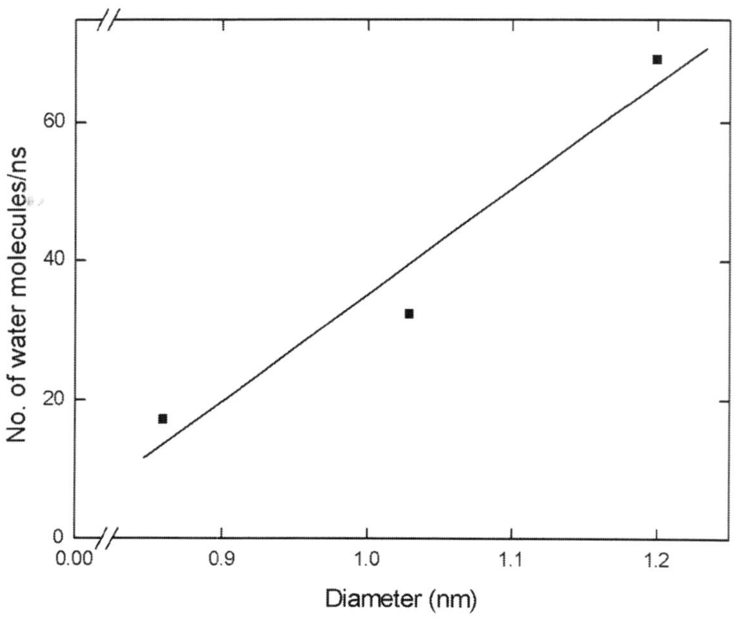

Figure 6.2 Water conduction through silicon carbide nanotubes as a function of diameter, determined using an applied pressure of 100 MPa, 500 mM salt concentration.

6.2.2 Ion-Selective Channels

We have successfully designed ion-selective carbon nanotubes (Hilder, Gordon, and Chung, 2010; Hilder and Chung, 2011; Hilder, Gordon, and Chung, 2011) that are selective to either positive or negative ions, depending on their chemical functionalization and nanotube diameter. For example, terminating a carbon nanotube 0.91 nm in diameter with carbonyl (C=O) groups (Hilder, Gordon, and Chung, 2010) or a carbon nanotube 1 nm in diameter with carboxylic acid (COOH) groups (Hilder, Gordon, and Chung, 2011) makes the nanotube selective to negative or positive ions, respectively. Each nanotube is embedded in a lipid bilayer separating two reservoirs, as illustrated in Fig. 6.3 for the carboxylic acid–terminated carbon nanotube.

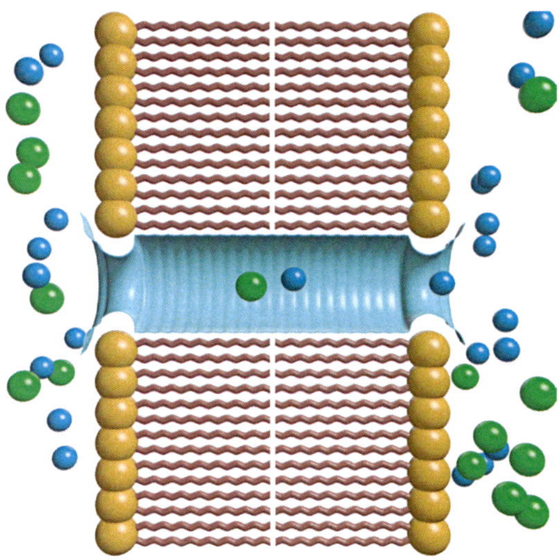

Figure 6.3 Schematic of the computational simulation assembly for the carbon nanotube terminated with carboxylic acid embedded in a lipid bilayer separating two reservoirs containing charged particles (blue and green balls). For clarity not all ions are shown. Figure reprinted in part with permission from Hilder, Gordon, and Chung (2011). Copyright 2011, American Institute of Physics.

We show (Hilder, Gordon, and Chung, 2010) that a carbon nanotube with an effective diameter of 0.91 nm and a length of 3.4 nm terminated with carbonyl (C=O) groups is stable in a lipid bilayer and is selective to chloride ions. Confirming the results of previous work (Lopez, Nielsen, Moore, and Klein, 2004; Nielsen, Ensing, Ortiz, Moore, and Klein, 2005), we demonstrate that the nanotube is stable in the lipid bilayer. The chloride selectivity is dependent on the carbonyl groups, which generate an energy well for chloride ions at the entrance and exit of the nanotube and a large energy barrier for sodium ions. The flow of chloride ions through this nanotube is larger than through biological chloride channels. This artificial channel shows a striking resemblance to a biological ion channel and has the potential to lead to numerous drugs designed to target ion channels, for example, for treatments for antibacterial and cystic fibrosis.

Alternatively, we show (Hilder, Gordon, and Chung, 2011) that a carbon nanotube with an effective diameter of 1 nm and a length of 3.6 nm terminated with carboxylic acid (COOH) is selective to sodium ions despite both sodium and chloride encountering an energy barrier to enter the nanotube. We find that chloride ions act as chaperones for sodium ions. Once inside the nanotube interior, resident chloride ions ferry sodium ions from the entrance to the exit. A large number of sodium ions flow through the nanotube, whereas the number of chloride ions is negligible. This artificial channel resembles the mutant glycine receptor, but the sodium conductance is seven times larger.

We also examine a carbon nanotube with an effective diameter of 0.91 nm and a length of 3.6 nm, hydrogen-terminated ends, and hydrogen attached to two short regions of the exterior surface (Hilder and Chung, 2011). This nanotube is selective to monovalent positive ions (+1), such as potassium; binds divalent positive ions (+2), such as calcium; and rejects monovalent negative ions (−1), such as chloride. The characteristics of this channel closely resemble the antibiotic gramicidin but with a potassium conductance six times larger.

6.3 Global Health and Environmental Implications

Using synthetic nanotubes that mimic biological ion channels to fabricate novel nanodevices may result in numerous innovative products that would positively impact global health and the environment. However, it will require significant research and development before these applications reach the market. The next section outlines applications that are likely to eventuate in the next 5 years and some others that are more idealistic and may or may not eventuate in the next 15 years. Although there are many positive impacts of these potential nanotube-based devices, the negative impacts of this technology are largely unknown due to a lack of fundamental research. Section 6.3.2 highlights some of the health and environmental issues that need to be addressed before nanotube-based devices can be realized.

6.3.1 Positive Impacts

Nanotube-based filtration membranes have the potential to progress toward the UN millennium development goal (MDG) to halve the number of people with access to safe drinking water by 2015. Approximately 25% of the world's population is affected by water shortage (Hilder, Gordon, and Chung, 2009a). With population growth and climate change limiting the world's freshwater stores, desalination and demineralization of seawater is fast becoming a possible solution. However, over 95% of salt ions must be removed from seawater for it to be fit for human consumption. There is an urgent and genuine need to make the process of desalination more effective and less power hungry. As discussed in Section 6.2.1, synthetic nanotubes have the ability to conduct water four to five times faster than currently available reverse osmosis membranes, while maintaining a salt rejection much greater than 95%. If nanotube-based water purification membranes can be fabricated at a reasonable cost, they could act as an alternative means of supplying safe drinking water. In recent experimental work, the feasibility of using carbon nanotubes as water purification devices was demonstrated (Fornasiero, Park, Holt et al., 2008).

Higher-permeability membranes have already contributed to reducing the energy consumption of desalination and may also help reduce capital costs by reducing the membrane area needed (Elimelech and Phillip, 2011). However, to reduce energy consumption further and reduce one of the environmental impacts of desalination, it will be necessary to design fouling-resistant membranes (Elimelech and Phillip, 2011). Biofouling is the growth of microbes on the membrane surface, which results in the need to regularly clean the membrane surface to maintain its efficiency. Either nanotube-based filtration membranes could be designed such that they are coated with antibiofouling designed polymers, or the nanotubes could be directly embedded in membranes with antibiofouling and antiviral properties, such as the membranes discussed by Botes and Cloete (2010). These antibiofouling nanotube-based filtration devices would not only exhibit enhanced water flow rates and ion selectivity but also reduce the chemical cleaning requirements due to their antimicrobial surfaces.

Water contamination, such as by heavy metal ions, is one of the leading causes of water shortage (Liang, Cao, Zhang et al., 2011). One-sixth of the world's population doesn't have access to safe water (Botes and Cloete, 2010). Due to the ability to design nanotubes for specific applications, it may also be possible to ensure the removal of heavy metals or other waterborne diseases. For example, the nanotube radius could be designed to selectively remove heavy metal ions or the nanotubes could be embedded in a membrane with antiviral properties to help prevent the contamination of drinking water by waterborne viruses and bacteria. Carbon nanotube filters have already been shown to efficiently remove *Escherichia coli* (*E. coli*), *Staphylococcus aureus* (*S. aureus*), and the polio-1 virus from water (Botes and Cloete, 2010).

The use of carbon nanotubes as single-ion detectors was recently confirmed experimentally (Lee, Choi, Han, and Strano, 2010). Therefore, in the next five years it is possible that ion-selective nanotubes will be used as ultrasensitive detection devices in much the same way that has been demonstrated with the biological ion channel gramicidin (Cornell, Braach-Maksvytis, King et al., 1997), whereby the flow of ions is prevented upon detection of a specific protein. Ion-selective nanotubes could be used to detect molecules such as trinitrotoluene (TNT) or glucose but at significantly lower concentrations than currently used detection devices. These molecules could be detected when they block the ion flow through the nanotube. As mentioned in Section 6.2.2, we have demonstrated the possibility of designing ion-selective nanotubes. In particular, we demonstrated the potential to mimic gramicidin with a potassium conductance six times larger.

A nanotube-based antibiotic that can selectively conduct either positive or negative ions across the cell membrane could be used for the purpose of killing, for example, bacteria causing meningitis and other gram-negative bacteria. It is possible to target gram-negative bacteria due to their thinner cell membranes compared to gram-positive bacteria and normal cells. The antibiotic gramicidin targets gram-negative bacteria and was one of the first antibiotics used in clinical practice. It exerts its antibacterial activity by allowing positive ions to selectively move across the thinner cell membranes of the bacteria and destroy the electric potential across the membrane, which then kills the bacterial cells (Kuyucak, Andersen, and Chung, 2001). We have demonstrated that synthetic nanotubes have the ability to be ion selective with ion conduction rates an order of magnitude larger than their biological counterparts. Moreover, synthetic nanotubes have a number of other qualities, listed in Section 6.1, that suggest they will present significant improvements as an antibacterial agent. It is unlikely that this type of nanotube-based device will be available in the next five years. However, a nanotube-based antibiotic may provide a more effective bacterial treatment in the future and, with the development of fabrication technology, will become increasingly affordable for developing countries.

Furthermore, it may one day be possible to use nanotubes to target specific cells, such as cancer cells, and destroy their membrane potential in a similar way to that described above for bacteria. Such nanotubes could have applications in the next generation of medical treatments, such as to treat cystic fibrosis, cancer, or other diseases. In addition, nanotubes that can mimic some of the functions of biological ion channels may be used in the treatment of diseases that result in ion channel mutations. For example, in cystic fibrosis, chloride channels malfunction. Thus, a nanotube such as the chloride-selective channel described in Section 6.2.2 could act as a substitute. However, these possible applications are very much futuristic ideas and much research needs to be achieved before devices such as these enter the marketplace.

6.3.2 Negative Impacts

There is a general consensus that pristine carbon nanotubes (i.e., with no functionalization) are toxic to cells (Cui, Tian, Ozkan, Wang, and Gao, 2005), whereas boron nitride nanotubes are considered to be nontoxic (Chen, Wu, Rousseas et al., 2009). The toxicity of pristine nanotubes is largely due to their insolubility in aqueous solutions, hydrophobicity (water-repelling nature), and tendency to clump together. However, with functionalization, carbon nanotubes can be made to be soluble (Bianco, Kostarelos, and Prato, 2005; Sayes, Liang, Hudson et al., 2006), thus reducing their cellular toxicity.

An obvious problem is how to responsibly dispose of used nanotube-based devices from both medical and engineering applications or from the excretion of nanotube-based pharmaceuticals. It is not known whether these devices may pose unwanted environmental impacts. Schierz and Zänker (2009) investigated carbon nanotubes as carriers of pollutants in the case of accidental release into the environment. They found that surface treatment of the nanotubes has a pronounced influence on their behavior in aqueous suspensions and that some treatments can enhance the binding of heavy metals such as uranium to the nanotubes. This could lead to the transport of heavy metals through natural aquatic systems and even into biological systems. On the other hand, Allen, Kichambare, Gou et al. (2008) were able to biodegrade single-walled carbon nanotubes using natural, enzymatic catalysis with natural horseradish peroxidase. Such a method could be used as a means of responsibly disposing of carbon nanotubes or as a chemical spill kit to help clean up carbon nanotubes in the environment, thus mitigating carbon nanotube environmental toxicity.

As mentioned in Section 6.3.1, one possible nanotube application is as a nanotube-based antibacterial agent. Obviously the main aim here is to kill bacteria cells, but it is not known for certain what will happen to the nanotubes once their mission is complete and the bacteria cells die. However, studies (Singh, Pantarotto, Lacerda et al., 2006) have shown that water-soluble single-walled carbon nanotubes are not retained in any of the reticuloendothelial organs (liver and spleen) and are rapidly cleared from systemic blood circulation. Moreover, the use of nanotubes as a possible substitute for malfunctioning ion channels in the body could have long-term health effects. Long-term toxicity in vitro and in vivo needs to be accurately determined before such a device could be developed.

There may also be risks associated with the fabrication of such devices to the scientists producing them, through inhalation or the skin. Some studies suggest nanotubes could be as dangerous as asbestos (Poland, Duffin, Kinloch et al., 2008). A hazardous fiber is defined by toxicologists to be thinner than 3 μm, longer than ~20 μm, and biopersistent in the lungs (Poland, Duffin, Kinloch et al., 2008). Although the nanotube-based devices proposed in this chapter are typically less than 1 nm in diameter, their lengths are much shorter (nanometer versus micrometer). However, further research needs to be done and care needs to be taken when creating nanotube-based devices.

6.4 Conclusions and Future Work

Through our computational studies we have shown that it is possible to design nanotubes that broadly mimic biological ion channels such that they either reject all charged particles or selectively reject positive or negative ions. Using ion-selective nanotubes in applications such as detection, desalination, and antibiotics will undoubtedly benefit global health and the environment in the future.

Nanotube-based filtration membranes have the potential to improve the efficiency of desalination by increasing water flow while maintaining salt rejection, making the water suitable for human consumption. It is possible that in the future these membranes are designed such that the nanotubes are embedded in antibiofouling and antimicrobial membranes to reduce the amount of chemical cleaning currently required during desalination. This would, in turn, reduce one of the negative environmental impacts of desalination and perhaps help it become a more accepted solution to water shortages. Moreover, nanotube-based membranes and devices may also help communities decontaminate water supplies by the removal of contaminants such as lead and *E. coli* and through detection of minute quantities of the same contaminants.

Unfortunately, there are many unknown risks associated with such devices that may result in negative impacts. To gain a better understanding of both positive and negative impacts of mimicking biological ion channels and for the successful development of such devices, it is vital that future experimental work be performed. First, it is essential that we conduct experiments to confirm our theoretical studies to illustrate the real possibility of these synthetic ion channels. Once their potential as nanodevices has been confirmed through experiment subsequent testing will be required to determine the associated risks, such as cellular and environmental toxicology. Moreover, significant advances need to occur in fabrication technology to reduce costs of nanotube-based devices so that they can be made readily available to local communities. There is endless research still to come, but the potential of these nanoscale devices is well worth the effort.

References

Allen, B. L., Kichambare, P. D., Gou, P., et al. (2008). Biodegradation of single-walled carbon nanotubes through enzymatic catalysis, *Nano Lett.*, **8**, 3899–3903.

Bianco, A., Kostarelos, K., and Prato, M. (2005). Applications of carbon nanotubes in drug delivery. *Curr. Opin. Chem. Biol.*, **9**, 674–679.

Botes, M., and Cloete, T. E. (2010). The potential of nanofibers and nanobiocides in water purification. *Crit. Rev. Microbiol.*, **36**, 68–81.

Chen, X., Wu, P., Rousseas, M., et al. (2009). Boron nitride nanotubes are noncytotoxic and can be functionalized for interaction with proteins and cells. *J. Am. Chem. Soc.*, **131**, 890–891.

Cornell, B. A., Braach-Maksvytis, V. L. B., King, L. G., et al. (1997). A biosensor that uses ion-channel switches. *Nature*, **387**, 580–583.

Corry, B. (2008). Designing carbon nanotube membranes for efficient water desalination. *J. Phys. Chem. B*, **112**, 1427–1434.

Cui, D., Tian, F., Ozkan, C. S., Wang, M., and Gao, H. (2005). Effect of single wall carbon nanotubes on human HEK293 cells. *Toxicol. Lett.*, **155**, 73–85.

Elimelech, M., and Phillip, W. A. (2011). The future of seawater desalination: energy, technology, and the environment. *Science*, **333**, 712–717.

Farhat, S., de La Chapelle, M. L., Loiseau, A., et al. (2001). Diameter control of single-walled carbon nanotubes using argon-helium mixture gases. *J. Chem. Phys.*, **115**, 6752–6758.

Fornasiero, F., Park, H. G., Holt, J. K., et al. (2008). Ion exclusion by sub-2-nm carbon nanotube pores. *Proc. Natl. Acad. Sci. U. S. A.*, **105**, 17250–17255.

Gordon, D., Krishnamurthy, V., and Chung, S. H. (2009). Generalized Langevin models of molecular dynamics simulations with applications to ion channels. *J. Chem. Phys.*, **131**, 134102-1–134102-11.

Hilder, T. A., and Chung, S. H. (2011). Carbon nanotube as a gramicidin analogue. *Chem. Phys. Lett.*, **501**, 423–426.

Hilder, T. A., Gordon, D., and Chung, S. H. (2009a). Salt rejection and water transport through boron nitride nanotubes. *Small*, **5**, 2183–2190.

Hilder, T. A., Gordon, D., and Chung, S. H. (2009b). Boron nitride nanotubes selectively permeable to cations or anions. *Small*, **5**, 2870–2875.

Hilder, T. A., Gordon, D., and Chung, S. H. (2010). Synthetic chloride-selective carbon nanotubes examined using molecular and stochastic dynamics. *Biophys. J.*, **99**, 1734–1742.

Hilder, T. A., Gordon, D., and Chung, S. H. (2011). Synthetic cation-selective nanotube: permeant cations chaperoned by anions. *J. Chem. Phys.*, **134**, 045103-1–045103-7.

Hilder, T. A., Yang, R., Gordon, D., Rendell, A. P., and Chung, S. H. (2012). Silicon carbide nanotube as a chloride-selective channel. *J. Phys. Chem. C*, **116**, 4465–4470.

Hilgenbrink, A. R., and Low, P. S. (2005). Folate receptor-mediated drug targeting: from therapeutics to diagnostics. *J. Pharm. Sci.*, **94**, 2135–2146.

Hille, B. (2001). *Ion Channels of Excitable Membranes*, 3rd ed. USA: Sinauer Associates.

Joseph, S., Mashl, R. J., Jakobsson, E., and Aluru, N. R. (2003). Electrolytic transport in modified carbon nanotubes. *Nano Lett.*, **3**, 1399–1403.

Kalra, A. S., Garde, S., and Hummer, G. (2003). Osmotic water transport through carbon nanotube membranes. *Proc. Natl. Acad. Sci. U. S. A.*, **100**, 10175–10180.

Khademi, M., and Sahimi, M. (2011). Molecular dynamics simulation of pressure-driven water flow in silicon-carbide nanotubes. *J. Chem. Phys.*, **135**, 204509-1–204509-7.

Kuyucak, S., Andersen, O. S., and Chung, S. H. (2001). Models of permeation in ion channels. *Rep. Prog. Phys.*, **64**, 1427–1472.

Lee, C. Y., Choi, W., Han, J. H., and Strano, M. S. (2010). Coherence resonance in a single-walled carbon nanotube ion channel. *Science*, **329**, 1320–1324.

Liang, H. W., Cao, X., Zhang, W. J., et al. (2011). Robust and highly efficient free-standing carbonaceous nanofiber membranes for water purification. *Adv. Funct. Mater.*, **21**, 3851–3858.

Lopez, C. F., Nielsen, S. O., Moore, P. B., and Klein, M. L. (2004). Understanding nature's design for a nanosyringe. *Proc. Natl. Acad. Sci. U. S. A.*, **101**, 4431–4434.

Majumder, M., Chopra, N., and Hinds, B. J. (2005). Effect of tip functionalization on transport through vertically oriented carbon nanotube membranes. *J. Am. Chem. Soc.*, **127**, 9062–9070.

Nielsen, S. O., Ensing, B., Ortiz, V., Moore, P. B., and Klein, M. L. (2005). Lipid bilayer perturbations around a transmembrane nanotube: a coarse grain molecular dynamics study. *Biophys. J.*, **88**, 3822–3828.

Nikolaev, P., Bronikowski, M. J., Bradley, R. K., et al. (1999). Gas-phase catalytic growth of single-walled carbon nanotubes from carbon monoxide. *Chem. Phys. Lett.*, **313**, 91–97.

Park, J. H., Sinnott, S. B., and Aluru, N. R. (2006). Ion separation using a Y-junction carbon nanotube. *Nanotechnology*, **17**, 895–900.

Poland, C. A., Duffin, R., Kinloch, I., et al. (2008). Carbon nanotubes introduced into the abdominal cavity of mice show asbestos-like pathogenicity in a pilot study. *Nat. Nanotechnol.*, **3**, 423–428.

Sayes, C. M., Liang, F., Hudson, J. L., et al. (2006). Functionalization density dependence of single-walled carbon nanotubes cytotoxicity in vitro. *Toxicol. Lett.*, **161**, 135–142.

Schierz, A., and Zänker, H. (2009). Aqueous suspensions of carbon nanotubes: surface oxidation, colloidal stability and uranium sorption. *Environ. Pollut.*, **157**, 1088–1094.

Singh, R., Pantarotto, D., Lacerda, L., et al. (2006). Tissue biodistribution and blood clearance rates of intravenously administered carbon nanotube radiotracers. *Proc. Natl. Acad. Sci. U. S. A.*, **103**, 3357–3362.

Suk, M. E., Raghunathan, A. V., and Aluru, N. R. (2008). Fast reverse osmosis using boron nitride and carbon nanotubes. *Appl. Phys. Lett.*, **92**, 133120-1–133120-3.

Wang, S., Liang, Z., Wang, B., Zhang, C., and Rahman, Z. (2007). Precise cutting of single-walled carbon nanotubes. *Nanotechnology*, **18**, 055301-1–055301-6.

Won, C. Y., and Aluru, N. R. (2007). Water permeation through a subnanometer boron nitride nanotube. *J. Am. Chem. Soc.*, **129**, 2748–2749.

Won, C. Y., and Aluru, N. R. (2009). A chloride ion-selective boron nitride nanotube. *Chem. Phys. Lett.*, **478**, 185–190.

Chapter 7

Nanostructured Materials: Implications for Information Technology

Dragomir N. Neshev
Nonlinear Physics Centre, Centre for Ultrahigh-Bandwidth Devices for Optical Systems, Research School of Physics and Engineering, Building 59, Le Couteur Building, Mills Road, Australian National University, Canberra, ACT 0200, Australia
Dragomir.Neshev@anu.edu.au

This chapter describes opportunities for the development of nanostructured metamaterials with unique light sensitivity for applications in information technologies. In particular we demonstrate experimentally the all-optical tuning of such metamaterials infiltrated with liquid crystals and show a fivefold enhancement of their nonlinear properties. Further applications of such materials in security or renewable energy are also discussed.

7.1 Introduction

Information technologies have been dramatically transformed in the last two decades with the development of the Internet and optical

---

Nanotechnology Toward the Sustainocene
Edited by Thomas A. Faunce

ISBN 978-981-4463-76-8 (Hardcover), 978-981-4463-77-5 (eBook)
www.panstanford.com

communications. Currently almost the entire communications technology relies on light for information transfer. Even our mobile phone conversations are eventually transmitted through an optical fiber after being received by mobile communication towers. However, something very important that is usually neglected is the fact that when information is being transmitted all-optically, eventually the vast amount of data hits a wall, often referred to as the "yellow wall." At this point all optical signals have to be converted into electrical in order for the information to be processed and routed to the desired directions.

Inevitably this conversion from optical to electrical signals results in large energy dissipation and indirectly to green-gas emission. As an example, in 2010 approximately 1% of the US energy consumption was attributed to electrical routers on the Internet. Hence, one can imagine that with the current exponential growth of the Internet, one can soon expect significant problems in the energy sector only because of the growth of Internet traffic.

The conversion of optical signals into electrical and back is considered the bottleneck of information transfer that slows down all our communications. And this is the problem that a large number of scientists aim to solve, including our research team at the Australian Research Council Centre of Excellence, Centre for Ultrahigh-Bandwidth Devices for Optical Systems (CUDOS). The goal of CUDOS researchers is to realize efficient signal processing of light entirely optically, thus avoiding the need for conversion of optical signals to electrical. The aim in such all-optical signal processing is to be able to perform the processing and routing of information on a smaller scale, much faster and more energy efficient.

However, to be able to process light all-optically, one needs to be able to control light with light, which is only possible through the use of the so-called optical nonlinearity. Importantly, nonlinear optical effects need to occur at low power in order to improve the efficiency of the signal processing.

The key in the development of all-optical control of light signals is to make them interact with each other. However, because light photons are bosonic particles they do not interact in free space. One photon can pass through another one without feeling it. Therefore, to enable interaction between the photons, one needs to introduce a "soft" material in which the photons propagate. In such a material the presence of photons will create an optical potential, which will act on all other photons, making them interact with each other.

The induction of an optical potential in the material is the essence of optical nonlinearity. Physically this is a result of the deformation of the electronic orbits in the atoms of the material under the action of the incident light intensity. As such, the oscillations of the electrons in the nuclei will be anharmonic, resulting in a nonlinear dependence of the material polarization on the external electric field.

This nonlinear dependence on an external electric field is the key to realizing *any kind of all-optical functionalities* for processing of information. For example, the effect of nonlinear refraction enables us to perform light switching, gating, and modulation. The ultrafast nonlinearities of atoms and molecules enable conversion of light from one color to another in a nonlinear frequency conversion process, or parametric nonlinear interactions allow for light amplification without the need of active gain materials. However, nonlinear processes usually happen at high laser intensities and require high-power laser sources that consume a lot of energy. Therefore the important question is, how do you achieve such all-optical functionalities at low power and with high efficiency?

A key solution to this fundamental power consumption problem comes from the use of photonic nanostructures. There are several ways by which the nanostructuring of optical materials can strongly enhance the nonlinear effects. The first possibility is to enable *light confinement* to the nanoscale. Because the nonlinear effects are proportional to the light intensity, the concentration of light to a very small volume will result in a multifold increase in the light intensity at the same laser powers. As such large research efforts have been focused on the increase of light confinement by using dielectric nanowire waveguides, photonic crystal waveguides, or even metal-dielectric-metal waveguides for subwavelength light concentration. Another possibility for utilizing nanostructuring is to *engineer the light dispersion* in nanoscale photonic waveguides in order to achieve phase matching of the desired nonlinear processes, including frequency light conversion.

However, something that is really at the frontier of the modern nonlinear optics is the new opportunities for *ultrastrong nonlinear response of materials with artificial magnetism* at optical frequencies. In such materials, the nonlinear interaction between photons arises not only due to the nonlinear response of the electric polarization of the material but also due to the nonlinearity in the magnetic polarizability of the medium. This magnetic-type nonlinear optics is reliant on the development of a new class of artificial optical materials—metamaterials—that exhibit a strong magnetic response at optical frequencies. Due to the strong magnetic nonlinearity of such materials it is possible to observe new nonlinear phenomena and apply them for power-stringent all-optical signal-processing applications.

7.2 Optical Metamaterials

Every material is composed of atoms and molecules, whose properties define the properties of the entire material. In nature, the properties of a material are constrained by the different chemical compositions of its compounds. However, with the development of nanotechnologies, it became possible to engineer artificial atoms with properties defined mostly by the geometry of the atom rather than by its chemical compounds. Such nanoscale artificial atoms are called *meta-atoms* and are typically much smaller than the wavelength of light. When many such meta-atoms are placed close together they form a composite material, called a *metamaterial*. Importantly, when the distance between the artificial atoms is much smaller than the wavelength of light, the material appears homogeneous for the incident light.

7.2.1 Split-Ring Resonator as an Artificial Meta-Atom

The key element for designing artificial atoms with a magnetic response is the *split-ring resonator* (Fig. 7.1a). The important feature of the split-ring resonator meta-atom is that when light is incident on the split ring, it can induce currents along the metal ring. These induced currents correspondingly cause an induced magnetic dipole moment. This dipole moment results in an induced magnetic field that is reversed with respect to the incident magnetic field at optical frequencies just above the resonant frequency of the split ring (see Fig. 7.1a,b), thus creating strong magnetic resonance. The split-ring resonator can be interpreted as a liquid crystal (LC) circuit (see Fig. 7.1b) with a resonant frequency $\omega_0 = 1/LC$ such that above this resonant frequency, the magnetic polarization of the meta-atom becomes negative. This negative magnetic polarization is the cause of the artificial magnetism of such meta-atoms.

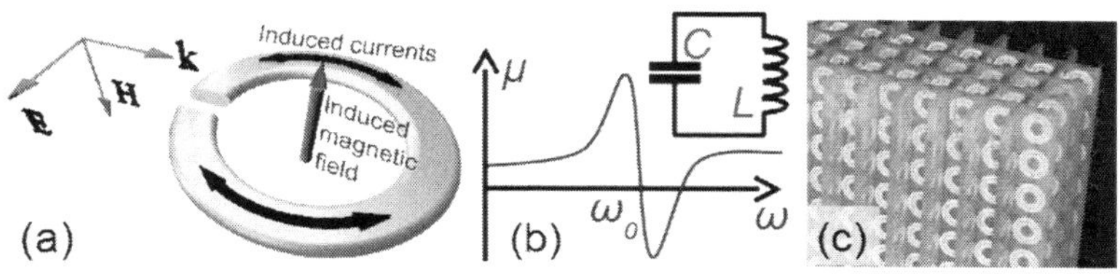

Figure 7.1 (a) Schematic of a split-ring resonator, illustrating the induction of currents and the corresponding magnetic dipole moment. (b) Equivalent rf circuit of the split ring, illustrating that above a resonant frequency ω_0, the effective permeability of the atom is negative over some frequency range. (c) Microwave metamaterial composed of closely spaced split-ring resonators. *Abbreviation*: rf, radio frequency.

Therefore, if we place many such split-ring resonators in an array, as shown in Fig. 7.1c, with lattice spacing much smaller than the excitation wavelength, we can assign to the composite material an effective permeability that is also negative over a certain frequency band. As mentioned above, this artificial magnetism calls for revisiting the field of nonlinear optics in order to account for nonlinearities that arise in the magnetic polarization of the medium.

To illustrate the opportunities coming from such magnetic nonlinearities, we can consider a split-ring resonator meta-atom where the slit of the ring is filled with a nonlinear dielectric (Fig. 7.2). Such a nonlinear dielectric changes its permittivity when a bias electric field is applied across the gap, therefore altering the capacitance C of the effective LC circuit. Therefore, according to our simplified picture of the effective meta-atom, the magnetic field H of the incident electromagnetic field will induce currents j along the ring, resulting in an electric field E_g inside the split-ring gap. This correspondingly will change the effective resonant frequency of the split ring and will therefore result in a nonlinear function of the magnetic permeability of the atom $\mu_{\text{eff}} = \mu_{\text{eff}}(H)$. In practice, such magnetic nonlinearity is achieved in microwave experiments by placing a varying capacitance diode across the gap of the split ring, where the effective capacitance of the diode is strongly affected by the flowing currents.

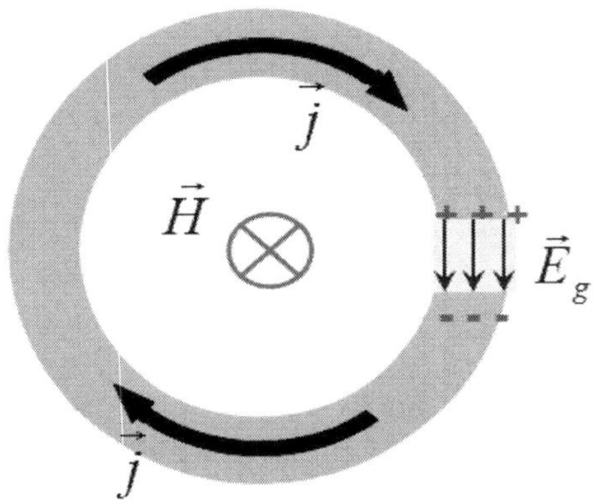

Figure 7.2 A split-ring resonator with a nonlinear dielectric in the gap of the split ring such that the magnetic permeability is a nonlinear function of the magnetic field H due to the magnetoelectric coupling in such meta-atoms (Zharov, Shadrivov, and Kivshar, 2003).

7.2.2 From Split Rings to Fishnet Optical Metamaterials

The big question, however, is how to transfer these split-ring resonator meta-atoms used in the microwave experiments to the optical domains, especially so that there are no diodes operating at optical frequencies. The answer comes from a simple continuous geometrical transformation of the split ring that is shown in Fig. 7.3 (Wegener and Linden, 2010). A single split ring in the left is transformed to a double split ring, which is rotated and stretched (Fig. 7.3). At the end of this transformation we obtain a cut-wire pair-metal-dielectric structure that can exhibit a negative magnetic permeability at optical frequencies. Combining this cut-wire-pair medium with the so-called metal wire medium, one finally ends up with the so-called fishnet metamaterials, as shown in Fig. 7.4.

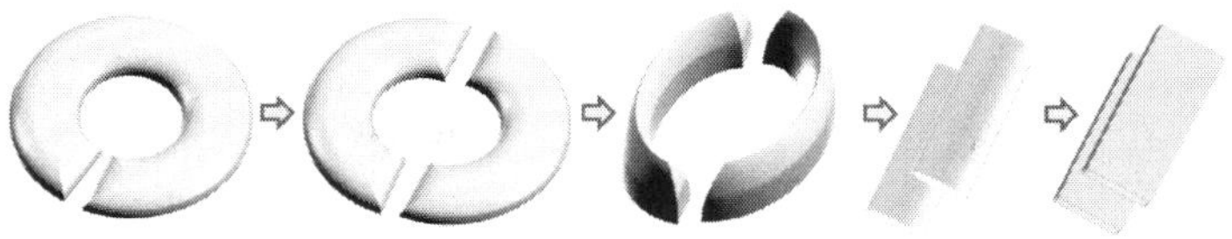

Figure 7.3 Set of images showing the transformation of the split-ring resonator to cut-wire pair metamaterials. Such transformations form the basis for artificial metamaterials at optical frequencies.

In Fig. 7.4, we show the development of metamaterials at the Australian National University (ANU). Whereas microwave metamaterials at the ANU were first designed more than six years ago, only recently it became possible to scale these designs down to fishnet metamaterials (right) to operate at telecommunication wavelengths (Minovich, Neshev, Powell, Shadrivov, Lapine et al., 2010). This breakthrough in development was possible due to the recently established state-of-the-art nanofabrication facilities at the ANU.

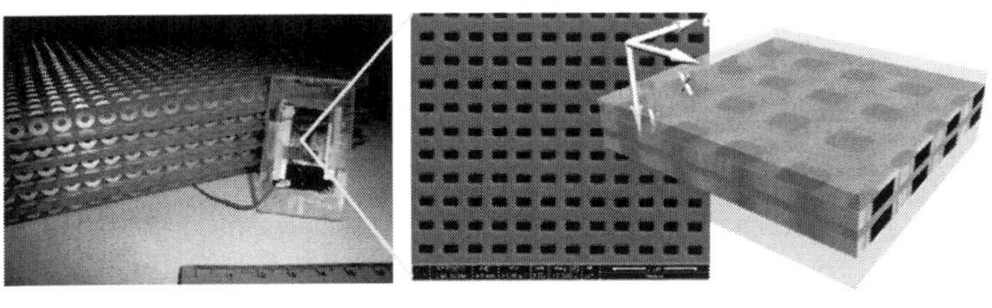

Figure 7.4 Transformation of the split-ring metamaterial operating at ~3 GHz (left) to a fishnet metamaterial at telecommunication wavelengths (right).

In general, fishnet metamaterials consist of two or more metal-dielectric layers that are milled through with rectangular-shaped holes (see Fig. 7.4 [right]). They were first demonstrated in 2005 (Zhang, Fan, Panoiu, Malloy, Osgood, and Brueck, 2005) and soon after used by a number of other groups around the world (Dolling, Enkrich, Wegener, Soukoulis, and Linden, 2006; Chettiar, Kildishev, Yuan et al., 2007; Li, Li, Wang et al., 2007; Valentine, Zhang, Zentgraf et al., 2008). In the southern hemisphere, this development has come only recently, when our group developed such materials in 2010

(Minovich, Neshev, Powell, Shadrivov, Lapine et al., 2010). Despite the strong international interest in fishnet metamaterials, we were the first to ask the question of how to make them highly nonlinear. And a clear answer to this question comes from the fact that there is strong electric field enhancement between the two metal plates of the fishnet, as well as inside the rectangular holes (Mary, Rodrigo, Garcia-Vidal, and Martin-Moreno, 2008).

7.3 Nonlinear Metamaterials

The optical properties of fishnet metamaterials are defined by their geometrical parameters and the relative permittivity of the materials composing the structure. Therefore, changing the dielectric permittivity of any of the structural compounds can lead to an overall change of the total refractive index of the fishnet metamaterial. Such index change can be induced by light itself (under the action of the enhanced electric field inside the dielectric layers of the fishnet) and with proper design of the fishnet structure can be amplified to lead to strong enhancement of the nonlinear response of the material.

7.3.1 Sensitivity of Fishnet Metamaterials

The occurrence of such enhanced sensitivity can be understood from the two-dimensional color plot of the overall metamaterial refractive index as a function of the index of the intermediate dielectric layer, as shown in Fig. 7.5a. The important features in this diagram are the two blue "fingers," representing the regions of negative refractive index of the metamaterial. Then, if we look at the dependence at a particular wavelength (see, e.g., the schematically drawn dependence for a wavelength of 1 μm in Fig. 7.5b), we can identify zones (red shading) where the overall index changes sharply with the change of the dielectric layer index. So if we tune the metamaterial parameter to one of these zones (e.g., using a dielectric layer of an appropriate refractive index), we can achieve very high sensitivity of the fishnet metamaterials to the induced index change in the dielectric layer between the two metal plates. Hence, if we use a nonlinear dielectric layer, we can achieve an overall enhancement of the nonlinear response of our metamaterial.

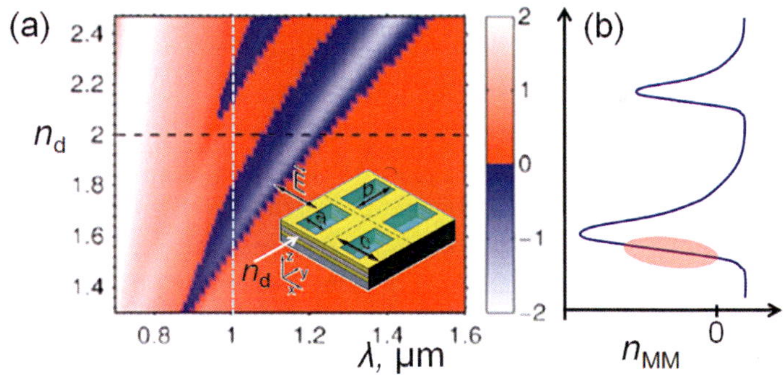

Figure 7.5 (a) The real part of the fishnet metamaterial effective refractive index as a function of the refractive index (n_d) of the middle dielectric layer (Minovich, Neshev, Powell, and Kivshar, 2010). (b) Schematic of the same dependence for a fixed wavelength of 1 μm. The structural parameters are a = 200 nm, b = 270 nm, and c = 410 nm. The metal and dielectric layer thicknesses are 30 nm and 50 nm, respectively.

However, because in reality the metamaterial thickness is very small, it is more practical to measure not the overall nonlinear index change but the change of the overall transmission through the fishnet material. Even this is not an easy task due to the broad spectral width of the plasmonic resonances at visible and near-infrared wavelengths. To be able to detect a substantial change in the transmission (corresponding to a substantial spectral shift of these resonances), the required index change needs to be of the order of $\Delta n \approx 0.1$ (Carretero-Palacios, Minovich, Neshev et al., 2010). While such a large change in the refractive index is doubtful to be achieved with fast electronic nonlinearity, it can be feasible by exploring the slow reorientational nonlinear response of LCs. Such reorientational nonlinearity is due to the strong dependence of the LC refractive index versus the orientation of the LC molecules, which is influenced by applied external electric fields. Therefore, the infiltration of the structure with LCs would provide an efficient route for tuning of its transmission and effective refractive index. Importantly, the LC reorientation can be achieved by the application of either external electric or optical fields.

We note that LCs have been utilized previously to tune electromagnetic transmission through structured metal surfaces for the control of optical transmission through metallic hole arrays (Dickson, Wurtz, Evans, Pollard, and Zayats, 2007; Evans, Wurtz, Hendren et al., 2007) and more recently for the control of transmission through fishnet metamaterials operating in the microwave spectral region (Zhang, Zhang, Zao et al., 2011). LC tenability of magnetic and negative index metamaterials has also been suggested (Khoo, Werner, Liang, Diaz, and Weiner, 2006; Wang, Kwon, Werner, Khoo, Kildishev, and Shalaev, 2007; Minovich, Neshev, Powell, Shadrivov, and Kivshar, 2010) and experimentally verified for thermal tunability of purely magnetic metamaterials (Xiao, Chettiar, Kildishev, Drachev, Khoo, and Shalaev, 2009). However, the transfer of these LC-tuning techniques to negative index fishnet metamaterials operating at optical frequencies has not been experimentally attempted to date, which is the task presented in this chapter.

7.3.2 Liquid Crystal–Infiltrated Fishnet Metamaterials

To achieve this goal first we perform numerical simulations to identify the best fishnet geometry for our experiments. The fishnet structure exhibits negative effective permittivity below the plasma frequency, which matches the cutoff frequency of the holes, that is, the transmission peak of the hole mode $\lambda_c >> 2b$, where b is the hole size along the direction of the magnetic field. The negative magnetic response is related to the excitation of surface plasmon polaritons (SPPs) modes in the dielectric gap between the two metal plates. If the SPPs are excited below the plasma frequency (at longer wavelengths), simultaneously negative effective permittivity and permeability are achieved. This results in an effective negative refractive index of the fishnet metamaterial (Mary, Rodrigo, Garcia-Vidal, and Martin-Moreno, 2008; Mary, Rodrigo, Martin-Moreno, and Garcia-Vidal, 2009).

The cutoff frequency of the hole modes depends strongly on the refractive index of the material inside the holes, namely, the infiltrated LCs (see Fig. 7.6a). Therefore, tuning its permittivity will effectively change the optical response of the entire structure. In addition, if the SPP resonance is located close to the cutoff frequency of the hole modes, the induced shift of its cutoff frequency can exchange the relative positions of both resonances, thus achieving switching of the effective index from positive to negative values. In our work (Minovich, Neshev, Powell, Shadrivov, and Kivshar, 2010), we have found that for the optimal geometrical parameters of gold-ZnO-gold trilayer fishnet metamaterials, infiltrated with *E*7 LCs ($\Delta n \approx 0.2$ in the near-infrared spectral range), there is indeed switching of the effective refractive index of the metamaterials from positive to negative over some spectral range at a central wavelength of about 1.3 μm (Minovich, Neshev, Powell, Shadrivov, and Kivshar, 2010). This switching is seen in our numerical results presented in Fig. 7.6b. Importantly, this is an overall change of the fishnet refractive index of $\Delta n \approx 1.1$, which is a fivefold enhancement of the nonlinear index change (at the resonant wavelength) in comparison to the change of the sole LC refractive index due to molecular reorientation. In practice, the LC molecules are reoriented by applying voltage U between the gold film and an indium tin oxide (ITO) transparent electrode placed above the sample, as shown in the bottom inset in Fig. 7.6(c).

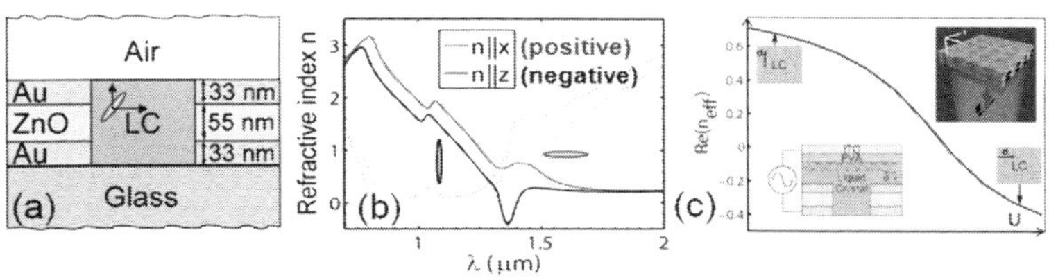

Figure 7.6 (a) Schematic of the LC-infiltrated fishnet metamaterial (side view). (b) The real (solid line) and imaginary (dotted line) parts of the effective refractive index for two orientations of the *E*7 LC molecules. The structural parameters are $a = 200$ nm, $b = 350$ nm, and $c = 420$ nm, and metal and dielectric layer thicknesses are as indicated in (a). (c) The change of the overall metamaterial refractive index at a wavelength of 1.3 μm.

7.3.3 Nonlinear Transmission through LC-Infiltrated Fishnet Metamaterials

To demonstrate our prediction experimentally, we fabricate several types of fishnet metamaterials using different dielectric layers sandwiched between two gold films (Fig. 7.7a). For nanostructuring of the metal-dielectric layers, we use focused ion beam milling, fabricating a typical fishnet structure, as shown in the scanning electron micrograph in Fig. 7.7b. The nanostructured material is infiltrated with an *E*7 LC layer and sandwiched on the top with a glass substrate coated with a transparent ITO electrode and a polyvinyl alcohol (PVA) layer for the prealignment of the LC molecules. Furthermore, a DC electric field can be applied between the top and bottom ITO electrodes for electrical control of the LC molecular alignment. The complete assembled LC cell is shown in Fig. 7.7c.

Figure 7.7 (a) Schematic of the LC-infiltrated fishnet metamaterial (side view). (b) A scanning electron micrograph image (top view) of the fabricated fishnet metamaterials. (c) Image of the final *E*7 LC-infiltrated metamaterial cell.

Below we show our results of the experimental testing of one particular sample—Au-MgF_2-Au trilayer fishnet—having the necessary geometrical parameters to achieve switching of the effective index from positive to negative values at telecommunication wavelengths. First we measure the linear transmission spectrum of the fishnet before infiltration (Fig. 7.8, solid red line) and after infiltration (Fig. 7.8, solid green line) to test our theoretical predictions (Minovich, Neshev, Powell, Shadrivov, and Kivshar, 2010). In our experimental setup we use excitation with a white-light-halogen linearly polarized light source. We focus the light onto the metamaterial surface and collect the transmitted light in the infrared spectral region onto an optical spectrum analyzer. The results are shown in Fig. 7.8. The solid curves correspond to our experimental measurements, while the dashed curves are based on finite-difference time domain numerical simulations. In our simulations, we use the six-term Drude–Lorentz expression for the dielectric permittivity of gold (Rakic, Djurišic, Elazar, and Majewski, 1998) and a constant refractive index of 1.38 for MgF_2. While there is a factor of two difference in the magnitude of transmission between our measurements (solid lines) and the numerical calculations (dashed lines), there is a good agreement in the position of the spectral lines. We attribute the lower transmission seen in the experiment to the inevitable roughness of the metal films and the higher ohmic losses of the sputtered gold. Importantly, from our measurements (Fig. 7.8) we can conclude that we have a complete LC infiltration of the nanoscale holes of the material and there are no air pockets formed during the infiltration process. The latter would be clearly visible in the spectral transmission, which would lie in between both peaks, as indicated by the results of the numerical simulations shown with blue lines in Fig. 7.8.

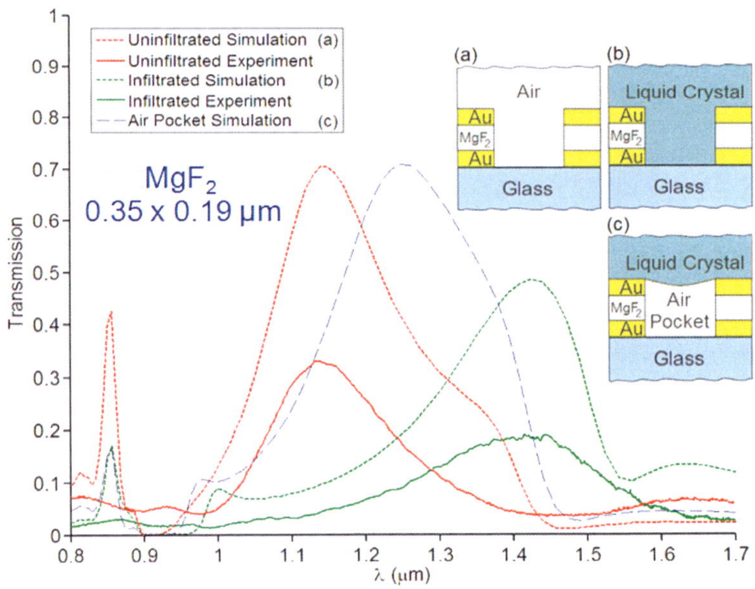

Figure 7.8 Linear transmission measurement of the LC cell, including Au-MgF_2-Au fishnet metamaterials. Solid lines—experimental results; dashed/dotted lines—numerical simulations; red—uninfiltrated fishnet metamaterials; green—fishnet metamaterials fully infiltrated with *E*7 LC; blue—calculations of hypothetical air pockets in the infiltrated metamaterials.

The next important point of our experiments is the test of the alignment of the LC molecules. The use of a PVA layer on the top glassed substrate forming our LC cell ensures that near this layer, the LC molecules are aligned parallel to the surface. However, we are unsure how this alignment propagates inside the cell and in particular inside the holes of the fishnet metamaterials. If this alignment horizontal is preserved, we should be able to use a DC electric field across the top and bottom electrodes of the LC cell to change it from horizontal to vertical. Such an external DC electric field is perpendicular to the initial alignment of the LC molecules and should therefore reorient them perpendicular to the top flat surface. However, after measuring the transmission spectra with and without an applied DC electric field, we saw no difference between the two. This is in contrast to the theoretically calculated transmission spectra that change significantly (of the order of 100 nm) for horizontal and vertical alignment of the LC molecules. The observed lack of change in the experimental transmission indicates that the LC molecules are already aligned vertically inside the holes of the fishnet. Such fixed alignment inside the nanoscale holes is possible due to surface anchoring of the molecules to the sidewalls. Therefore, for the initial vertical alignment of the LC molecules inside the holes, their reorientation would only be possible through an external x-polarized electric field, such as optical fields.

To test this dependence of the optical transmission on the light intensity, we measure the transmission through our infiltrated metamaterials versus an incident laser power at a fixed wavelength of 1550 nm. Our pump source is a fiber laser operating at telecom wavelengths. The particular pump wavelength is positioned on the long wavelength side of the transmission maxima (see Fig. 7.8) and near the wavelength of the gap plasmon resonance. The laser is mildly focused onto the metamaterial, and the transmitted light is detected on an InGaAs detector. First, we performed a reference measurement by testing the transmission of the entire stack of layers in the LC cell but without the metamaterial. In this test experiment we obtain a perfectly linear dependence of the transmission with power regardless of the applied DC electric field.

However, this linear dependence dramatically changes once we have the fishnet metamaterials inserted into the beam path. In this case, we observe a drop in the transmission (sublinear dependence) with an increase in the incident laser power. At low input powers the transmission through the infiltrated structure is ~7%. As the power is increased, the transmission is reduced by approximately 30%. The normalized dependence of the transmission is shown in Fig. 7.9 for the power range from 0 mW to 40 mW. Above these power levels, the thermal effects start to dominate the transmission dependence, which is reflected in the increase of the error bars in our transmission measurements. As seen in Fig. 7.9, the drop in transmission saturates at about 20 mW, which is an indication for complete LC reorientation at these laser powers. We also observe that the measured transmission drop is dependent on the application of the bias DC electric field, indicating again that we have molecular reorientation inside the holes of the fishnet metamaterials. However, further experimental studies are needed to determine the exact mechanisms behind the observed effects, as well as to find ways to limit the adverse LC surface anchoring inside the holes of the fishnet metamaterials.

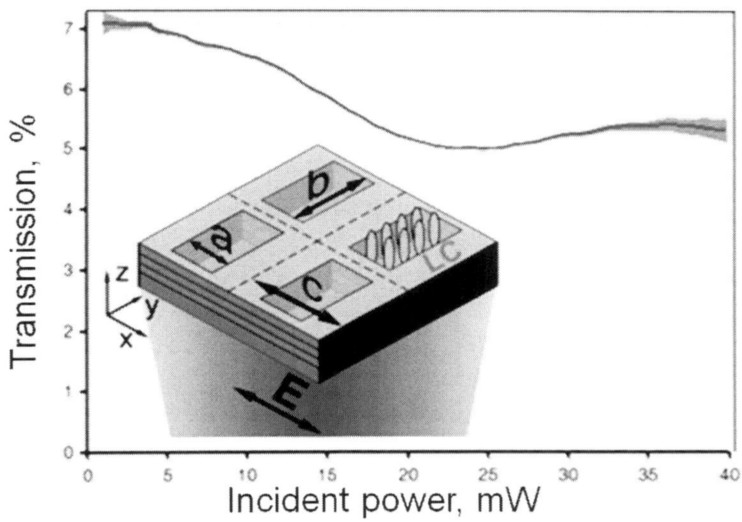

Figure 7.9 Measured transmission at 1550 nm versus input power. (Inset) Experimental geometry.

7.4 Discussions and Conclusions

Our results to dynamically tune the refractive index of fishnet metamaterials with the incident light itself represent an important milestone in the development of metamaterials with an enhanced nonlinear response. We believe that further studies on nonlinear metamaterials will lead to the discovery of entirely new fundamental physical phenomena, thus opening novel applications for adaptive control of metamaterial parameters.

The results presented in this chapter only scratch the surface of possibilities for future development. Obviously there is a lot more work to be done to understand deeper the underlying physics of the observed effects. For example, it is important to understand the specific anchoring of the LC molecules to the wall of the fishnet materials. Because of the complex composite metal-dielectric structure and nanoscale dimensions of the sidewall, it is hard to predict the exact surface chemistry of LC molecules. Possible solutions are the application of monolayers that modify the surface tension on the sidewall, thus removing the anchoring of the LC molecules. The use of negative LCs could also show particular promises for different reorientation. Finally, the electric fields inside the fishnets materials consist of both transverse and longitudinal components, resulting in complex vector fields inside the fishnets and hence a complex structure of the molecular orientation.

The possible future implementation of metamaterials in signal-processing devices further requires the implementation of ultrafast nonlinear processes. Therefore in future research it is important to utilize entirely solid-state materials rather than LCs with a slow orientation nonlinear response. Our group is currently working on metamaterials with semiconductor or piezoelectric dielectric layers, thus opening new opportunities for metamaterials tuning. In general, there are vast opportunities for the future communication networks that will come from the development of novel nanomaterials (Islam and Logeeswaran, 2010). A combination of metamaterials with other nanotechnologies, such as quantum dot–based, DNA-based, or carbon-based nanomaterials, will surely open new avenues for research and technologies that will contribute to the sustainable development of communication technologies, feeding progressively to more secure and greener communications.

In all these future efforts, one needs to identify the true opportunities for nanostructured metamaterials without restriction to specific geometries or constituent materials. While the research to follow will most likely have entirely exploratory character, we believe that the impact will go far beyond optical communications and will influence other related areas, including photovoltaic technologies and homeland security. Indeed metamaterials have promised to provide an avenue for the development of perfect light absorbers (Liu, Starr, Starr, and Padilla, 2010) that can absorb light in an ultrathin nanoscale layer, enabling the development of ultrathin solar cells. In homeland security, metamaterials have really inspired the public with the ability to create invisibility cloaks (Schurig, Mock, Justice et al., 2006) and hide objects or create the illusion that objects are invisible—an opportunity that has certain appeal in military or security applications.

We are certainly looking to a bright and dynamic future!

Acknowledgments

I would like to acknowledge a number of research students and collaborators for their substantial and valuable contributions to the results summarized in this chapter. Most of all I would like to thank A. Minovich, J. Farnell, I. Shadrivov, D. Powell, H. H. Tan, C. Jagadish, and Y. S. Kivshar. This work was supported by the Australian Research Council under the Centre of Excellence research program. I acknowledge the support of the Australian National Computational Infrastructure and the ACT Node of Australian Nanofabrication Facility.

References

Carretero-Palacios, S., Minovich, A., Neshev, D. N., et al. (2010). Optical switching in metal-slit arrays on nonlinear dielectric substrates. *Opt. Lett.*, **35**(24), 4211–4213.

Chettiar, U. K., Kildishev, A. V., Yuan, H.-K., et al. (2007). Dual-band negative index metamaterial: double negative at 813 nm and single negative at 772 nm. *Opt. Lett.*, **32**(12), 1671–1673.

Dickson, W., Wurtz, G. A., Evans, P. R., Pollard, R. J., and Zayats, A. V. (2007). Electronically controlled surface plasmon dispersion and optical transmission through metallic hole arrays using liquid crystal. *Nano Lett.*, **8**(1), 281–286.

Dolling, G., Enkrich, C., Wegener, M., Soukoulis, C. M., and Linden, S. (2006), Simultaneous negative phase and group velocity of light in a metamaterial. *Science*, **312**(5775), 892–894.

Evans, P. R., Wurtz, G. A., Hendren, W. R., et al. (2007). Electrically switchable nonreciprocal transmission of plasmonic nanorods with liquid crystal. *Appl. Phys. Lett.,* **91**(4), 043101–043103.

Islam, M. S., and Logeeswaran, V. J. (2010). Nanoscale materials and devices for future communication networks. *IEEE Commun. Mag.,* **48**(6), 112–120.

Khoo, I. C., Werner, D. H., Liang, X., Diaz, A., and Weiner, B. (2006). Nanosphere dispersed liquid crystals for tunable negative-zero-positive index of refraction in the optical and terahertz regimes. *Opt. Lett.,* **31**(17), 2592–2594.

Li, T., Li, J., Wang, F., et al. (2007). Exploring magnetic plasmon polaritons in optical transmission through hole arrays perforated in trilayer structures. *Appl. Phys. Lett.,* **90**(25), 251112.

Liu, X., Starr, T., Starr, A. F., and Padilla, W. J. (2010). Infrared spatial and frequency selective metamaterial with near-unity absorbance. *Phys. Rev. Lett.,* **104**(20), 207403.

Mary, A., Rodrigo, S. G., Garcia-Vidal, F. J., and Martin-Moreno, L. (2008). Theory of negative-refractive-index response of double-fishnet structures. *Phys. Rev. Lett.,* **101**(10), 103902.

Mary, A., Rodrigo, S. G., Martin-Moreno, L., and Garcia-Vidal, F. J. (2009). Holey metal films: from extraordinary transmission to negative-index behavior. *Phys. Rev. B,* **80**(16), 165431.

Minovich, A., Neshev, D. N., Powell, D. A., and Kivshar, Y. S. (2010). Influence of the substrate on negative index fishnet metamaterials. *Opt. Commun.,* **283**(23), 4770–4774.

Minovich, A., Neshev, D. N., Powell, D. A., Shadrivov, I. V., and Kivshar, Y. S. (2010). Tunable fishnet metamaterials infiltrated by liquid crystals. *Appl. Phys. Lett.,* **96**(19), 193103.

Minovich, A., Neshev, D. N., Powell, D. A., Shadrivov, I. V., Lapine, M., et al. (2010). Tilted response of fishnet metamaterials at near-infrared optical wavelengths. *Phys. Rev. B,* **81**(11), 115109.

Rakic, A. D., Djurišic, A. B., Elazar, J. M., and Majewski, M. L. (1998). Optical properties of metallic films for vertical-cavity optoelectronic devices. *Appl. Opt.,* **37**(22), 5271–5283.

Schurig, D., Mock, J. J., Justice, B. J., et al. (2006). Metamaterial electromagnetic cloak at microwave frequencies. *Science,* **314**(5801), 977–980.

Valentine, J., Zhang, S., Zentgraf, T., et al. (2008). Three-dimensional optical metamaterial with a negative refractive index. *Nature,* **455**(7211), 376–379.

Wang, X., Kwon, D.-H., Werner, D. H., Khoo, I.-C., Kildishev, A. V., and Shalaev, V. M. (2007). Tunable optical negative-index metamaterials employing anisotropic liquid crystals. *Appl. Phys. Lett.,* **91**(14), 143122.

Wegener, M., and Linden, S. (2010). Shaping optical space with metamaterials. *Phys. Today,* **63**(10), 32–36.

Xiao, S., Chettiar, U. K., Kildishev, A. V., Drachev, V., Khoo, I. C., and Shalaev, V. M. (2009). Tunable magnetic response of metamaterials. *Appl. Phys. Lett.,* **95**(3), 033115–033117.

Zhang, F., Zhang, W., Zhao, Q., et al. (2011). Electrically controllable fishnet metamaterial based on nematic liquid crystal. *Opt. Express,* **19**(2), 1563–1568.

Zhang, S., Fan, W., Panoiu, N. C., Malloy, K. J., Osgood, R. M., and Brueck, S. R. J. (2005). Experimental demonstration of near-infrared negative-index metamaterials. *Phys. Rev. Lett.,* **95**(13), 137404.

Zharov, A. A., Shadrivov, I. V., and Kivshar, Y. S. (2003). Nonlinear properties of left-handed metamaterials. *Phys. Rev. Lett.,* **91**(3), 037401.

Chapter 8

Laser Trapping of Nanoparticle Agglomerates in Air

Andrei V. Rode, Vladlen G. Shvedov, Cyril Hnatovsky, and Weislaw Krolikowski
Laser Physics Centre, Research School of Physics and Engineering, Australian National University, Canberra, ACT 0200, Australia
avr111@physics.anu.edu.au

This chapter presents a review of a recently developed technique of trapping nanoparticles in air using optical vortices. The technique is based upon thermal forces induced by uneven heating of light-absorbing particles with a laser beam. The laser vortices with a doughnut-shaped intensity profile provide a means to trap particles and push them along the zero-intensity beam axis to a desirable location. A brief overview will cover recent results of trapping, guiding, and pinpoint-positioning agglomerates of carbon nanoparticles in air, as well as simultaneous trapping of thousands of particles with a single laser beam. Optical vortices can be applied to touch-free transport of containers holding gases, ultrapure or dangerous substances, and biological macromolecules. Laser-trapping and agglomerating nanoparticles can be applied for

Nanotechnology Toward the Sustainocene
Edited by Thomas A. Faunce

ISBN 978-981-4463-76-8 (Hardcover), 978-981-4463-77-5 (eBook)
www.panstanford.com

monitoring and removal of nanopollutants from air, offering a new avenue toward environmental protection in the workplace of the nanotechnology industry.

8.1 Introduction

New generations of nanomaterials evolving with the dynamic developments in nanotechnology hold great promise for the creation of new products with enhanced properties and attributes. Nanomaterials characteristics such as greater catalytic efficiency or improved hardness and strength are highly desirable for applications in commercial, medical, and environmental sectors. The explosive growth of nanotechnology and the rapid expansion of nanoparticle-based products in many industrial sectors in the first decade of the new millennium have raised new, unresolved issues of occupational exposure and potential environmental risks unique to artificially produced engineered nanoparticles. The fast-developing field of nanotechnology presents us with serious new challenges in occupational health and safety in the workplace as the adverse effects of nanoparticles on human health are potentially significant and mostly unknown (Sahoo, Parveen, and Panda, 2007; Nel, Xia, Mädler, and Li, 2006; Singh and Nalwa, 2007; Faunce, Walters, Williams, Bryant, Jennings, and Musk, 2006; Maynard, Aitken, Butz et al., 2006; Shatkin, 2008). Until recently, the spectacular developments in nanotechnology have been with little regard to the potential effects of nanoparticles on human health and the environment. Some of the unique properties of nanoparticles may pose hazards to humans as they are able to pass through cells membranes or cross the blood-brain barrier (Maynard, Aitken, Butz et al., 2006; Shatkin, 2008)

A serious challenge regarding environmental protection is to develop new methods for effective removal of nanoparticle contamination from air. There is a growing demand for techniques enabling efficient removal of nanocluster air contaminants, which cause serious health risks. Methods for effective capture and removal of nanoparticle contamination from air need to be developed both quickly and cost effectively.

While optical trapping in liquids at the microscopic scale is well documented with various reviews (Dholakia and Reece, 2006;

Dholakia, Reece, and Gu, 2008), little attention has been paid to the trapping and manipulation of nanometric-size particles, which fall largely between the size region of cells and that of atomic ensembles. In particular the trapping of particles at the nanometer scale and direct manipulation of biological macromolecules have gained momentum. In this chapter, we present a brief summary of recent developments in laser trapping, guiding, and manipulation of agglomerated nanoparticles suspended in air with optical vortices. The new concept of touch-free transport and pinpoint positioning of objects in air opens up diverse and rich, practical opportunities for handling objects in dangerous or hard-to-reach areas. The concept can be applied to touch-free trapping and agglomeration of nanosize particles, viruses, and living cells.

8.2 Nanoparticles and Their Properties

8.2.1 What Makes Nanoparticles So Special?

The term "nanoparticle" is usually applied to particles with dimensions between the atomic size and the submicron size of a bulk material, that is, between approximately 1 nm and 100 nm. The main reason for setting apart the particles in this size range is the unique properties and behavior of matter at the nanometer scale: while the properties of a bulk material are constant, physical properties of nanoparticles are size dependent. For example, gold nanoparticles below 100 nm change their color to red; their melting point reduces from the bulk value of 1064°C for particles below 10 nm and goes quickly down to ~300°C for particle sizes from 5 nm to ~2.5 nm (Buffat and Borel, 1976; Wang, Chen, Wang, Wang, Lu, and Zhao, 2002). Moreover, gold nanoparticles become chemically reactive and behave as a semiconductor at a particle size below approximately 5 nm (Pyykkö, 2007; Link and El-Sayed, 2003). Size-dependent photoluminescence of silicon nanoclusters has been observed experimentally (Link and El-Sayed, 2003). When the cluster size was decreased from 3.5 nm to 1 nm, the peak in the luminescence spectrum shifted from 750 nm to 300 nm (Marine, Patrone, Luk'yanchuk, and Sentis, 2000). Another example is magnetic properties of carbon: while graphite, diamond, and fullerenes have diamagnetic properties, polymerized fullerenes (Makarova, Han, Esquinazi et al., 2003) and laser-deposited carbon nanofoams (Rode, Gamaly, Christy et al., 2004; Arčon, Jagličič, Zorko et al., 2006), which appear to contain schwarzite structures with hyperbolically curved graphitic-like sheets, demonstrate paramagnetic behavior.

The underlying foundation for the size-dependent properties of nanoparticles is in quantum size effects, which are exhibited by atoms on the surface. While the number of surface atoms is reduced in proportion to R^{-2} with the decreasing radius R of a particle, the number of atoms in the bulk is reduced much faster, in proportion to $\sim R^{-3}$, so the relative number of surface atoms grows with the decreasing size. In the bulk, the interference of electronic wave functions results in a band structure. The surface atoms differ slightly from their bulk counterparts because they do not have their symmetry in interaction with neighboring atoms. Their electronic orbits are slightly energetically shifted; thus the dielectric function, which is a key factor dictating the material behavior, is different. When the size of the system decreases to the nanometer scale, the energy bands split into energy levels. The difference in energy between the electronic levels ΔE in a small cluster increases in inverse proportion to R as $\Delta E \propto v_F/R$, where v_F is the electron Fermi velocity (Gorkov and Eliashberg, 1965; Kreibig and Vollmer, 1995; Prasad, 2004; Gamaly and Rode, 2004). For this reason, surface atoms, and as a consequence the surface area density (m^2/g) of industrial components, play a more and more significant role in determining the nanomaterial physical properties and chemical activity.

8.2.2 Health Implications

The dependence of material characteristics on the particle size opens up many industrial benefits. By tuning a physical characteristic to a required value, nanotechnology promises a new dimension and a new level of complexity in the manufacture of the existing and developing novel materials with new features. It paves the way for a new technological revolution. The absence of complete information precludes, however, a full assessment of possible health effects, both short term and long term. Considerable uncertainty exists regarding health risks from nanoscale materials, such as the adverse effects of nanoscale particles emitted as air pollutants (Faunce, Walters, Williams, Bryant, Jennings, and Musk, 2006; Maynard, Aitken, Butz et al., 2006; Shatkin, 2008).

Human exposure to these nanomaterials is inevitable, as they can enter the body through inhalation, food consumption, or absorption through skin and affect different organs and tissues, such as the brain, liver, kidney, heart, colon, spleen, bone, and blood, with possible cytotoxic effects, for example, the deformation and inhibition of cell growth, leading to various diseases in humans (Sahoo, Parveen, and Panda, 2007; Nel, Xia, Mädler, and Li, 2006; Singh and Nalwa, 2007; Dreher, 2004). Nanomaterials that are most likely to present health risks are nanoparticles, agglomerates of nanoparticles, and particles of nanostructured materials in air. On the cellular level, an ability to act as a gene vector has been demonstrated for nanoparticles. Carbon black nanoparticles have been implicated in interfering with cell signaling. Researchers in Germany have observed nanomaterials pass the blood–brain barrier and concentrate in the cerebellum, responsible for controlling balance and voluntary movement (Tolstoshev, 2006).

Given the rapid rate of development, it is not surprising that concerns have been raised relating to the safety of nanomaterials in a variety of products. German authorities recently recalled a bathroom-cleansing product MagicNano that was purported to contain nanosize particles. After it was on the market for only three days, more than 100 people suffered from severe respiratory problems, while 6 of them were hospitalized with pulmonary edema (Weiss, 2006). The exact cause of the respiratory effects had not been determined, and it was not clear whether the product actually contained nanoparticles or whether the reported effects were due to nanoparticles (ICON press release, 2006). Another example is TiO_2 nanoparticles used in some sun creams; they have the potential to cause neurological damage (Ball, 2006). It is too early to know the scale of how worrying these findings are.

Sufficient scientific evidence exists to conclude that many nanomaterials are highly likely to be toxic to human health and the environment. Issues regarding safe handling of potentially toxic nanomaterials, including questions of whether personal protective equipment is effective for protection against nanoparticle exposures, have not yet been solved. Below we present an overview of a new concept recently developed at the Australian National University in Canberra for capturing and handling of airborne nanoparticles, potentially dangerous for human health, using optical vortices (Shvedov, Desyatnikov, Rode, Krolikowski, and Kivshar, 2009; Desyatnikov, Shvedov, Rode, Krolikowski, and Kivshar, 2009; Shvedov, Rode, Izdebskaya, Desyatnikov, Krolikowski, and Kivshar, 2010a, 2010b; Shvedov, Rode, Izdebskaya et al., 2010).

8.3 Laser Trapping of Airborne Particles

8.3.1 Photophoretic Force

The mechanical effect of light on matter was first noticed by Johannes Kepler about four centuries ago, when he was intrigued by his observation that the tail of a comet pointed away from the sun at all times, which he attributed to solar pressure. The fact that electromagnetic radiation exerts a pressure was first predicted by Maxwell and proven experimentally by Lebedev (1901) and Nichols and Hull (1901). The ability to apply forces through radiation pressure gave rise to the optical tweezers technique, a noncontact method to trap and manipulate particles with a laser beam tightly focused by a high-numerical aperture microscope objective. The strong-intensity gradients in the converging beam cause polarization of particles and draw these particles toward the focus, while the radiation pressure pushes them along the optical axis. If the gradient force dominates, the particles can be trapped near the focal point. Optical tweezers led to a large number of unique opportunities to probe particle dynamics and enabled a wide range of studies in physical, chemical, and biological sciences. Optical trapping of nanometric objects is more challenging than trapping of micron-size objects due to a number of fundamental factors. First of all, there is a limitation on the amount of force one can possibly exert on a particle of nanodimensions by light. Another challenge relates to the different optical properties of nanoparticles when compared to the bulk material. A comprehensive review of forces in optical trapping of particles, particularly metallic nanoparticles, is presented by Dienerowitz, Mazilu, and Dholakia (2008).

A vast majority of optical-trapping and manipulation experiments, however, were carried out with particles immersed in liquids, where the buoyancy of the particles simplifies the trapping and suppresses the particle motion. Little work has been carried out on airborne particles due to the difficulties associated with trapping particles in air. A very limited number of pilot studies with airborne particles have been carried out, some with liquid aerosols, and are related to the chemical reactions in atmospheric aerosols (Tong, 1975; Pluchino, 1983; McGloin, Burnham, Summers, Rudd, Dewar, and Anand, 2008).

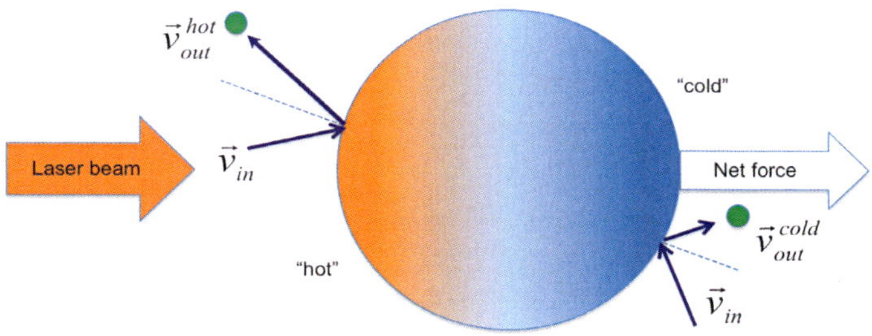

Figure 8.1 Illustration of the concept of photophoresis and thermal forces induced by heating a particle nonuniformly.

The main difficulty with optical trapping of absorbing particles in air is the presence of an additional, when compared to trapping in liquids, and strong influence of a photophoretic force due to nonuniform temperature distribution across a particle. Photophoresis results from the momentum transfer from gas molecules and occurs only if the surface of a particle is nonuniformly heated by an incident light beam. Under anisotropic heating, gas molecules rebound off the surface with different velocities, creating an integrated force on the particle. This effect was discovered by Ehrenhaft and termed *photophoresis* (Ehrenhaft, 1917; Preining, 1966; Allen, Beijersbergen, Spreeuw, and Woerdman, 1992). This force depends on the particle size relative to the mean free path of gas molecules, absorption of laser light, and thermal conductivity rather than the refractive index of the particle. In the conditions of a temperature difference across the particle, gas molecules with mass m_{gas}, which are much smaller than the particle, acquire different momentum depending on the point of impact (Melzer, 2001):

$$v_{\text{in}} \propto \sqrt{\frac{k_{\text{B}}T_{\text{gas}}}{m_{\text{gas}}}}; \quad v_{\text{out}}^{\text{hot}} \propto \sqrt{\frac{k_{\text{B}}T_{\text{p}}^{\text{hot}}}{m_{\text{gas}}}} > v_{\text{out}}^{\text{cold}} \propto \sqrt{\frac{k_{\text{B}}T_{\text{p}}^{\text{cold}}}{m_{\text{gas}}}} \tag{8.1}$$

where v_{in} and v_{out} are the molecule velocities before and after the impact, $k_{\text{B}}T_{\text{gas}}$ is the gas temperature, and $k_{\text{B}}T_{\text{p}}$ is the particle local temperature at the point of impact (Fig. 8.1). Particles scattered from a hotter area gain higher momentum; therefore, the resulting force is always directed away from the higher temperature and thus the maximum laser intensity. This is the reason that absorbing particles cannot be trapped using a Gaussian beam profile. Semitransparent particles move away (positive photophoresis) or toward the light source (negative photophoresis), depending on whether the front or the back surface is hotter.

The interaction of light with airborne particles is characterized by a Knudsen number, *Kn*, which is the ratio of the mean free path of the gas molecules to the size of the particle. The mean free path in air at atmospheric pressure is about 100 nm; thus very different physical mechanisms are involved when trapping particles of a few nanometers compared to particles of a few microns in air. For the small particles (large $Kn > 10$ range) the photophoretic forces can be estimated by kinetic theory. This force is always positive, which means the force is pointing in the same direction as the laser beam and comparable to the radiation pressure for both weakly and strongly absorbing particles. The radiation pressure is directed toward the lower intensity, which helps to trap particles in the minimal intensity in the center of the optical vortices. At the same time, the small particles develop an electric dipole moment in response to the laser electric field, which draws the particles toward the gradient of electric field, that is, to the focus with higher intensity.

For a low $Kn \ll 1$ (larger particles) the nature of the interaction is quite different—the particle's motion is due to the thermal creep phenomenon. Gas molecules slide over the particle surface, and the corresponding tangential stress tends to move the gas from colder to hotter parts of the surface. The temperature difference, in turn, depends on the particle transmission for the laser light, which is wavelength dependent, and thermal conductivity; thus the photophoretic force can be either positive or negative (i.e., directed toward the laser source or away from the source). To make things even more complicated, large particles, larger than the laser wavelength, act as lenses refracting the rays of light and redirecting the momentum of the photons, driving the particle toward the higher flux, which can have a substantial influence on mesoscopic agglomerates of nanoparticles.

In spite of well-supported skepticism, the first trapping of highly absorbing particles in air was demonstrated by levitation of carbon microspheres of 0.75 μm to 4.0 μm near a focal beam by Pluchino in 1983. It was established that the mechanism of trapping was entirely dependent on the particle's thermal properties, laser beam power, and the type of surrounding gas molecules. It was not possible to trap weakly absorbing particles in these conditions. The explanation of trapping by photophoretic forces was based on a close examination of the nonparaxial focal region and calculations of the energy distribution in the vicinity of the focal plane, which showed minima in energy density and that the flow of energy was actually whirling around at the focal plane. It was also observed that the number of particles trapped increases after one was suspended, suggesting that the electromagnetic field is further disturbed by the particle to form more intensity minima and swirling in the energy flow, which, in turn, caused more particles to be suspended. To date, there has been no theory developed unifying laser trapping by radiation force and photophoretic force and there is no united approach that would cover this difficult range ($0.1 \leq Kn \leq 10$) of particular interest for nanoparticle trapping and manipulation.

We demonstrate below that counterpropagating optical vortices create a radially symmetrical and azimuthally homogeneous intensity distribution in a large volume of the trap, enabling efficient particle trapping.

8.3.2 Optical Vortex

In 1992 it was found theoretically that Laguerre–Gaussian (LG) light beams possess an orbital angular momentum (OAM) (Allen, Beijersbergen, Spreeuw, and Woerdman, 1992). These beams are examples of light beams with a doughnut-like intensity structure that is symmetric about the beam's axis and a phase structure of intertwined phase fronts (Fig. 8.2). Since then, several groups have demonstrated the mechanical effects of the OAM in such beams:

when microscopic particles absorb such light, they begin to rotate. Specifically, this has been demonstrated very elegantly in optical tweezers, where the same light beam traps the particle, resulting in a microscopic tool that can not only hold microscopic particles still and translate them in the *x*-, *y*-, and *z*-directions but also rotate them about the beam's axis (Rubinsztein-Dunlop, Nieminen, Friese, and Heckenberg, 1998). Optical vortices offer better stability and robustness. There has been no previous attempt, to the best of our knowledge, to employ optical vortices for trapping and guiding of absorbing airborne particles.

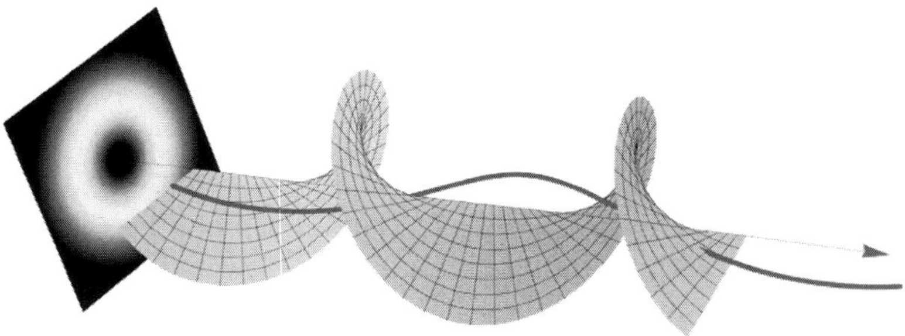

Figure 8.2 An optical vortex. The helical wave front represents OAM of the vortex beam. The spiraling curve is a pointing vector of a linearly polarized LG mode. The projection on the left shows a doughnut-shaped intensity distribution with zero intensity at the beam axis in the transverse cross section of the vortex beam.

8.3.3 Optical Trap with Counterpropagating Vortex Beams

Aiming to construct a trapping volume with a localized minimum of intensity we turned to the original scheme of a dual-beam trap with two counterpropagating coaxial beams, first suggested by Ashkin (1970) in his first paper on optical trapping (Fig. 8.3a). The advantages of a dual-beam trap formed between the focal planes are

- better stability of trapping;
- the ability to trap a wide range of particle sizes in a larger volume not accessible to the tight focusing of optical tweezers;

- the increased probability of trap loading due to increased trapping volume; and
- a more robust platform to trap particles in a hostile environment, outside the safe, well-controlled optical lab.

The dual-beam trap was modified in two ways. First, two Gaussian beams were replaced with counterpropagating vortices, and second, the trapping area was formed with beams of opposite topological charge and linear polarizations rotated by 90° relative to each other to avoid interference. The phases of these two beams are in the same direction; thus the momentum longitudinal projections of both beams have the same direction in the trapping area. The intensity of the electric and magnetic fields of the counterpropagating beams retains the axial distribution symmetry with zero intensity on the common axis and hence forms a stable trap. The doughnut-like vortex beam shape provides two-dimensional confinement in the radial directions, while the displacement of focal planes ensures stability in the longitudinal direction. The distinct feature of this trapping geometry is that the particles are trapped in the minimum intensity of the beam, which ensures their minimal heating and perturbation by light, which can be important for studies of properties of particle and/or for trapping live cells in air.

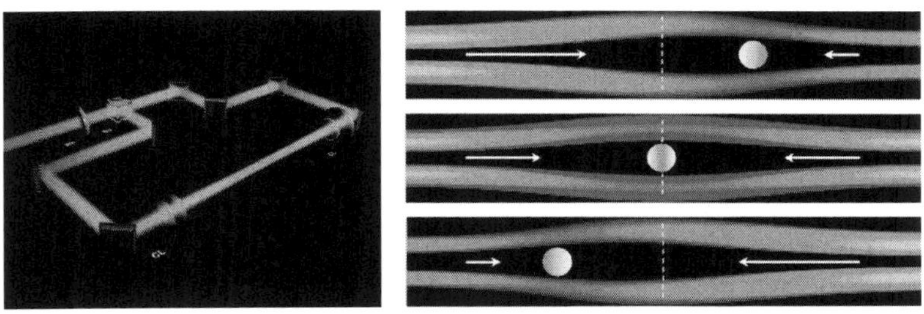

Figure 8.3 (a) A scheme of a dual-beam optical trap produced by counterpropagating coaxial vortex beams (WP: a half-waveplate; BS: a polarizing beam splitter cube; CL: collimators). (b) The illustration of a dual-beam trap formed between the focal planes of focusing lenses, first suggested by Ashkin (1970). The central positioning is at the equal laser intensity in the counterpropagating beams; if the balance is disturbed, the particle moves to a new equilibrium position toward the lower-intensity beam.

A detailed description of the experimental setup is presented elsewhere (Shvedov, Desyatnikov, Rode, Krolikowski, and Kivshar, 2009; Desyatnikov, Shvedov, Rode, Krolikowski, and Kivshar, 2009). Briefly, a linearly polarized Gaussian beam from a *cw* laser (Verdi-5, l = 532 nm) passed through the diffraction "fork"-type hologram, where it was transformed into a LG_{01} vortex beam with a topological charge l = 1. A half-waveplate adjusted the polarization of the beam so that a polarizing beam splitter cube divided the beam into two beams with equal power but with a 90° relative polarization orientation. The round-trip scheme was formed with an odd number of reflectors so that the axial symmetry of the intensity distribution was preserved for any polarization state of counterpropagating beams. The scheme was designed to ensure both beams had only a single round trip; thus unwanted interference of the beams on the following paths had been prevented. Lenses with the same focal length focused the coaxially aligned forward- and backward-propagating beams to the focal planes spaced apart along the axis to form a trapping volume confined between the planes.

Carbon nanoclusters produced by a high-repetition-rate laser ablation technique were used to illustrate the trapping of nanoparticles in air (Fig. 8.4) (Rode, Gamaly, and Luther-Davies, 2000; Gamaly, Madsen, Golberg, and Rode, 2009). The average size of carbon nanoclusters could be varied in the range from 3 nm to 8 nm by changing the Ar gas pressure in the formation chamber from 50 torr to atmospheric pressure (Gamaly, Madsen, Golberg, and Rode, 2009). The record low bulk density and low thermal conductivity make agglomerates of carbon nanoparticles a nearly ideal substance for experiments with photophoretic trapping (Rode, Gamaly, and Luther-Davies, 2000). Moreover, the optical absorption of nanostructured carbon is high, while reflectivity is extremely low, which makes such particles approach the properties of a black body.

The nanoclusters were trapped and agglomerated in larger aggregates and collected from the trap for diagnostics of the content with scanning electron microscopy (SEM). The size of the aggregates of the nanoclusters was varied from 0.2 μm to 10 μm in the experiments with 60 mm focal lenses and focal planes separated by ~7 mm, mainly depending on the laser power in the trap and the accuracy of the vortex alignment. The waist of the vortex beams was ~8 μm, and the operating power in the trapping volume at which the trapping of particles was observed was in the range of 0.3–30 mW per beam. Stability of trapping in both transverse and longitudinal cross sections was estimated from the fluctuations of the particle position; a typical value was within ±2 μm and was improving with the increase of the laser power.

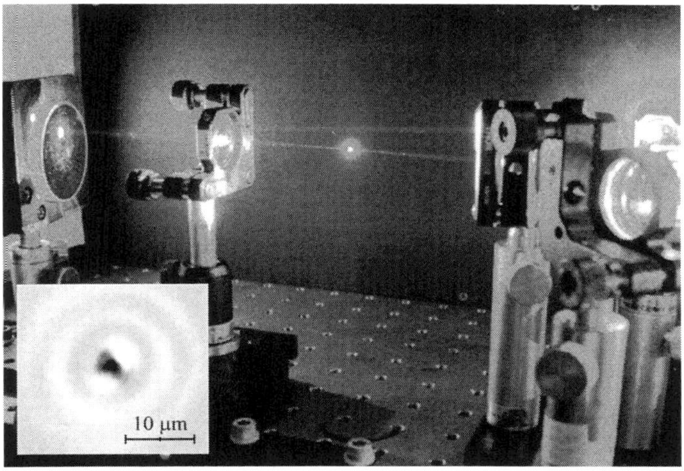

Figure 8.4 Trapping of nanoparticles in air with optical vortices. The bright spot seen between the lenses is the light scattered from the agglomerated nanoclusters. An insert shows a view through the notch filter acting as a mirror; the dark spot in the middle is the trapped agglomerated nanoparticles.

The new trapping scheme is very robust; the capture of nanocluster aggregates in open air could withstand air movement in the room due to the air-conditioning and people's loud and excited conversations while watching the trapping process. The level of *cw* laser energy in the trap was of the order of 1–10 mW, within the energy range of a conventional laser pointer. The level of force exerted on the trapped particle was conservatively estimated to be in the order of 1–10 pN (Shvedov, Desyatnikov, Rode, Krolikowski, and Kivshar, 2009; Desyatnikov, Shvedov, Rode, Krolikowski, and Kivshar, 2009).

The scheme provides two principal degrees of freedom for optical manipulation of trapped particles, depending upon (i) changing the polarization of the optical vortex and (ii) changing the distance between the focal planes. Both ways influence the structure of the trapping potential and provide control over the transverse dynamics of the trapped particles. The trapped particles could be moved backward and forward over 2.5 mm along the axis. These experiments paved the way for upscaling the vortex-based trapping scheme to a larger number of particles and for transporting them over much longer distances. These will be presented in the following sections.

8.3.4 Optical Trapping of Multiple Particles with a Speckle Field

In this section, we discuss a new strategy for the multiple, three-dimensional (3D) trapping of absorbing particles in air with a unidirectional coherent laser beam by taking advantage of trapping sites provided by a speckle pattern (Shvedov, Rode, Izdebskaya, Desyatnikov, Krolikowski, and Kivshar, 2010a; Shvedov, Rode, Izdebskaya et al., 2010). In particular, recent development of the holographic traps (Polin, Ladavac, Lee, Roichman, and Grier, 2005) and the so-called dark-core bottle beams with an axial region of zero intensity surrounded by light in all three dimensions (Arlt and Padgett, 2000; Bokor and Davidson, 2007; Shvedov, Izdebskaya, Rode, Desyatnikov, Krolikowski, and Kivshar, 2008) have attracted a lot of attention as their hollow areas enable the simultaneous trapping of a large number of particles (Shvedov, Rode, Izdebskaya, Desyatnikov, Krolikowski, and Kivshar, 2010a; Shvedov, Rode, Izdebskaya et al., 2010).

A speckle pattern is a wave intensity pattern produced by the interference of random coherent wave fronts formed by reflectance off or transmission through an irregular surface of the incident wave. If the size of the scattering area is sufficiently small, each of the scattering irregularities can be considered an independent point source generating an individual elementary wave of coherent light. The speckle pattern is well pronounced when the phase shifts between these waves are in the range from 0 to 2π. In case of several coherent point sources each of them generates an individual diffraction pattern in the image plane. The superposition of these patterns results in a field of isolated spots, with the smallest size approximately equal to the diameter of the first diffraction disk formed by the focusing lens aperture from a point source, representing a resolution of the imaging system. Assuming the large number of irregularities within the resolution limit, the speckle pattern can be considered as generated by an aperture-limited imaging system and can be calculated using the imaging system parameters only. A detailed description of the trapping volume formed by a speckle pattern is presented by Shvedov, Rode, Izdebskaya et al. (2010).

To conduct experiments with speckle trapping we used a ground glass diffuser as a speckle source, irradiated it with a *cw* laser (Verdi-5, λ = 532 nm), and imaged the diffuser with a 19 mm focal length lens into a glass cell filled with carbon particles (Shvedov, Rode, Izdebskaya, Desyatnikov, Krolikowski, and Kivshar, 2010a; Shvedov, Rode, Izdebskaya et al., 2010). The average speckle size in the transverse cross section $<\varepsilon_{\perp}>$ is determined by (Goodman, 2007):

$$\langle \varepsilon_{\perp} \rangle \approx 2.44 \lambda z_{\mathrm{a}} / D_{\mathrm{a}} \tag{8.2}$$

where z_a is the imaging distance and D_a is the diameter of the lens aperture. The length of an individual speckle $<\varepsilon_{||}>$ is estimated by the following relation (O'Holleran, Dennis, and Padgett, 2009):

$$\langle \varepsilon_{||} \rangle \approx 16 \lambda z_{\mathrm{a}}^{2} / D_{\mathrm{a}}^{2} \tag{8.3}$$

which coincides with the distance between two intensity minima in the longitudinal cross section of the Fresnel diffraction pattern generated by a circular aperture of diameter D_a (Goodman, 2007).

The speckle pattern of a coherent beam presents an ideal 3D trapping field with multiple traps for particles with sizes sufficiently smaller than the individual speckle size. Indeed, a randomly positioned single speckle has a phase singularity—a point of perfect destructive interference with zero intensity. The energy circulates in the closed loops, and the phase varies between 0 and 2π around this point, forming a 3D optical vortex (O'Holleran, Dennis, and Padgett, 2009). Many individual dark regions in the speckle field act as multiple traps for light-absorbing particles (Fig. 8.5); the size of the area as well as the average speckle size could be varied by using lenses with different numerical apertures. With the laser beam switched on, the particles suspended in air were entering the trapping area through Brownian motion and were reliably captured in the random speckle traps. The particles could be held in the traps potentially for unlimited time—in our experiments for up to 24 hours without any visible change in the number of the trapped particles.

The efficiency of trapping in these experiments was estimated by measuring the laser power in the image plane of ~100 mW, estimating the number of trapping centers, which was typically in the range from 1000 to 2000, and by building a size distribution and evaluating the mass of the trapped agglomerated nanoparticles using SEM. As an example, the total estimated mass of particles trapped using 125 mW laser power was $\sim 8 \times 10^{-10}$ g; thus the efficiency of trapping in our experiments was $\sim 7 \times 10^{-9}$ g/W.

Aiming to test the dependence of the average speckle size on the size of the trapped particles, we compared the size distribution of the trapped particles in experiments by focusing optics of different numerical apertures. While generating different average speckle sizes of 2.0 μm and 6.4 μm in a transverse cross section, which is determined by the z_a/D_a ratio—see Eqs. 8.2 and 8.3 above, we kept the same total laser intensity in both experiments. The trapped particles were then collected on a silicon wafer and analyzed under a scanning electron microscope. The image of particles suspended in a speckle field is presented in Fig. 8.5a, and the histograms of the particles' size distribution for both traps are shown in Fig. 8.5b; a total of 460 trapped particles were measured. The Poisson fittings show the average size of agglomerates $>d_{p1}< = 1.14$ μm for the trap with an average speckle size $<\varepsilon_{\perp 1}> = 2.0$ μm and $>d_{p2}< = 1.67$ μm for $<\varepsilon_{\perp 2}> = 6.4$ μm. The results indicate a tendency for agglomeration and trapping of particles of a larger, in average, size with a larger speckle size.

The studies presented in this section demonstrate the ability of reliable 3D multiple trapping of absorbing particles in air using a monodirectional laser beam. The multiple trapping volume was induced by a speckle pattern, which led to a multitude of microscopic-size bottle beams determined by optical singularities in each individual speckle. Multiple trapping, holding, and sorting of a large scale of light-absorbing particles and aerosols in air and other gases had so far been beyond the abilities of the standard laser-trapping systems. The results also mark the emergence of a possibility for selective trapping and separation of suspended particles in air by contactless optical means, which opens up diverse and rich practical opportunities for laser trapping of matter in air or any gaseous environment.

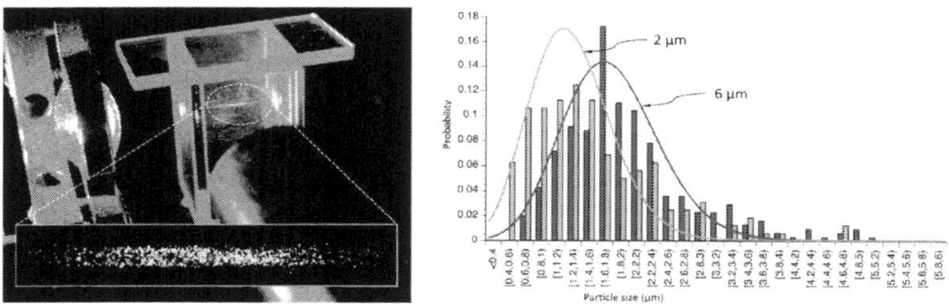

Figure 8.5 (a) Trapping of multiple particles with a speckle field. The lens on the left images a diffuser scattering surface (not shown) into a long line inside the rectangular cavity. The trapping volume manifests itself by the scattered laser light from the agglomerates of nanoparticles trapped and suspended in air by the speckle pattern. The inset shows a magnified side view of the trapping volume. The trapping area is ~7.5 mm long and 570 μm in diameter. (b) Size histograms for the particles collected from the traps with average speckle sizes of 2.0 μm and 6.4 μm. The size intervals are 0.2 μm (data from Shvedov, Rode, Izdebskaya et al., 2010). Both histograms were fitted with a Poisson distribution function (solid lines).

8.4 Long-Range Transport and 3D Manipulation

8.4.1 Optical Pipeline

The experimental results of trapping micron-size particles with counterpropagating vortex beams and demonstrated ability to move particles within the trap over millimeter distances, described in Section 8.3.3, laid the groundwork for developing a long-distance optical transport of particles. Below we describe how ~100 μm particles can be transferred over meter-long distances with a collimated vortex beam, which we termed the "optical vortex pipeline" (Shvedov Rode, Izdebskaya, Desyatnikov, Krolikowski, and Kivshar, 2010b; Smart, 2010). Recently, photophoretic trapping and back-and-forth transportation of light-absorbing aerosols have been shown inside an optical vortex bottle beam generated with a spatial light modulator (Zhang, Zhang, Prakash et al., 2011).

The horizontally oriented pipeline was formed from a vortex beam generated with an optical phase plate from Holo/Or Ltd., which converted 532 nm *cw* Verdi-5 laser radiation with a TEM_{00} mode into an LG_{01} doughnut-shaped mode with topological charge $l = 1$. The beam was collimated to form a waist w_0 in the range from 150 μm to 600 μm, which could be controlled by a pair of collimating lenses to form a long, slow-diverging optical duct with a Rayleigh length of 120 mm to 1000 mm (Shvedov, Rode, Izdebskaya et al., 2010). Both carbon nanofoam–agglomerated particles and graphite-coated glass microshells were used in transport experiments. These particles were chosen due to their low mass of the order of 10^{-9}–10^{-8} g, low thermal conductivity, and high absorption at the laser beam wavelength, to maximize the photophoretic force; the size of the particles was varied in the range from 50 μm to 100 μm. The glass shells with spherical surfaces were selected for further accurate evaluation of forces acting on a particle transported by a vortex beam. The particles were loaded into the pipeline from a coaxially located glass cylinder and, being trapped in the low-intensity core of the beam, were transported along the beam axis. The bright ring of maximum light intensity acted as a repelling "pipe wall" on particles, counterbalancing the gravitation force and keeping them trapped in the dark region on axis, while the axial *z* component F_z of thermal force pushed particles in a horizontal direction along the pipeline. The power in the beam was varied from 0.1 W to a maximum value of 2 W. The trapped particles could be transported within the pipeline over a very long distance—up to 1.5 m in our demonstration experiments—limited only by the divergence of the vortex beam.

The transport velocity was determined by the equilibrium between two forces, namely, the longitudinal component of thermal force F_z from illuminating the particle and the Stokes drag force $F_z = F_s = -6\pi\mu r v$, where μ is the viscosity of air($\mu = 1.73 \times 10^{-5}$ Nsm^{-2}), r is the radius of a spherical particle, and v is the particle's setting velocity. A typical transport speed recorded in the experiments was in the range 3 mm/s to 25 mm/s for the particle sizes 30–100 μm, where the larger particles were moving faster due to larger cross sections and thus a higher beam illumination, like sailing along the beam. The dependence of particle velocity on the particle radius measured with the pipeline waist w_0 = 75 μm and the total beam power of 0.8 W are presented in Fig. 8.6; the large scattering of data is due to a variation of the wall thickness and the resulting variation of mass of particles equal in size.

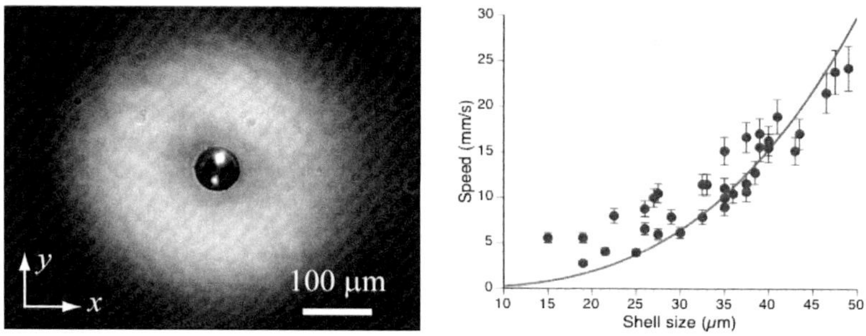

Figure 8.6 (a) A snapshot of a graphite-coated glass microshell with a radius of 33.5 μm trapped in a 125 μm vortex beam. (b) The transport velocity of particles in a pipeline with a waist radius of 75 μm and a total beam power of 0.8 W. The solid line is a fitting curve $v_s \propto (r/w_0)^3$—see Shvedov, Rode, Izdebskaya et al. (2010) for details.

The ability to transport particles along the pipeline axis opened up the way to pinpoint remote positioning of the particles in a desirable location. Indeed, by steering the beam during the particle travel with a mirror located 0.5 m from a substrate we placed each transported particle with the accuracy ±10 μm—see Fig. 8.7. This presents the angular precision of $\pm 2 \times 10^{-5}$, limited by the beam point stability. This precision is an analogue of note that the level of precision achieved with an optical vortex pipeline is higher than the beam divergence of $\sim 10^{-4}$. By varying the vortex beam diameter it was also possible to influence the size of the trapped and transported particles. This was experimentally tested by positioning the 100 μm shells on top of the letter "A" and 60 μm shells at the bottom of the letter by varying the vortex beam diameters from 250 μm to 150 μm—see Fig 8.7b.

Fixing the position of particles in air in a desirable location on the z-axis was tested in a dual-beam scheme similar to that described in Section 8.3.3 above. A 30 cm long transport line was formed with two carefully adjusted counterpropagating pipelines to guarantee that both beam axes coincide better than ±5 μm over the whole transport distance. This was done with the help of a cell filled with a liquid suspension of fluorescent nanoparticles by controlling the beam propagation with a microscope along the whole path. The position of the particle was guided by rotating a waveplate before the polarizing beam splitter (see Fig 8.3b), which changed the relative intensities but preserved the total laser power of the counterpropagating beams.

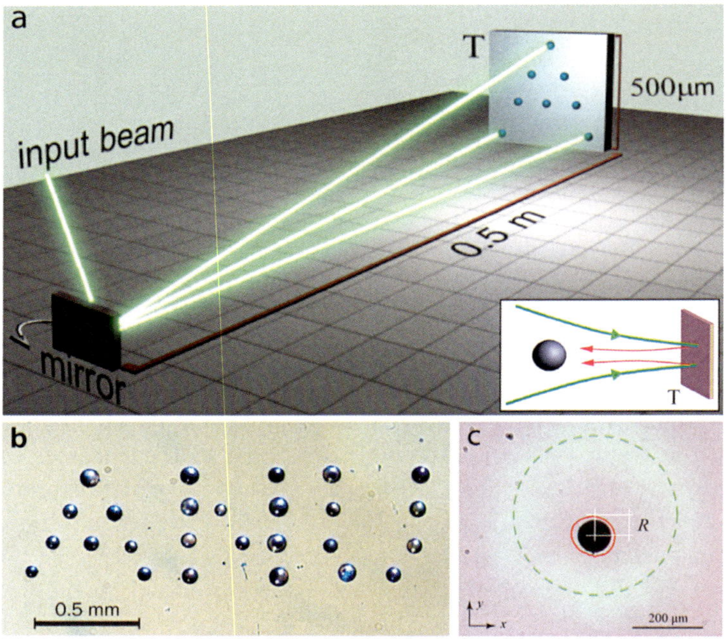

Figure 8.7 (a) Remote positioning of particles transported by the optical vortex pipeline. The inset shows the possibility of fixing the particle position at the target substrate (T) by placing a mirror in the optical path. (b) An example of the remote deposition of glass microspheres from the distance of 0.5 m. The letters "ANU" stand for the Australian National University, The positioning accuracy is ±10 μm, and the scale bar is 0.5 mm. (c) Analysis of transverse stability of trapping. The solid (red) line encircles the area of particle localization. The radius of the sphere is 50 μm. The dashed green line shows the vortex intensity maximum with a radius w_0 = 236 μm. The white lines show the transverse displacement of the particle ~61 μm from the vortex axis due to gravity. Reprinted from Shvedov, Rode, Izdebskaya et al. (2010b), with permission from American Physical Society.

The results of experiments on fixing the particle position at a desirable point along the pipeline are presented in Fig. 8.8a. The particle position depends on the balance of power $\Delta P = (P_{\text{left}}-P_{\text{right}})$ between the counterpropagating beams in the duct $\Delta P/P_{\text{tot}}$, the dependence fits a linear approximation within ±4%. The disturbance of balance introduced an excess of thermal force counterbalanced by viscosity of air; the resulting force, which varies along the pipeline due to the beam divergence, moved the particle to a new equilibrium position (Fig. 8.8b). The balanced position of particles was also dependent on the total power in the pipeline (Fig. 8.8c). By changing the total power but keeping the balance of the power

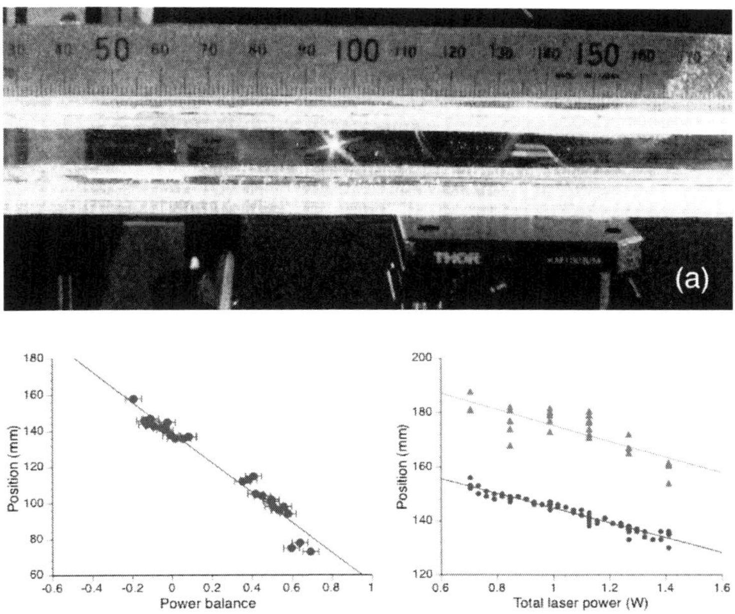

(b) (c)

Figure 8.8 (a) Stable positioning of a particle in the optical pipeline. (b) Equilibrium positions of a 67 μm glass shell guided backward and forward along the 300 mm optical duct by changing the power balance $\Delta P/P_{\text{tot}}$ between the counterpropagating beams; the total power P_{tot} in the beam was kept constant at 1 W; the solid line is a linear fit to the data. (c) Changes in the equilibrium positions of the 90 μm (triangles) and 103 μm shells (dots), depending on the total power in two equal power beams.

fixed at the equilibrium, the particle's position was changed. We speculate that the change of total power changes the particle illumination in the beam, lifting the particle in the vertical direction with an increase of power to counterbalance gravity and lowering the particle when the power decreases. The minute change in the position relative to the beam axis slightly changes the balance in the longitudinal direction, shifting the particle to a new balanced point in the pipeline. This hypothesis was confirmed by the observation of two particles of different sizes of 90 μm and 103 μm, changing their equilibrium positions, depending on the total laser power before the beam splitter. The shift of the equilibrium point is due to the beam divergence. More detailed studies are required to fully uncover the particle illumination conditions in various positions relative to the beam axis.

Observation of the dynamics of a particle with known size and mass offers another insight into the value of the laser-induced force involved. Figure 8.9a presents the dynamics of a carbon-coated glass shell 67 μm in diameter in a 300 mm pipeline; the ratio of the particle-to-vortex radius is $r/w_0 = 0.15$. The related acceleration at the initial "push" from the laser beam was determined as the average acceleration between the moment when the balance of power was disturbed and the moment of maximum speed $>a< = \Delta V/\Delta t$—see Fig. 8.9b. The estimated mass of a particle of this size with a typical 0.25–0.5 μm wall thickness is within the range of 1.9×10^{-9}–1.6×10^{-8} g. Taking the acceleration of particles from Fig. 8.9b yields the driving force of the order of $\sim(0.2–3.2) \times 10^{-12}$ N at 1 W of the total laser power. This value is in agreement with previous estimates, taking into account that less than 1% of the total laser power illuminates the particles with the $r/w_0 = 0.15$ ratio located close to the minimum intensity of the doughnut-shaped beam. These experiments show that the force that controls the position of particle in the pipeline is linearly dependent on laser power.

Practically, an optical vortex pipeline constitutes a versatile toolset able to trap and move the particle two-dimensionally touch free from one place to another, to stop it in a desirable position along the path, and to position a particle on a substrate with very high precision. In the next section we will consider a technique of full-scale 3D manipulation of particles in air.

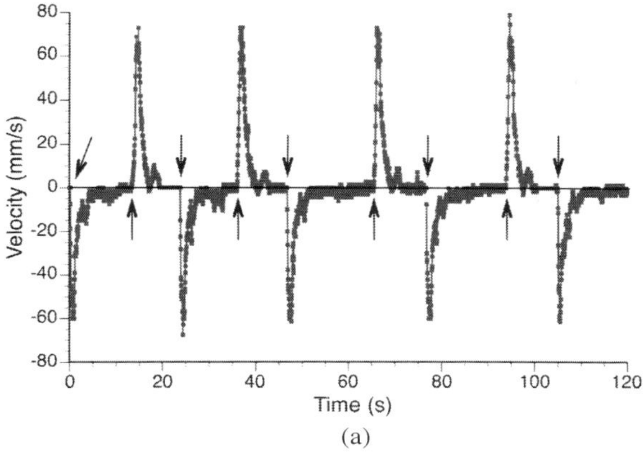

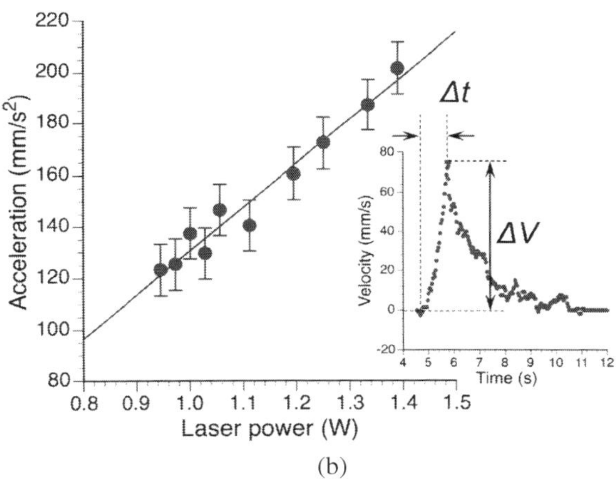

Figure 8.9 Dynamics of the 67 μm glass shell from Fig. 8.8a,b moved between the equilibrium points in the optical pipeline. (a) Velocity trace after the balance was disturbed by changing the relative intensities of the beams; the balance positions are at zero axis; positive and negative values are the speed data of the particle moving back and forth between the equilibrium positions; the moments when power balance was changed are shown by arrows. (b) Acceleration of that shell from an equilibrium position to a maximum speed after the balance of power was disturbed, at various levels of total laser power in the trap. The solid line is a linear fit to the data. The inlet shows a blown-up section of the velocity graph between the two equilibrium positions.

8.4.2 Full-Scale 3D Manipulation

To ensure full-scale 3D manipulation of particles a trapping enclosure of zero or lower intensity should be surrounded by a high-intensity barrier, ideally of a spherical shape. Optical bottle beams (Shvedov, Rode, Izdebskaya et al., 2010; Singh and Nalwa, 2007; Smart, 2010; Steuernagel, 2005; Tolstoshev, 2006) are well suited to form such a trap. They present the nearest approach to the best-possible trapping geometry. In this section we demonstrate one of the ways to form a deep 3D potential well formed in the focal volume by strong spherical aberrations (Shvedov, Hnatovsky, Rode, and Krolikowski, 2011). A combination of focusing optics with an optical fiber as a source of diverging coherent radiation provides a fully controllable approach to form a 3D optical bottle beam for manipulation of particles in air.

The experimental setup for the trap is schematically shown in Fig. 8.10a. The output TEM_{00} mode from a *cw* solid-state laser operating at λ = 532 nm is coupled into a single-mode optical fiber, collimated with an objective O, and then focused in free space with a planoconvex lens L with a focal distance f = 25.4 mm. The experimental plot was obtained by scanning the imaging optics along the beam propagation direction z. The entrance aperture of the lens has to be significantly underfilled to achieve the desired intensity distribution in its focal region; it can be adjusted by changing the distance between the fiber output and the front focal plane of O.

The aberration function φ, which defines the deviation of the distorted wave front from the ideal convergent spherical wave front, is well approximated by (Mahajan, 1991):

$$\phi = -\frac{(a\rho)^4 n^2}{8f^3(n-1)^2} \tag{8.4}$$

where a is the aperture of lens L; $\rho = r/a$, where r is a variable denoting the radial distance from the optical axis; and n = 1.5 is the lens refractive index. The spatial intensity distribution I in the focal region of a lens illuminated with a Gaussian beam can be found by evaluating a scalar Debye integral (Born and Wolf, 2003):

$$I(u,v) = \frac{8\pi a^4}{\lambda^2 f^2 w_0^2} P \left| \int_0^1 \exp\left(-\frac{a^2\rho^2}{w_0^2}\right) \exp\left[-i\left(k\phi - \frac{u\rho^2}{2}\right)\right] J_0(v\rho)\rho d\rho \right|^2 \tag{8.5}$$

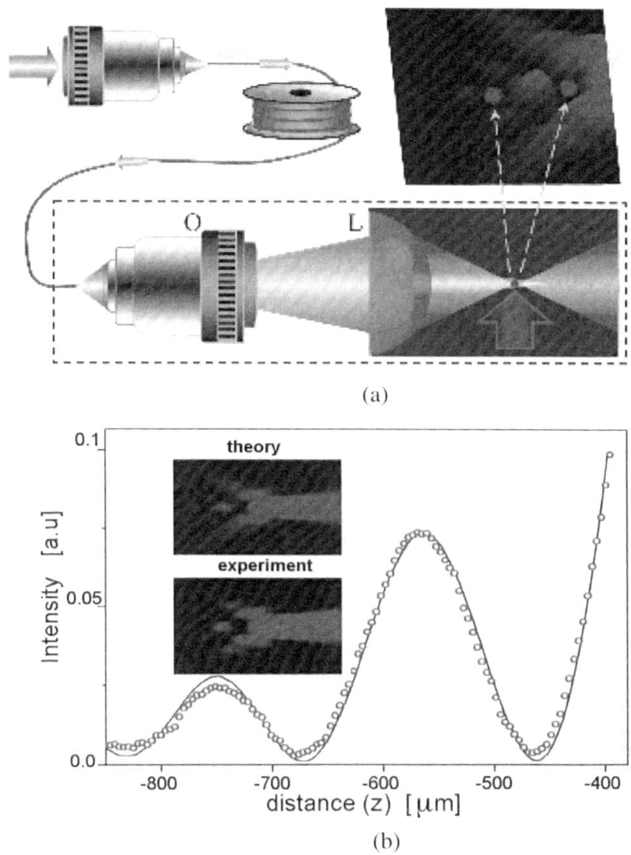

Figure 8.10 (a) A bottle beam formed in the focal volume by spherical aberrations. Particles are trapped inside the low-intensity regions of the bottle trap formed by the spherical aberrations of lens L and objective O. (b) Experimental intensity distribution along the z axis of the focal area (circles) before the maximum intensity. An aberration-free Gaussian focus is at z = 0. The insets represent the simulated and experimentally obtained intensity distributions in the xz cross section. The intensity has rotation symmetry relative to the beam axis. Reprinted from Shvedov, Hnatovsky, Rode, and Krolikowski (2011), with permission from American Physical Society.

where u and v are dimensionless variables defined as $u = kz(a/f)^2$ and $v = kr(a/f)$, $k = 2\pi/\lambda$ is the wave number, w_0 is the lens L Gaussian beam waist, $J_0(v\rho)$ is the Bessel function of 0 order, and p

is the beam power. The intensity distribution calculated along the axis is presented as a graph, while the distribution in the x–z plane is presented as a top inset in Fig. 8.10b. As one can see from Fig. 8.10, the theoretically calculated intensity distribution $I(u,v)$, with the aberration function given by Eq. 8.4, follows closely the experimental results. The intensity minima in the graph represent an array of optical bottles to be used as 3D particle traps (Arlt and Padgett, 2000). Taking the experimental beam aperture a = 12 mm, the zero intensity in the first minimum with the highest-intensity barrier was achieved for w_0 = 1.1 mm. The intensity profiles for other w_0 values can be found in Shvedov, Hnatovsky, Rode, and Krolikowski (2011).

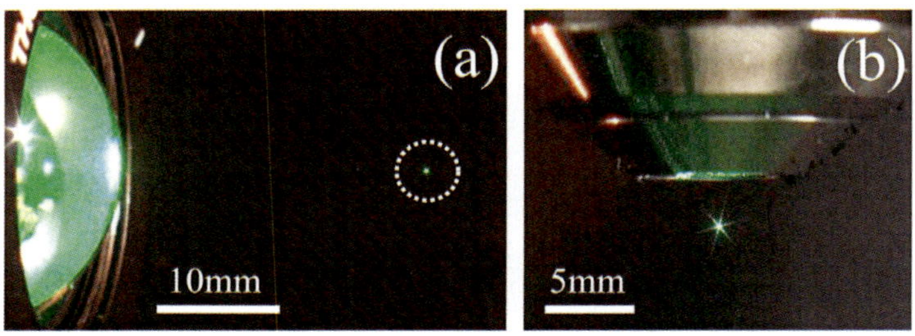

Figure 8.11 3D manipulation of absorbing particles in air with a bottle beam setup. (a) A solid graphite particle (encircled) with an estimated of 5 μm size (~10^{-10} g) trapped inside the aberrated focus of the lens L. The trapping power is 25 mW. (b) An agglomerate of carbon nanoparticles inside the focus of a 60x microscope objective NA = 0.7. The radius of the beam waist is 0.8 mm, the radius of the entrance aperture of the objective is 2.5 mm, and the estimated radius of the bottle for these parameters is 0.7 μm. The trapping power is 1 mW. Reprinted from Shvedov, Hnatovsky, Rode, and Krolikowski (2011), with permission from American Physical Society.

The described bottle beam–generating setup combined with an optical fiber demonstrates the robust trapping ability of particles in air with limitless 3D manipulation of trapped particles along any trajectory (see Fig. 8.11). With a trapping power of only 0.5–30 mW, the aberration-based trap was used in open air as it could easily withstand air currents estimated at 10–50 cm/s, depending on the focusing geometry and the type and size of the particles. The efficiency of trapping was estimated by trapping a graphitic particle of ~5 μm size with 25 mW of laser power, yielding ~4×10^{-9} g/W, which is close to the efficiency in a speckle-based bottle beam. For a fixed trapping power a tighter focusing provides a more stable trap because the intensity minima are surrounded by a higher-intensity barrier. The size of the trap can be varied by tuning the effective *NA* of the focusing optics.

In general, the obvious simplicity of this scheme allows one to use a bottle trap in a wide range of geometries. If the light beam is delivered through an optical fiber, the 3D manipulation is restricted only by the length of the fiber and the maximum laser power in the system. Polychromatic light in combination with achromatic focusing optics can also be used for trapping as the focal intensity pattern itself depends only weakly on the spectral bandwidth; the spectral range of the trapping light is limited solely by light absorption in the fiber core and the lens, that is, $\lambda \sim 0.3$–3 μm for glass, for example.

8.5 Future Directions

The new concept of optical trapping, manipulation, transport, and pinpoint positioning of objects using vortex beams opens up diverse and rich practical opportunities for handling objects in a clean environment or hard-to-reach areas (Omori and Suzuki, 1998). The method can be applied, for example, in biomedical fields requiring touch-free transport of biological samples and ultrapure or dangerous substances. The nanoparticles agglomerated and collected in the trap can be further investigated in terms of their chemical activity, unique toxicity (as compared to the bulk), and health risk assessment. Better localization of nanoparticles will allow one to suggest appropriate engineering controls and personal protective equipment in the manufacturing processes. The outcomes are of fundamental importance for a wide range of other fields of science, such as interstellar dusty plasmas and atmospheric physics. The bottle beam traps allow one to monitor the presence and to study the behavior of aerosol particles, such as atmospheric black carbon soot, which is widely expected to lead to global warming, in a very controlled way.

For a successful application, a number of factors need to be improved. Transport stability has to be increased, and delivery may need to be timed to a periodic or triggering signal. Key to refining the methods is an understanding of the underlying physical forces responsible for the transport phenomenon, namely, the interplay between the radiation pressure and the photophoretic force, and their values relative to gravity and buoyancy forces, in a wide range of gaseous and vacuum conditions.

The force dominating in gaseous environments is the photophoretic force, as has been confirmed by the levitation experiments (Tong, 1975; Pluchino, 1983; McGloin, Burnham, Summers, Rudd, Dewar, and Anand, 2008). Particle mass can be determined by suspending the particles in a trap formed by a diverging vortex beam and by measuring the particle diameter, beam power, and beam size in the cross section at the particle suspension position. This is a concept for a new type of an optical scale working in the range from a picogram to hundreds of nanograms and is easily extendable to lower or higher ranges. This opens up an easy and convenient way to weigh particles, as the setup itself is simple and easy to realize.

Nanoscale X-ray morphology of proteins and nucleic acids is a principal factor that helps define chemical reactivity, aggregation, and how alterations in their structures affect their functions. The primary aim for the foreseeable future is to develop an efficient and stable trapping device to freely maneuver agglomerates of molecules such as proteins or viruses to the focus of a free electron laser (FEL) spectrometer. The preeminent aim is to identify the nature of the forces acting upon macromolecules in air and helium at pressures varying from several atmospheres to high vacuum. Secondly, a novel concept of loading the trap sites is required. Envision synchronized delivery of particles with FEL pulses, a twofold setup trapping thousands of particles in a first holding chamber suspended with a speckle-based trap, stable enough to be moved around in air without disturbing the trapped particles. The transport stage of the trap will then expel a single target from the speckle trap and use an optical pipeline to deliver it to the focal spot of the FEL with accuracy down to a few microns. The trapping environment and parameters to be studied extensively and delivery speed stability, trapping efficiency, and spatial stability will be optimized in order to synchronize the arrival of targets with FEL pulses. One of the benefits of this approach is the ability to recapture particles missing the FEL interaction region at the far end of the pipeline, thus allowing for an overall increase in delivery efficiency.

Temperature measurements of the molecules trapped in the pipeline and the levitation setup are of paramount importance as biomolecules may be damaged at temperatures exceeding 60°C. Direct temperature measurement of microscopic particles is an elaborative task itself. However, our preliminary studies indicate that the maximum temperature of trapped particles did not exceed 55°C, indicating potential applicability of the scheme for trapping of biomolecules. We also plan a series of modeling experiments aiming to evaluate heating of the target by measuring the temperature over much a larger area at known laser. Finally, we aim to integrate the optical delivery pipeline into the experimental chamber at FLASH at DESY in Hamburg. The system design and engineering should be flexible enough to be adapted to the X-ray diffraction apparatus as well as other diagnostics, such as florescence, Raman spectra, and electrical and magnetic properties, without any detrimental influence to other diagnostics.

Summing up, in many respects, laser trapping of airborne particles, and particularly nanometric-scale particles, is in its infancy. There is ample room for further experimental studies to confirm existing theoretical predictions and develop new theoretical models for forces involved. There are good grounds to believe that the future of manipulation of airborne particles with optical vortex beams looks very bright indeed.

Acknowledgments

We are grateful to A. S. Desyatnikov, Y. V. Izdebskaya, and N. Eckerskorn for many contributions to these studies and to Y. S. Kivshar, P. Reece, D. McGloin, and M. Padgett for fruitful and critical discussions. This work was supported by the National Health and Medical Research Council, project #525456, and the Australian Research Council through the Discovery scheme, project #110100975.

References

Allen, L., Beijersbergen, M. W., Spreeuw, R. J. C., and Woerdman, J. P. (1992). Orbital angular momentum of light and the transformation of Laguerre-Gaussian modes. *Phys. Rev. A*, **45**, 8185–8189.

Arčon, D., Jagličič, Z., Zorko, A., et al. (2006). Origin of magnetic moments in carbon nanofoam. *Phys. Rev. B*, **74**, 014438.

Arlt, J., and Padgett, M. J. (2000). Generation of a beam with a dark focus surrounded by regions of higher intensity: the optical bottle beam. *Opt. Lett.*, **25**, 191–193.

Ashkin, A. (1970). Acceleration and trapping particles by radiation pressure. *Phys. Rev. Lett.*, **24**, 156–159.

Ball, P. (2006). Nanoparticles in sun creams can stress brain cells. *Nature News* 16 June 2006, http://www.nature.com/news/archive/date/2006/06.html?num=160.

Bokor, N., and Davidson, N. (2007). A three dimensional dark focal spot uniformly surrounded by light. *Opt. Commun.*, **279**, 229–234.

Born, M., and Wolf, E. (2003). *Principles of Optics*, 7th ed. Cambridge, UK: Cambridge University Press.

Buffat, P., and Borel, J.-P. (1976). Size effect on the melting temperature of gold particles. *Phys. Rev. A*, **13**, 2287–2298.

Desyatnikov, A. S., Shvedov, V. G., Rode, A. V., Krolikowski, W. Z., and Kivshar, Y. S. (2009). Photophoretic manipulation of absorbing aerosol particles with vortex beams: theory versus experiment. *Opt. Express*, **17**, 8201–8211.

Dholakia, K., and Reece, P. (2006). Optical manipulation takes hold. *Nano Today*, **1**, 18–27.

Dholakia, K., Reece, P., and Gu, M. (2008). Optical micromanipulation. *Chem. Soc. Rev.*, **37**, 42–55.

Dienerowitz, M., Mazilu, M., and Dholakia, K. (2008). Optical manipulation of nanoparticles: a review. *J. Nanophoton.*, **2**, 021875.

Dreher, K. L. (2004). Health and environmental impact of nanotechnology: toxicological assessment of manufactured nanoparticles. *Toxicol. Sci.*, **77**, 3–5.

Ehrenhaft, F. (1917). On the physics of millionths of centimeters. *Physik. Z.*, **18**, 352–368.

Faunce, T. A., Walters, H., Williams, T., Bryant, D., Jennings, M., and Musk, B. (2006). Policy challenges from the "White" Senate inquiry into workplace-related health impacts of toxic dusts and nanoparticles. *Aust. N.Z. Health Policy*, **17**, 7.

Gamaly, E. G., and Rode, A. V. (2004). Nanostructures created by lasers. In *Encyclopaedia of Nanoscience and Nanotechnology* 7 (Nalwa, H. S., ed.). Stevenson Range: American Scientific, 783–809.

Gamaly, E. G., Madsen, N. R., Golberg, D., and Rode, A. V. (2009). Expansion-limited aggregation of nanoclusters in a single-pulse laser-produced plume. *Phys. Rev. B*, **80**, 184113.

Goodman, J. W. (2007). *Speckle Phenomena in Optics* (Ben Roberts & Co., Greenwood Village).

Gorkov, L. P., and Eliashberg, G. P. (1965). Minute metallic particles in an electromagnetic field. *Sov. Phys. JETP*, **21**, 940.

Grier, D. G. (2003). A revolution in optical manipulation. *Nature*, **424**, 810–816.

ICON press release (9 April 2006). *Workshop on Engineered Nanomaterials and Human Health Hazards, International Council on Nanotechnology ICON*, at Woodrow Wilson International Center for Scholars, Rice University, on 6–7 April 2006, http://www.wilsoncenter.org/article/.

Kreibig, U., and Vollmer, M. (1995). *Optical Properties of Metal Clusters*. Berlin, Heidelberg: Springer-Verlag.

Lebedev, P. (1901). Untersuchungen über die druckkräfte des lichtes. *Ann. Phys.*, **311**(11), 433–458.

Link, S., and El-Sayed, M. A. (2003). Optical properties and ultrafast dynamics of metallic nanocrystals. *Annu. Rev. Phys. Chem.*, **54**, 331–366.

Mahajan, V. N. (1991). *Aberration Theory Made Simple*. Washington, DC: SPIE.

Makarova, T. L., Han, K. H., Esquinazi, P., et al. (2003). Magnetism in photopolymerised fullerenes. *Carbon*, **41**, 1575–1584.

Marine, W., Patrone, L., Luk'yanchuk, B., and Sentis, M. (2000). Strategy of nanocluster and nanostructure synthesis by conventional pulsed laser ablation. *Appl. Surf. Sci.*, **154–155**, 345–352.

Maynard, A. D., Aitken, R. J., Butz, T., et al. (2006). Safe handling of nanotechnology. *Nature*, **444**, 267–269.

McGloin, D., Burnham, D. R., Summers, M. D., Rudd, D., Dewar, N., and Anand, S. (2008). Optical manipulation of airborne particles. *Faraday Discuss.*, **137**, 335–350.

Melzer, A. (2001). Laser manipulation of particles in dusty plasmas. *Plasma Sources Sci. Technol.*, **10**, 303–310.

Nel, A., Xia, T., Mädler, L., and Li, N. (2006). Toxic potential of materials at the nanolevel. *Science*, **311**, 622–627.

Nichols, E. F., and Hull, G. F. (1901). A preliminary communication on the pressure of heat and light radiation. *Phys. Rev.*, **13**, 307–320.

O'Holleran, K., Dennis, M. R., and Padgett, M. J. (2009). Topology of light's darkness. *Phys. Rev. Lett.*, **102**, 143902.

Omori, R., and Suzuki, A. (1998). Uranium dioxide particles collection using radiation pressure of a laser light in air. *J. Nucl. Sci. Technol.*, **35**, 830–832.

Pyykkö, P. (2007). Magic nanoclustures of gold. *Nat. Nanotechnol.*, **2**, 273–274.

Pluchino, A. B. (1983). Radiometric levitation of spherical carbon particles. *Appl. Opt.*, **22**, 1861–1866.

Polin, M., Ladavac, K., Lee, S. H., Roichman, Y., and Grier, D. G. (2005). Optimized holographic optical traps. *Opt. Express*, **13**, 5831–5845.

Prasad, P. N. (2004). *Nanophotonics.* Hoboken, NJ: John Wiley & Sons.

Preining, O. (1966). Photophoresis. In *Aerosol Sciences* (Davies, C. N., ed.). New York: Academic Press, 111–135.

Rode, A. V., Gamaly, E. G., and Luther-Davies, B. (2000). Formation of cluster-assembled carbon nano-foam by high-repetition-rate laser ablation. *Appl. Phys. A*, **70**, 135–144.

Rode, A. V., Gamaly, E. G., Christy, A. G., et al. (2004). Unconventional magnetism in all-carbon nanofoam. *Phys. Rev. B*, **70**, 054407.

Rubinsztein-Dunlop, H., Nieminen, T. A., Friese, M. E. J., and Heckenberg, N. R. (1998). Optical trapping of absorbing particles. *Adv. Quantum Chem.*, **30**, 469–492.

Sahoo, S. K., Parveen, S., and Panda, J. J. (2007). The present and future nanotechnology in human health care. *Nanomed.: Nanotechnol., Biol., Med.*, **3**, 20–31.

Shatkin, J. O. (2008). *Nanotechnology: Health and Environmental Risk*. Boca Raton, FL: CRC Press.

Shvedov, V. G., Desyatnikov, A. S., Rode, A. V., Krolikowski, W. Z., and Kivshar, Y. S. (2009). Optical guiding of absorbing nanoclusters in air. *Opt. Express*, **17**, 5743– 5757.

Shvedov, V. G., Hnatovsky, C., Rode, A., and Krolikowski, W. (2011). Robust trapping and manipulation of airborne particles with a bottle beam. *Opt. Express*, **19**, 17350–17356.

Shvedov, V. G., Izdebskaya, Y. V., Rode, A. V., Desyatnikov, A. S., Krolikowski, W. Z., and Kivshar, Y. S. (2008). Generation of optical bottle beams by incoherent white-light vortices. *Opt. Express*, **16**, 20902–20907.

Shvedov, V. G., Rode, A. V., Izdebskaya, Y. V., Desyatnikov, A. S., Krolikowski, W. Z., and Kivshar, Y. S. (2010a). Selective trapping of multiple particles by volume speckle field. *Opt. Express*, **18**, 3137–3142.

Shvedov, V. G., Rode, A. V., Izdebskaya, Y. V., Desyatnikov, A. S., Krolikowski, W. Z., and Kivshar, Y. S. (2010b). Giant optical manipulation. *Phys. Rev. Lett.*, **105**, 118103.

Shvedov, V. G., Rode, A. V., Izdebskaya, Y. V., et al. (2010). Laser speckle field as a multiple particle trap. *J. Opt.*, **12**, 124003.

Singh, S., and Nalwa, H. S. (2007). Nanotechnology and health safety: toxicity and risk assessments of nanostructured materials on human health. *J. Nanosci. Nanotechnol.*, **7**, 3048–3070.

Smart, A. G. (2010). Optical manipulation of light-absorbing particles takes to the air. *Phys. Today*, **63**(11), 13–14.

Steuernagel, O. (2005). Coherent transport and concentration of particles in optical traps using varying transverse beam profiles. *J. Opt. A: Pure Appl. Opt.*, **7**, S392–S398.

Tolstoshev, A. (2006). *Nanotechnology: Assessing the Environmental Risk for Australia.* Australia: Earth Policy Centre, University of Melbourne.

Tong, N. T. (1975). Experiments on photophoresis and thermophoresis. *J. Colloid Interface Sci.*, **51**, 143–151.

Wang, J., Chen, X., Wang, G., Wang, B., Lu, W., and Zhao, J. (2002). Melting behavior in ultrathin metallic nanowires. *Phys. Rev. B*, **66**, 085408.

Weiss, R. (2006). Has all the magic gone? *The Economist.* http://www.economist.com/node/6795430

Zhang, P., Zhang, Z., Prakash, J., et al. (2011). Trapping and transporting aerosols with a single optical bottle beam generated by Moir techniques. *Opt. Lett.*, **36**, 1491–1493.

Chapter 9

The Bhopal Disaster and Peroxide Bombs: Nanoscale Aspects of Oscillatory Thermal Instability

Rowena Ball
Mathematical Sciences Institute, John Dedman Building 27, Australian National University, Canberra, ACT 0200, Australia
Rowena.Ball@anu.edu.au

Thermal runaway reactions were involved in the Bhopal disaster of 1984, in which methyl isocyanate was vented from a storage tank, and occur in liquid peroxide explosions; yet to date there have been few investigations into the mechanism of thermal runaway in such liquid thermoreactive systems. Consequently protocols for storing thermally unstable liquids and deactivating liquid peroxide bombs may be suboptimal. This contribution presents simulation results for hydrolysis of methyl isocyanate and thermal decomposition of triacetonetriperoxide, using a gradientless, continuous-flow reactor paradigm. This approximation enabled stability analyses of steady-state solutions of the dynamical mass and enthalpy equations. Results show that thermal runaway in both systems is due to the onset of a

Nanotechnology Toward the Sustainocene
Edited by Thomas A. Faunce

ISBN 978-981-4463-76-8 (Hardcover), 978-981-4463-77-5 (eBook)
www.panstanford.com

large amplitude, hard thermal oscillation initiated at a subcritical Hopf bifurcation. This type of thermal misbehavior cannot be predicted using classical ignition theory and may be typical of liquid thermoreactive systems. The mechanism of oscillatory thermal instability on the nanoscale is elucidated, and the potential for exploiting this instability in nanoscale thermal analysis for detection of peroxide explosives is explored.

9.1 Introduction

The causes of the world's worst industrial disaster at Bhopal, India, on December 2–3, 1984, are still being debated in the international media, more than 25 years after the event. Was it caused by neglect, parsimony, or procrastination by Union Carbide on training, safety, and maintenance? Corruption, sabotage, and cover-up? Inadequate government regulation? Any or all of the above may well have helped set up the worst possible scenario—for it could not have been any worse—but they are contributing factors rather than causes. (A brief account of the disaster is given in the appendix.)

The primary *cause* of the thermal runaway that led to the venting of a lethal mist of methyl isocyanate (MIC) over Bhopal City was physicochemical. In this chapter, I present a stability analysis of a simple dynamical model for the MIC–H_2O reacting system, revealing oscillatory thermal misbehavior that cannot be predicted using classical ignition theory. A similar instability is shown to govern the explosive thermal decomposition of triacetonetriperoxide (TATP) in liquid solution. The provenance of oscillatory thermal instability on the nanoscale is elucidated and shown to lie in the ability of the reactant molecules to store energy in the internal molecular motions—in other words, the heat capacity. The significant potential for exploiting oscillatory thermal instability is explored as a diagnostic in nanoscale thermal analysis for the detection of explosives.

Despite the enormity of the Bhopal disaster little or no research has been published that elucidates the fundamental physicochemical cause of thermal runaway in liquid reactive systems such as MIC hydrolysis. In terms of achieving the millennium development goals (MDGs) (United Nations, 2011) it is important that thermoreactive processes in liquids be thoroughly investigated at the nanoscale and the knowledge disseminated widely. Given the horrific legacy of the disaster, the long-term adverse health effects in children and adverse reproductive effects in women of MIC exposure that have been well documented (Mishra, Samarth, Pathak, Jain, Banerjee, and Maudar, 2009), such knowledge is relevant to the MDGs of child health and maternal health. More generally people have a right to expect that thermally unstable and hazardous liquids be stored safely.

The *Millennium Development Goals Report* (United Nations, 2011) highlights the challenges posed by conflicts and armed violence to human security and MDG achievements. A growing threat to security is the use of liquid peroxide explosives by terrorists. The ingredients for making such bombs are cheap and widely available, and they cannot be detected by metal detectors and nitrogenous explosives detectors or distinguished from hand lotion by X-ray machines. TATP and other peroxide-based explosives were used in the suicide attacks on the London transit system in 2005, which killed 56 people, and the terrorists convicted of the 2006 transatlantic aircraft conspiracy had planned to blow up a number of planes using TATP and other peroxide explosives. (Many more accounts of peroxide misuse incidents are easily found on the web.) To this day there are severe restrictions on carrying liquids through security barriers at most airports. It seems grimly inevitable that the use of peroxide explosives as mass murder weapons will increase. Knowledge of their fundamental mechanism of action at the nanoscale may help to counter their use by informing recent efforts to develop nanoscale thermal analysis for explosives' detection (Zuck, Greenblatt, Zifman et al., 2008; Carreto-Vazquez, Wojcik, Liu, Bukur, and Mannan, 2010).

The literature on thermal runaway criteria and parametric sensitivity in batch reactors and storage tanks was summarized by Velo, Bosch, and Recasens (1996). In defining critical conditions they, along with other authors cited therein, begin with the assumption that storage tanks can be modeled as well-stirred batch reactors with linear thermal coupling to the environment. However, batch reactors have no nontrivial steady states, and there is no general theory for determining whether a thermal excursion will grow or decay. It is shown in this work that a simple model with nonequilibrium steady states that is also spatially homogeneous—the continuous-flow stirred tank reactor (CSTR) paradigm—can provide great insight into thermoreactive instabilities in liquid systems and provide fundamental causative information that cannot easily be extracted from detailed numerical simulations that include convective motions.

In Section 9.2, the chemical reactions are described and the relevant data for the physical properties of the reactants and thermodynamic and kinetic parameters is provided. The CSTR paradigm is described in Section 9.3, and the dynamical equations are given. In Section 9.4 the results of numerical stability analyses of the equations are shown graphically and discussed where values of the parameters for MIC hydrolysis and for TATP thermal decomposition in solution were used in the equations. Some points regarding the applicability of the CSTR paradigm are discussed in Section 9.5; the nanoscale aspects of oscillatory thermal runaway are elucidated through examining the behavior in the limits of the two timescales of the relaxation oscillation, and the potential these results offer for improved microthermal analysis in explosives detection is discussed. A summary of the conclusions is given in Section 9.6.

9.2 Chemistry and Data

9.2.1 MIC Hydrolysis

Isocyanates hydrolyze exothermically to the corresponding amine and carbon dioxide. In excess water isocyanates react exothermically with the hydrolysis product amine to form the disubstituted urea (Saunders and Slocombe, 1948; D'Silva, Lopes, Jones, Singhawangcha, and Chan, 1986). With MIC the product is *N,N*-dimethyl urea, and the reaction sequence is as follows:

$$CH_3NCO_{(l)} + H_2O_{(l)} \xrightarrow{K_1(T)} CH_3NH_{2(aq)} + CO_{2(aq)} \qquad \text{(R9.1)}$$

$$CH_3NCO_{(l)} + CH_3NH_{2(aq)} \xrightarrow{K_2(T)} CH_3NHCONHCH_{3(aq)} \qquad \text{(R9.2)}$$

For reaction R9.2 $\Delta H_2(298\text{ K}) = -174.6$ kJ/mol, and for the sequence overall $\Delta H_{tot}(298\text{ K}) = -255$ kJ/mol (Lide, 2008–2009).

A chemical analysis of the residue in the MIC storage tank (Tank 610) at the Union Carbide plant at Bhopal, sampled 17 days after the event, found a variety of MIC condensation products, mainly the cyclic trimer (D'Silva, Lopes, Jones, Singhawangcha, and Chan,

1986). However, experiments indicated that these condensations must have been initiated at temperatures and pressures well above the normal boiling point of MIC. So for modeling the initial thermal runaway these reactions need not be considered. No kinetic data are available for reaction R9.2. So only reaction R9.1 is used in the model. It will be seen from the results that reaction R9.1 alone is sufficient to induce thermal runaway. Relevant physicochemical data is given in Table 9.1.

Table 9.1 Physical, kinetic, and thermochemical parameters for MIC hydrolysis

Molecular weight MIC	57.05 g/mol	
Specific heat capacity C_p° (298) MIC	1959 J/(kg K)	Poling, Thomson, Friend, Rowley, and Wilding (2008)
Specific heat capacity C_p° (298) H_2O	4181 J/(kgK)	
Boiling point MIC at 1 atm	38.3°C	Lide (2008–2009)
Density MIC at 25°C	0.9588 g/cm^3	Lide (2008–2009)
R1 reaction enthalpy	80.4 kJ/mol	Lide (2008–2009)†
R1 activation energy	65.4 kJ/mol	Castro, Moodie, and Sansom (1985)
R1 pseudo first-order frequency factor	4.13×10^8/s	Castro, Moodie, and Sansom (1985)

†From standard enthalpies of formation at 298 K

9.2.2 TATP Thermal Decomposition

TATP, a cyclic trimer, is an explosive made by mixing acetone and hydrogen peroxide, both of which are legal, cheap, and readily available over the counter. Pure TATP is a white crystalline powder that looks like sugar and is soluble in organic solvents. The thermal decomposition of TATP does not involve combustion; the main reaction products are acetone, some carbon dioxide, and ozone (Eyler, Mateo, Alvarez, and Canizo, 2000; Oxley, Smith, and Chen, 2002). Its high explosive power is in part due to the large entropy increase of the formation of four gas molecules from one condensed-phase molecule (Dubnikova, Kosloff, Almog et al., 2005). Relevant parameters for the thermal decomposition of TATP in toluene are given in Table 9.2.

Table 9.2 Physical, kinetic, and thermochemical parameters for thermal decomposition of TATP in toluene

Molecular weight	222.2356 g/mol	
Specific heat capacity C_p° (298) toluene	1698.25 J/(kgK)	Poling, Thomson, Friend, Rowley, and Wilding (2008)
Boiling point toluene at 1 atm	110.8°C	
Density of toluene at 298 K	866.9 kg/m^3	
Reaction enthalpy	330–420 kJ/mol†	Dubnikova, Kosloff, Almog et al. (2005)
Activation energy	178.52 kJ/mol	Eyler, Mateo, Alvarez, and Canizo (2000)
Frequency factor	9.57×10^{16}/s	Eyler, Mateo, Alvarez, and Canizo (2000)
Feed concentration of TATP	2 mol/kg	

†Depending on reaction products

9.3 The CSTR Paradigm

The spatially homogeneous flow reactor, or reacting mass or volume, in which a reactant undergoes a first-order, exothermic conversion is a simple but elucidatory model for thermoreactive systems when it is appropriate to ignore convection, because as a dynamical system it has nontrivial steady states that can be analyzed for stability. The dynamical mass and enthalpy equations may be written as follows:

$$M\frac{\mathrm{d}c}{\mathrm{d}t} = MAe^{-E/RT}c + F(c_\mathrm{f} - c) \tag{9.1}$$

$$MC_\mathrm{r}\frac{\mathrm{d}T}{\mathrm{d}t} = (-\Delta H)\, MAe^{-E/RT}c + F(C_\mathrm{f}T_\mathrm{a} - C_\mathrm{r}T) - L(T - T_\mathrm{a}). \tag{9.2}$$

Notation and quantities are defined in Table 9.3. For numerical and comparative reasons it is more convenient to work with the following dimensionless system corresponding to Eqs. 9.1 and 9.2:

$$\frac{dx}{d\tau} = -xe^{-1/u} + f(1 - x) \tag{9.3}$$

$$\varepsilon\frac{du}{d\tau} = xe^{-1/u} + \varepsilon f(\gamma u_a - u) - l(u - u_a), \tag{9.4}$$

where the dimensionless groups are defined in Table 9.3.

Table 9.3 Quantities, definitions, and units

A	reaction frequency (s^{-1})		
c	$c(t)$, concentration of reactant (mol/kg)		
c_f	inflow reactant concentration (mol/kg)		
C_r	specific heat capacity of the reaction mixture (J/(kg K))		
C_f	specific heat capacity of the inflow stream (J/(kg K))		
E	activation energy (J/mol)		
F	flow through rate (kg/s)		
ΔH	reaction enthalpy (J/mol)		
M	mass of reaction mixture (kg)		
R	gas constant (J/(mol K))		
t	time (s)		
T	$T(t)$ reaction temperature (K)		
T_a	ambient temperature (K)		
L	heat loss coefficient (W/K)		
ε	$C_rE/c_f(-\Delta H)R$	τ	tA
f	F/MA	u	RT/E
γ	C_f/C_r	u_a	RT_a/E
l	$LE/c_fMA(-\Delta H)R$	x	c/c_f

Use of the CSTR paradigm, defined by Eqs. 9.3 and 9.4, is appropriate on a timescale over which the reacting volume remains relatively constant and gradientless relative to the much faster rate of reaction. In this analysis the focus is on the dynamics and we circumscribe a reacting volume, perhaps a small volume inside a large vessel or storage tank, or a virtual CSTR, where the spatial gradients are insignificant in comparison to the time evolution and therefore can be neglected. If this approximation does not hold, then we are free to reduce the circumscribed volume until it does.

The advantage is that the powerful tools of stability and bifurcation theory can be applied to a CSTR model.

A numerical analysis of Eqs. 9.3 and 9.4 was carried out using values of the dimensionless groups calculated from the data in Table 9.1 and Table 9.2 and assigned values of the inverse residence time f, heat loss coefficient l, and inflow concentration.

9.4 Results

9.4.1 Onset of Thermal Runaway in MIC Hydrolysis

From Eqs. 9.3 and 9.4, in the steady state we can define the rate of reactive heat generation as the nonlinear term

$$r_g \equiv fe^{-1/u}/(e^{-1/u} + f)$$

and the rate of nonreactive heat loss, or cooling, as the linear term

$$r_l \equiv -u(\varepsilon f + l) + u_a(\varepsilon\gamma f + l).$$

According to classical ignition theory the reacting mixture self-heats if r_g exceeds r_l. Thermal runaway occurs if r_g exceeds r_l beyond a system-specific threshold; for the hydrolysis of MIC this is taken as the boiling point of MIC. These rates were computed using data from Table 9.1 and are plotted in Fig. 9.1 against the reaction temperature, labeled on the plot in dimensional units. We see that the system self-heats until the reaction temperature T reaches the steady state of ~305 K, at which the heating and cooling rates are balanced. Since the boiling point of MIC is 312 K, according to this diagram the Bhopal disaster did not happen. On the basis of this diagram we would not expect thermal runaway to develop, even when the ambient temperature is allowed to drift slowly up to 292 K.

However, thermal balance diagrams such as that in Fig. 9.1 can be dangerously misleading because they infer stability rather than assess stability rigorously, although such diagrams are often used in chemical reactor engineering. The dangers inherent in what is apparently a recent and alarming trend to completely ignore stability in prescribing thermal runaway criteria were bluntly pointed out by Ball and Gray (2013). The steady states, periodic solutions, and stability analysis of Eqs. 9.3 and 9.4 were computed numerically and yield a dramatically different picture of the thermal stability of

MIC hydrolysis. Figure 9.2 shows a bifurcation diagram in which the steady-state temperature and the temperature amplitude envelope of periodic solutions are plotted as a function of T_a.

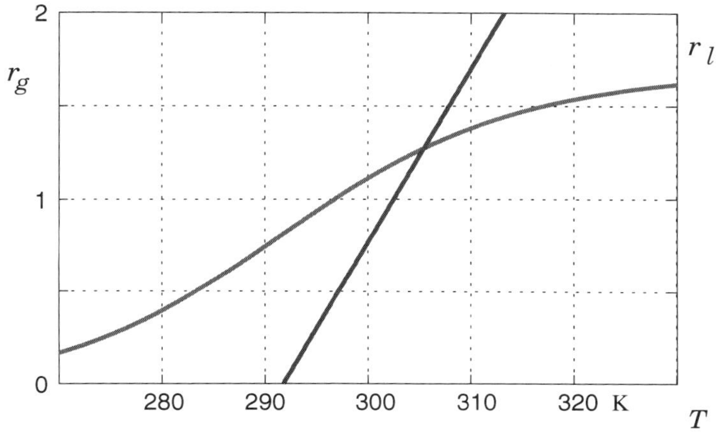

Figure 9.1 Rates of reactive heat generation r_g (red) and heat loss r_l (blue) versus T from Eqs. 9.3 and 9.4. $u_a = 0.0379$ (corresponding to $T_a = 292$ K), $f = 1.7$, $l = 700$, and $\varepsilon = 10$.

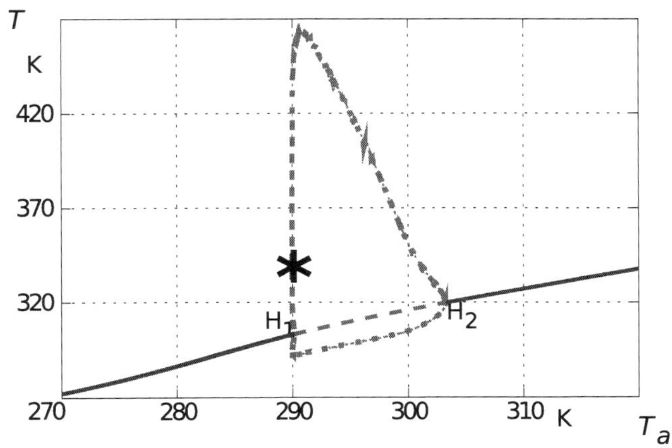

Figure 9.2 A bifurcation diagram of Eqs. 9.3 and 9.4. Stable steady states are plotted with a solid blue line, unstable steady states are plotted with a dashed red line, and the amplitude envelope of periodic solutions is marked with a dotted magenta line. H_1 and H_2 label the Hopf bifurcations, and the large * marks the change in stability of the limit cycles. $f = 1.7$, $l = 700$, and $\varepsilon = 10$.

The steady state is stable at $T_a \approx 286$ K, the temperature at which the tank of MIC had been held for several months. As T_a is allowed to drift upward quasistatically the reaction temperature T increases slowly, but at $T_a = 290.15$ K the stability analysis flags an abrupt change in the nature of the solutions. At this point the steady-state solution becomes unstable at a Hopf bifurcation and the hard onset of a high-amplitude thermal oscillation ensues. Clearly, at $T_a = 292$ K we have catastrophic thermal runaway, contrary to the prediction given by Fig. 9.1. (In the resulting superheated fluid exothermic condensation reactions would increase the temperature even further.)

This is quite different from classical ignition of a thermoreactive system, which occurs at a steady-state turning point. The dynamics of oscillatory thermal runaway can be understood by studying the close-up of the region around the lower Hopf bifurcation H_1 shown in Fig. 9.3. H_1 is subcritical, and the emergent limit cycle is unstable. The amplitude envelope of the unstable limit cycles is marked with a thin dotted line; they grow as T_a is decreased. At the turning point * of the periodic solution branch, the limit cycles become stable. Thermal runaway *may* occur if there are significant perturbations while T_a is within the regime *–H_1, and it *must* occur when T_a drifts above H_1. In principle the rapid thermal excursion takes the system to the amplitude maximum of the stable limit cycle. In reality reactants and products would vaporize, the pressure would soar, and the system would vent or explode. But it must be emphasized that the thermal runaway is initiated at the oscillatory instability rather than at a classical turning point.

The presence of oscillatory instability is all-pervasive and dominant in this system. This can be appreciated by the inspection of Fig. 9.4, a plot of the loci of the steady-state turning points and the Hopf bifurcations of Eqs. 9.3 and 9.4 over the two parameters u_a and the inverse residence time f. In the filled region thermal runaway will always be oscillatory. The bistable regime, indicated by the dashed line, occurs at very high flow rates (short residence times). However, classical thermal runaway at a steady-state turning point does not occur because the oscillatory instability is still present and dominant.

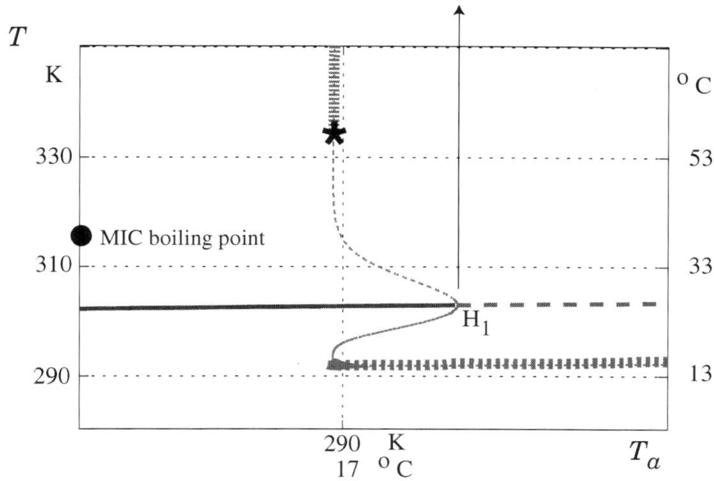

Figure 9.3 A close-up of the region around the Hopf bifurcation H_1 in Fig. 9.2.

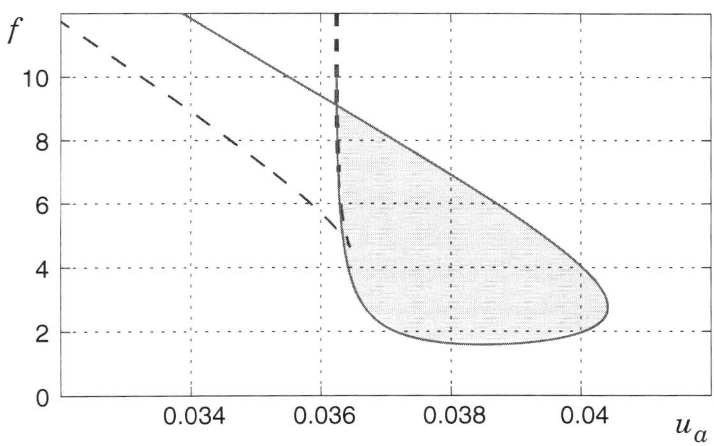

Figure 9.4 The locus of Hopf bifurcations is marked with a solid line, and the locus of steady-state turning points is marked with a dashed line. $l = 700$, $\varepsilon = 10$.

Two computed time series are compared in Fig. 9.5. In the upper plot $T_a = 308.4$ K and the transient perturbation decays to a stable steady state. However, the onset of thermal instability is violent: in the lower plot $T_a = 308.5$ K, a difference of only 0.1 K, and the transient does not decay but swings into sustained high-amplitude relaxation oscillations with a period of about 166 seconds.

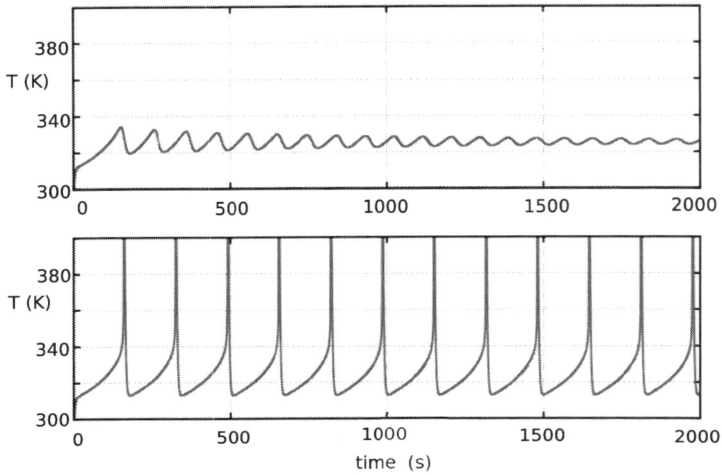

Figure 9.5 Computed time series for MIC hydrolysis, with F = 0.0016 kg/s and L = 560 W/K. Upper plot: T_a = 308.4 K; lower plot: T_a = 308.5 K.

The behavior of the system under a slow upward drift of the ambient temperature can be simulated easily; a time series with T_a = 0.02°C/s is shown in Fig. 9.6, which confirms the abrupt onset of the instability. Of note is the decay in amplitude of the oscillations as the thermostat temperature *increases;* physically this occurs because the reactant is consumed faster than it is supplied as the temperature increases.

9.4.2 Onset of Thermal Runaway in TATP Thermal Decomposition

The steady states, periodic solutions, and stability analysis of Eqs. 9.3 and 9.4 were computed using the data for TATP thermal decomposition in Table 9.2, and results are shown in Fig. 9.7 and Fig. 9.8.

In Fig. 9.7, the loci of classical ignition points (turning points) and Hopf bifurcations are plotted over T_a and F—they are qualitatively similar to those in Fig. 9.4, and the same interpretation applies. A point within the oscillatory region was selected to compute the time series in Fig. 9.8. The system exhibits violent relaxation oscillations; explosive thermal decomposition of TATP is initiated at the onset of this oscillatory behavior rather than by classical ignition.

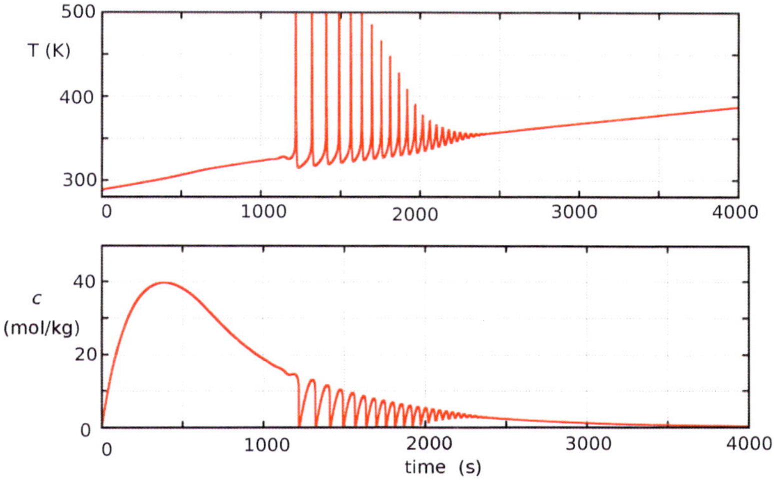

Figure 9.6 Computed time series for MIC hydrolysis with a drift in thermostat temperature T_a = 0.02 C/s. F = 0.0016 kg/s, and L = 560 W/K.

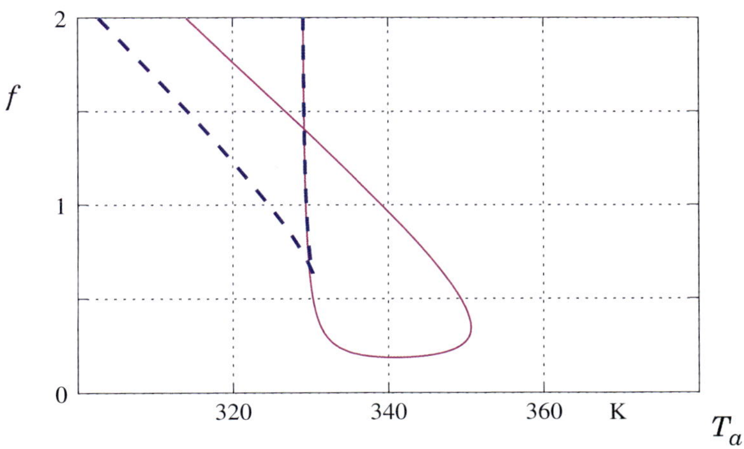

Figure 9.7 The locus of Hopf bifurcations (solid magenta line) and the locus of turning points (classical ignition points) (dashed blue line) for the TATP system.

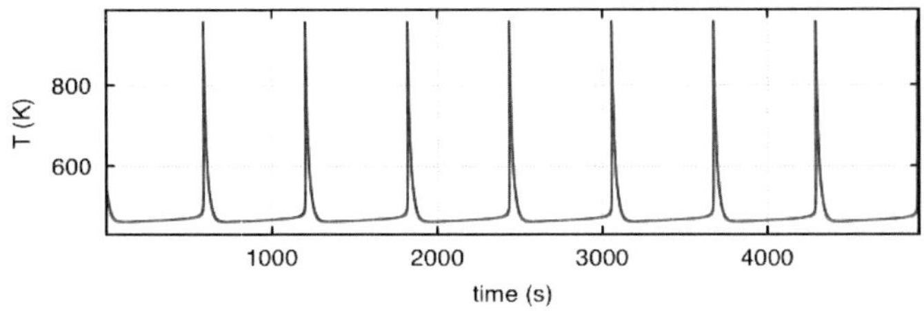

Figure 9.8 Strong relaxation oscillations in the TATP thermal decomposition. The time series was computed for T_a = 460 K and $F = 5 \times 10^{-4}$ kg/s.

9.5 Discussion

The tendency to undergo oscillatory, rather than classical, thermal runaway is typical of exothermically reactive organic liquids. In Ball and Gray (1995) the hydration of 2,3-epoxy-1-propanol in a CSTR was found to exhibit similar nonclassical thermal misbehavior. Here it is shown that the thermal runaway that led to the Bhopal disaster may have been initiated at an oscillatory instability and that liquid peroxide explosions may be initiated at an oscillatory instability rather than by classical thermal ignition. Comparable results have been obtained using parameters for the decomposition of cumenehydroperoxide in Eqs. 9.3 and 9.4 (Ball, 2011).

9.5.1 Nanoscale Aspects of Oscillatory Thermal Instability

Since this is a book about all things nano, it is pertinent to elucidate the mechanism of oscillatory thermal instability on the nanoscale. But before we zoom into nanospatial scales we need to discuss the characteristics of relaxation oscillators in general and examine the two timescales of the relaxation oscillation solutions of Eqs. 9.3 and 9.4.

Relaxation oscillators are often and easily implemented in electrical and electronic circuits, but any dissipative dynamical system with nonequilibrium steady states and two or more dynamical state variables has the potential to exhibit relaxation oscillations. In general they are not smoothly sinusoidal in form. Instead, relaxation oscillators are characterized by energy dissipation and energy accumulation occurring on different timescales. There may be relatively slow dissipation until the system reaches a threshold state at which the internal energy rapidly and nonlinearly increases or rapid dissipation is followed by slow but accelerating increase in internal energy. This latter, two-time dynamics occurs with the thermochemical oscillator in Eqs. 9.3 and 9.4.

The two timescales in each period evident in Fig. 9.5 are shown in Fig. 9.9. On the "fast" timescale (left), during the few seconds between the temperature maximum and about 990 seconds, the concentration remains close to zero. On the "slow" timescale (right), the system barely heats at all over the long interval, although evidently some reaction takes place because the reactant accumulates nonlinearly, reaches a maximum, and then declines slowly before the exponential spike.

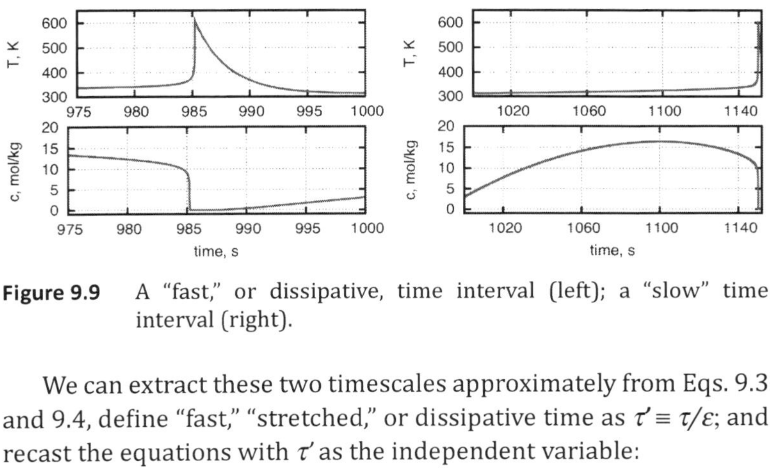

Figure 9.9 A "fast," or dissipative, time interval (left); a "slow" time interval (right).

We can extract these two timescales approximately from Eqs. 9.3 and 9.4, define "fast," "stretched," or dissipative time as $\tau' \equiv \tau/\varepsilon$; and recast the equations with τ' as the independent variable:

$$\frac{\mathrm{d}x}{\mathrm{d}\tau'} = \varepsilon[-xe^{-1/u} + f(1-x)] \tag{9.5}$$

$$\frac{\mathrm{d}u}{\mathrm{d}\tau'} = xe^{-1/u} + \varepsilon f(\gamma u_a - u) - l(u - u_a) \tag{9.6}$$

For ε sufficiently small, we have $\mathrm{d}x/\mathrm{d}\tau' \approx 0$ and $x \approx x_0 \approx 0$ since the reactant is almost fully depleted and $\mathrm{d}\boldsymbol{u}/\mathrm{d}\tau' \approx -l(u - u_a)$, so $\boldsymbol{u}$ decays, as $u \approx \exp(-l\tau')(u_0 - u_a) + u_a = \exp(-l\tau')(u_{max} - u_a) + u_a$ since the temperature amplitude is at a maximum when $x = 0$ at $\tau'_{max} = \tau'_0$

= 0. As τ' becomes "large," $u \to u_a$; in Fig. 9.9 left this occurs around ~1000 s. However, the reactant is gradually accumulating and the system evolves in "slow" time.

Over much of the "slow" time interval the system behavior can be understood by dividing Eq. 9.4 through by ε. At low temperatures and with a large ε, we have $du/d\tau \approx -f(u - u_a)$, which gives $u \approx \exp(-f\tau)(u_0 - u_a) + u_a$, and since $u_0 = u_a$ on this timescale, we have $u \approx u_a$. In Fig. 9.9 this approximation holds over much of the "slow" time interval: the internal energy of the system increases, but the temperature remains almost constant, although slowly increasing. During this time the reactant accumulates but the reaction does take place: x evolves as $x \approx f[1 - \exp(-b\tau)]/b$, where $b = \exp(-1/u_a)$.

If the reaction takes place in "slow" time but the system is barely heating up, where does the heat of the reaction go? Now we focus on the nanoscale.

The answer is that the heat is stored in the internal motions of the reactant mixture molecules. Referring to the definition of the dimensionless group e in Table 9.3 we see that e is large if the specific heat capacity C_r of the reaction mixture is large. The specific heat capacity of a substance is a function of the number of degrees of freedom of motion available to its constituent particles. A monoatomic perfect gas has only the kinetic energy of each atom. So the translational motion in three dimensions is the only motion it can undergo. The equipartition theorem tells us that the three translational contributions to the constant volume molar heat capacity $C_V^{tr} = (3/2)R = 12.47$ J/(mol K); thus the kinetic energy contribution does manifest as temperature change.

Polyatomic molecules have many additional degrees of freedom though, and physical properties of liquids are governed much more by the potential energy of the system than the kinetic energy. Potential energy is stored in intramolecular rotational and vibrational degrees of freedom and in the vibrational intermolecular force potential. The mean translational energy of each liquid molecule is the same as that for gases, $(3/2)kT$, where k is Boltzmann's constant, and since $C_p = C_v + R$ we can take the molecular translational contribution to the molar heat capacity of a liquid as $C_p^{tr} = 20.78$ J mol^{-1}K^{-1}.

Since the molar heat capacity of MIC is 111.76 J mol^{-1}K^{-1} and that of water is 75.29 J mol^{-1}K^{-1}, on a molar basis the MIC–H_2O liquid system can store a large amount of heat in the internal rotational modes and in the vibrational intermolecular potentials. (The internal vibrational modes are not excited at this "slow" temperature.)

The specific heat capacity of MIC, as opposed to the molar heat capacity, is not particularly large, but that of water is anomalously high due to strong intermolecular forces. For the reaction mixture it is large enough to depress the first and third terms on the right hand side of Eq. 9.4 relative to the second term at low temperature. The rapid temperature spike and transition to "fast" time occurs because the activation energy is high enough to make the reaction very temperature sensitive. In other words, the Arrhenius term in Eq. 9.4 can take over very rapidly once the reaction zone reaches a certain temperature. At the temperature peak the system enters the "fast" dissipative time interval since the reactant is fully depleted. The specific heat capacity cannot affect the dynamics on this timescale.

This analysis suggests that containment systems with large heat capacities for thermoreactive liquids may not suppress the oscillatory instability, although they would certainly lengthen the "slow" timescale. This may or may not be a good thing. A lengthy "slow" interval may provide enough time before the thermal runaway to deactivate or quench the system in other ways. On the other hand, it might give a false sense of security: *Nothing has happened over the last four hours; we might as well leave it.*

9.5.2 Opportunities for Nanoscale Thermal Analysis

The ubiquitous presence of oscillatory thermal instability in liquid peroxide explosives suggests that it may be exploited to great advantage in new techniques of nanoscale calorimetry for explosives' detection. The state of the art of nanocalorimetry was reviewed by Carreto-Vazquez, Wojcik, Liu, Bukur, and Mannan (2010), who concluded that although there are still technical challenges in reducing uncertainties and refining reproducability, such chip-scale calorimeters will provide valuable fast measurements of exothermic behaviors of suspect substances in trace quantities using low-cost, reliable, handheld devices in the field.

Zuck, Greenblatt, Zifman et al. (2008) performed comparative differential scanning microcalorimetry on a number of explosive and nonexplosive materials to demonstrate that the exotherms obtained for the explosives are diagnostic. Several of their explosives' thermograms showed multiple exotherms declining in amplitude, which Zuck, Greenblatt, Zifman et al. (2008) ascribed to inhomogeneous heating of the melted sample. An alternative explanation is that the explosive undergoes thermal decomposition via oscillatory thermal instability and the scan passes through the oscillatory region.

Figure 9.11a shows an approximately redrawn thermogram from Zuck, Greenblatt, Zifman et al. (2008) for the thermal decomposition of trinitrotoluene. The tunable heater voltage is proportional to a tunable temperature. The endotherm is due to melting, and the exotherms are due to thermal decomposition of the TNT sample. In Fig. 9.11b the dataset that was computed to produce Fig. 9.10 is plotted with T against T_a—this simulates a thermogram without, of course, the melting endotherm and expression of the temperature in terms of departure from a baseline. Although the heating rates in Fig. 9.11a and Fig. 9.11b are vastly different the qualitative similarity of the plots is striking. With intensive experimental workup and optimization, the presence of oscillatory thermal instability with multiple peaks diminishing in amplitude could be definitively diagnostic of liquid explosives. Furthermore, in standardized apparatus, features such as the temperature of onset and the number of peaks and their amplitude could identify the explosive substance.

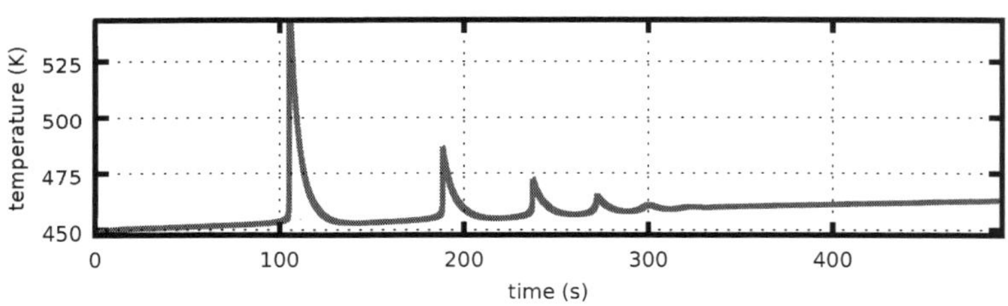

Figure 9.10 Time series for TATP thermal decomposition with a thermostat drift T_a = 0.43°C/s and F = 1.8 x 10^{-3} kg/s.

9.6 Conclusions

1. The CSTR paradigm was applied to investigate the thermal stability of MIC hydrolysis and the thermal decomposition of triacetone triperoxide in solution.

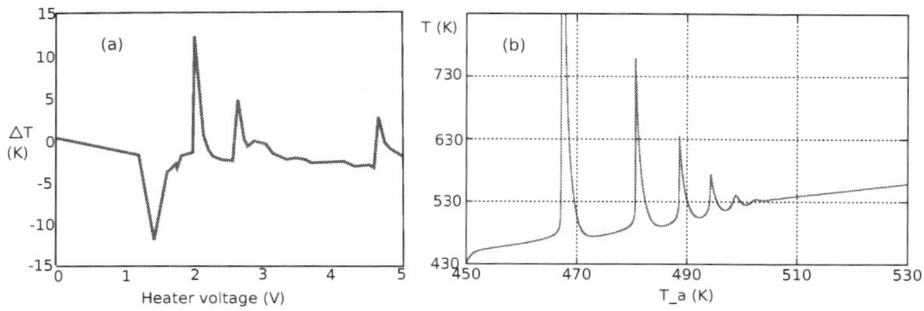

Figure 9.11 (a) The thermogram from Zuck, Greenblatt, Zifman et al. (2008) (approximately redrawn) for the thermal decomposition of TNT. (b) A simulated thermogram for the thermal decomposition of TATP.

2. Stability analyses of the steady-state solutions of the dynamical model found that in both cases, thermal runaway occurs due to the hard onset of a thermal oscillation at a subcritical Hopf bifurcation. Classical thermal ignition at a steady-state turning point does not occur in these systems, and over the thermal regime of interest, they are dominated by oscillatory instability.
3. Analysis of the cause of oscillatory thermal instability on the nanoscale shows that this nonclassical thermal misbehavior predominates in liquid thermoreactive systems where the specific heat capacity and activation energy are high. These results provide new information about the cause of thermal runaway that may inform improved designs of storage systems for thermally unstable liquids and better management of organic peroxide–based explosives.
4. The thermal decomposition of liquid explosives via oscillatory thermal instability suggests that this may be developed as a new and valuable diagnostic and identifier of explosives in the emerging field of nanocalorimetry.

Appendix A

The following brief account of the Bhopal disaster has been compiled from the following sources: Weir (1987), Shrivastava (1987), Lepkowski (1994), and Abbasi and Abbasi (2005).

The Union Carbide plant at Bhopal carried out the production of carbaryl, an agricultural insecticide that has been used widely throughout the world since 1945. MIC, a low-boiling-point, highly reactive, and extremely toxic liquid, used in the synthesis of carbaryl, was stored in an underground stainless steel tank (Tank 610), which was encased in a concrete shell. The temperature of the 41 tons of MIC in Tank 610 was 12°C–14°C rather than the recommended 0°C–4°C because the refrigeration unit had been nonoperational for several months. On the evening of December 2, 1984, a worker had been sent to hose out a nearby tank. The hose was left running unattended, and it is believed that a faulty valve allowed entry of water into the connected Tank 610. (Union Carbide disputes this, asserts that nothing was wrong with its equipment and procedures, and argues that sabotage by a disaffected employee must have led to the disaster.) By 11:30 p.m., when workers detected lachrymose whiffs of leaking MIC, water had been running into Tank 610 for several hours. Although a slow rise in temperature and pressure in the tank had been noted, the early signs of trouble were not acted upon. Shortly after 11:30 p.m., the contents of the tank reached thermal criticality and began escaping as vapor from the flare tower.

Downwind of the flare tower lay the crowded suburbs and shantytowns. Most of the fluid in the tank streamed from the tower, then drifted low over the city, and sank and seeped in deathly mist in lungs and eyes, a period to sleep and swift arrest of retreat. About 4000 lives were claimed immediately, and many tens of thousands, through the subsequent days and months and years, lost their lives or their health to the poison's effects, and the dead are still being counted.

Acknowledgment

The author acknowledges the support of Australian Research Council Future Fellowship FT0991007.

References

Abbasi, T., and Abbasi, S. (2005). The expertise and the practice of loss prevention in the Indian process industry: some pointers for the Third World. *Process Saf. Environ. Prot.*, **83,** 413–420.

Ball, R. (2011). Oscillatory thermal instability and the Bhopal disaster. *Process Saf. Environ. Prot.*, **89,** 317–322.

Ball, R., and Gray, B. F. (1995). Transient thermal behavior of the hydration of 2,3-epoxy-1-propanol in a continuously stirred tank reactor reactor. *Ind. Eng. Chem. Res.*, **34,** 3726–3736.

Ball, R., and Gray, B. F. (2013). Thermal instability and runaway criteria: the dangers of disregarding dynamics. *Process Saf. Environ. Prot.*, **91**, 221–226.

Carreto-Vazquez, V. H., Wojcik, A. K., Liu, Y.-S., Bukur, D. B., and Mannan, M. S. (2010). Miniaturized calorimeter for thermal screening of energetic materials. *Microelectron. J.*, **41,** 874–881.

Castro, E., Moodie, R., and Sansom, P. (1985). The kinetics of hydrolysis of methyl and phenyl isocyanates. *J. Chem. Soc., Perkin Trans. 2*, 737–742.

D'Silva, D., Lopes, A., Jones, R., Singhawangcha, S., and Chan, J. (1986). Studies of methyl isocyanate chemistry in the Bhopal incident. *J. Org. Chem.*, **51,** 3781–3788.

Dubnikova, F., Kosloff, R., Almog, J., et al. (2005). Decomposition of triacetonetriperoxide is an entropic explosion. *J. Am. Chem. Soc.*, **127,** 1146–1159.

Eyler, G., Mateo, C., Alvarez, E., and Canizo, A. (2000). Thermal decomposition reaction of acetone triperoxide in toluene solution. *J. Org. Chem.*, **65,** 2319–2321.

Lepkowski, W. (1994). Bhopal ten years later. *Chem. Eng. News*, **December 19**, 8–18.

Lide, D. R. (ed.) (2008–2009). *CRC Handbook of Chemistry and Physics*, 89th ed. CRC Press, Taylor & Francis Group.

Mishra, P., Samarth, R., Pathak, N., Jain, S., Banerjee, S., and Maudar, K. (2009). Bhopal gas tragedy: review of clinical and experimental findings after 25 years. *Int. J. Occup. Med. Environ. Health*, **22,** 193–202.

Oxley, J., Smith, J., and Chen, H. (2002). Decomposition of a multi-peroxidic compound: triacetonetriperoxide (TATP). *Propellants, Explos., Pyrotech.*, **27,** 209–216.

Poling, B., Thomson, G., Friend, D., Rowley, R., and Wilding, W. (2008). *Perry's Chemical Engineers Handbook*, 8th ed. McGraw-Hill.

Saunders, J., and Slocombe, R. (1948). The chemistry of the organic isocyanates. *Chem. Rev.*, **43**(2), 203–218.

Shrivastava, P. (1987). *Bhopal: Anatomy of a Crisis*. Cambridge, MA: Ballinger.

United Nations (2011). *The Millennium Development Goals Report 2011*. New York: United Nations, www.un.org/millenniumgoals.

Velo, E., Bosch, C., and Recasens, F. (1996). Thermal safety of batch reactors and storage tanks: development and validation of runaway boundaries. *Ind. Eng. Chem. Res.*, **35,** 1288–1299.

Weir, D. (1987). *The Bhopal Syndrome*. San Francisco: Sierra Club Books.

Zuck, A., Greenblatt, J., Zifman, A., et al. (2008). Explosive detection by microthermal analysis. *J. Energ. Mater.*, **26,** 163–180.

Chapter 10

Fusion Power and Nanoscience Challenges for Extreme Materials

Matthew J. Hole and Cormac S. Corr
Plasma Research Laboratory, Australian National University, Canberra, ACT 0200, Australia
matthew.hole@anu.edu.au, cormac.corr@anu.edu.au

Fusion is the energy process that powers the sun and the stars. If harnessed on the earth, it could produce millions of years of clean base-load power. The process is free of CO_2 emission, is intrinsically safe, and has no weapon potential. Radioactive waste is very low level and indirect, arising from neutron activation of the first wall. With current technology, a fusion power plant could be completely recycled within 100 years of shutdown.

An offset to the attractive features of fusion power is that the reaction is difficult to initiate and sustain: burn necessitates very high temperatures (~100 million K), strong confinement magnetic fields, and a large plasma confinement volume. In addition to the physics confinement challenges cast by a "burning" plasma, the high more fuel-abundant deuterium-deuterium for next-generation experiments, such as the ITER. In a commercial power plant, the

Nanotechnology Toward the Sustainocene
Edited by Thomas A. Faunce

ISBN 978-981-4463-76-8 (Hardcover), 978-981-4463-77-5 (eBook)
www.panstanford.com

neutron and heat flux challenge is even more extreme: the first wall has to withstand neutron damage of 100 displacements per atom over its lifetime. For the most part, nanoscience challenges for fusion lie in the field of extreme materials. In this chapter, we review the science, engineering, and economics of fusion power and highlight some of the nanoscience challenges for extreme materials.

10.1 The Basis of Fusion Power

Nuclear force is the strongest of the four fundamental forces of nature: gravitational, weak electromagnetic, strong electromagnetic, and nuclear. Exothermic chemical reactions, such as the combustion of coal, access the strong electromagnetic force. The difference between the internal energies of reactants and products is converted to heat. Accessing the nuclear force involves reactions of the nucleus of atoms. Nuclear reactions convert the binding energy of nuclei to kinetic energy.

Per reaction, nuclear reactions offer millions of times greater energy yield compared to chemical reactions. To illustrate, reaction R10.1 is the combustion of pure anthracite coal (dry mass). Reaction R10.2 is the fission of uranium into energetic products xenon and strontium and a thermal neutron. The easiest fusion reaction to initiate is the fusion of deuterium and tritium, isotopes of hydrogen, to form helium and an energetic neutron (Wesson, 2004). Per unit mass of fuel, a deuterium–tritium (D–T) reaction yields four times as much heat as the fission of uranium and millions of times more heat than the combustion of pure anthracite coal (dry mass).

$$C_6H_2 + 6.5O_2 \rightarrow 6CO_2 + H_2O + 30\text{ eV} \quad \text{(R10.1)}$$

$$U^{235} + n \rightarrow Xe^{134} + Sr^{100} + n + 200\text{ MeV} \quad \text{(R10.2)}$$

$$D^2 + T^3 \rightarrow He^4 + n^1 + 17.6\text{ MeV} \quad \text{(R10.3)}$$

$$D^2 + He^3 \rightarrow He^4 + p + 18.3\text{ MeV} \quad \text{(R10.4)}$$

$$D^2 + D^2 \rightarrow He^3 + n^1 + 3.27\text{ MeV} \quad \text{(R10.5)}$$

$$D^2 + D^2 \rightarrow T^3 + H^1 + 4.03\text{ MeV} \quad \text{(R10.6)}$$

Fusion power research is focused on exploitation of the D-T cycle, codiscovered by Australian Sir Mark Oliphant in 1933 (Oliphant, Harteck, and Rutherford, 1934). The next generation of fusion cycles based on the aneutronic deuterium–helium (D–He) reaction and the more fuel-abundant deuterium–deuterium reaction will be based on the same technology but will involve more extreme conditions.

Unlike the fission of U^{235}, which can be initiated by a neutron, the D-T and D-He cycles require the fuel nuclei to collide with sufficient kinetic energy (in the center of the mass frame) to overcome their electrostatic repulsion. For a given collision, the probability of a fusion reaction σ is a function of the relative velocity v between the two reactant nuclei. The reaction rate (fusion reactions per volume per time) is σv times the product of the reactant number densities. Then, in a gas with a distribution of velocities, the velocity average must be taken, $<\sigma v>$ (Wesson, 2004).

Figure 10.1 shows the reaction rate $<\sigma v>$ as a function of the ion temperature. Although the collision cross section of the D-T reaction is a maximum at around 60 keV, the temperature need not be that high, because the required reactions occur in the very high-energy tail of the velocity distribution function. The necessary temperature is around 10 keV, or 100,000,000°C. At these extreme temperatures, which are six to seven times hotter than the core of the sun, the fuel exists as ions in the plasma state. That is, the fuel atoms are ruptured into their component electrons and nuclei.

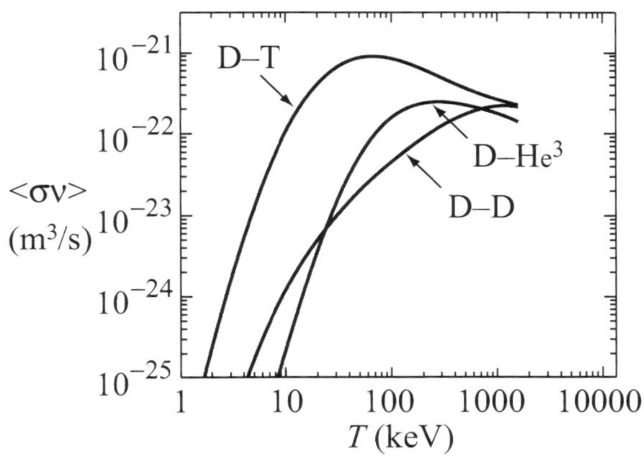

Figure 10.1 Velocity averaged cross section for the D–T, D–He, and D–D fusion reactions as a function of temperature. Reprinted from Jeffrey P. Freidberg (2007). *Plasma Physics and Fusion Energy,* with permission from Cambridge University Press.

To reach such high temperatures, a D–T plasma must be heated. To be useful, the heated plasma must be confined and controlled. From the power balance relationship in a confined plasma, a condition for plasma ignition (when the input heating power P_{H} approaches zero) can be obtained. The confined thermonuclear power output per unit volume in a D–T plasma of equal parts deuterium and tritium is given by:

$$P_{\alpha} = \frac{1}{4} n^2 \langle \sigma v \rangle \varepsilon \qquad \text{(R10.7)}$$

where n = the density of fuel ions and ε = the kinetic energy of the confined He^4 ions (see reaction R10.3). The rate of energy loss $P_{\mathrm{L}} = W/\tau_{\mathrm{E}}$, where W = the stored energy and τ_{E} = the energy confinement time. Conservation of energy yields the overall power balance relation:

$$P_{\mathrm{H}} + P_{\alpha} = P_{\mathrm{L}} \qquad \text{(R10.8)}$$

Finally, substituting expressions for P_{α}, P_{L}, and $<\sigma v>$ yields the important self-sustaining ignition condition or Lawson criterion:

$$nT\tau_{\mathrm{E}} \geq 3 \times 10^{21}\ \mathrm{m^{-3} keVs} \qquad \text{(R10.9)}$$

The importance of the triple product $nT\tau_{\mathrm{E}}$ is that it provides a figure of performance merit for fusion plasmas and a reactor performance threshold. A sample reactor relevant condition is $n = 10^{20}\ \mathrm{m^{-3}}$, $T = 10$ keV, and $\tau_{\mathrm{E}} = 3$ seconds.

By treating the fusion plasma as a heat source, concept realization of fusion power as an energy source can be described independent of the details of the plasma. Figure 10.2 shows such a schematic of concept, showing the D–T plasma as a heat source, which is used (in this case) to generate electricity. As with fossil-fuel-based large-scale power systems, electricity is generated by the production of steam, which drives a turbine-generator combination. The plasma is a heat source, which could also be used for industrial processes such as desalination or hydrogen production.

A distinctive feature of D–T fusion is the fuel handling and processing cycle. One of the fuel ions, tritium, is radioactive, with a half-life of 12.3 years, and so does not occur naturally. Instead, it needs to be manufactured, by neutron transmutation of lithium. The two processes are:

$$\mathrm{n + Li^7 \rightarrow n + T + He^4 - 2.5\ MeV} \qquad \text{(R10.10)}$$

$$\mathrm{n + Li^6 \rightarrow n + T + He^4 + 4.8\ MeV} \qquad \text{(R10.11)}$$

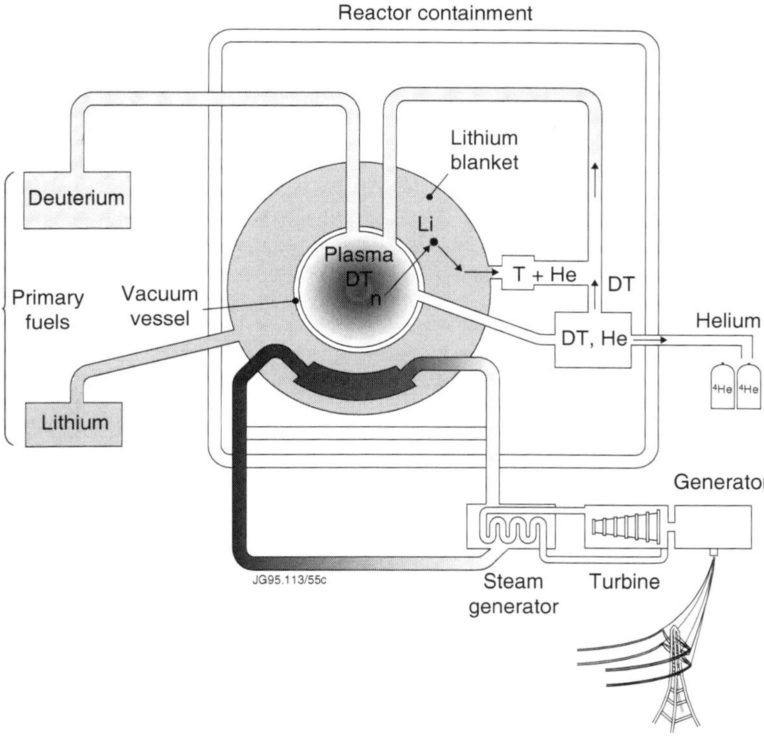

Figure 10.2 Schematic of a fusion power plant. Photograph/figure supplied courtesy of JET-EFDA publications copyright Euratom.

In a working fusion power plant, tritium will mostly be generated in situ by the neutron activation of lithium in a blanket surrounding the vessel walls. The D-T reaction releases a 14.1 MeV neutron, which carries 80% of the reaction kinetic energy out of the plasma. When the energetic neutron impacts lithium within the blanket, it generates tritium, via reactions R10.8 and R10.9. The blanket cannot be engineered so that all neutrons undergo such a reaction, and so a neutron multiplier such as beryllium or lead will be used. By processing the lithium, tritium can be separated and used for injection into the plasma, while the helium gas can be removed as a waste product. Heat exchangers will remove the heat to produce steam for the turbine. To initiate the fusion cycle, some tritium will have to be manufactured separately.

10.2 Fusion Fuel Abundance

The fuels for fusion are isotopes of hydrogen, the most abundant element in the universe. Of normal matter, hydrogen accounts for over 75% of all mass in the universe. Normally, abundance of the elements is quoted with respect to hydrogen. For deuterium and lithium the abundance ratios are 1D:6500H (1 deuterium atom per 6500 hydrogen atoms), 1Li:10^6H (earth), and 1Li:1000H (solar system).

Deuterium is chemically indistinguishable from hydrogen and is found commonly in water. Enrichment of deuterated or heavy water (HDO) is usually accomplished by distillation, electrolysis, or isotopic exchange. Deuterium itself is extracted from heavy water by electrolysis. Any nation with access to water (sea or fresh) thus immediately has access to deuterium. The total mass of the hydrosphere is 1.4×10^{21} kilograms, giving an estimated terrestrial deuterium reserve of 23×10^{10} kilotons.

Excluding the oceans, lithium is found as both a mineral salt and a brine solution. Estimated economically demonstrated resources of lithium total 4.1 million tons (McKay, McKay, Miezitis et al., 2005), while the world's estimated reserves total 11 million tons (Norton and Leahy, 2006). Lithium may also be extracted from seawater (0.17 g/m^3), yielding a further potential 230,000 million metric tons.

Like any resource, the duration for which D-T fuel can power civilization depends on the rate of use. Exploitation of the D-T cycle is limited only by the world's lithium reserves. Using a D-T fuel cycle, the complete fusion burn of the earth's 11-million-ton lithium reserves would produce 3×10^{24} joules. According to US Department of Energy statistics (2012), the world electricity consumption in 2010 was 15 TW-yr. Even assuming a pessimistic efficiency of 30% of a fusion power plant efficiency and total reliance on fusion power for all of the earth's electricity needs, the burn duration of the D-T cycle is 2000 years. The burn duration is a lower estimate as it assumes no improvement in the efficiency of power use and a pessimistic fusion efficiency calculation. The use of lithium in seawater increases this limit to several million years. Exploitation of the D-D cycle offers billions of years of energy.

Compared to coal and fission, fusion power offers very high energy density. A 1 GW (electric) power station operating at 30%

conversion efficiency will require 430 kg of lithium and 650 kg of deuterium in one year. In contrast, a coal power plant of the same size requires 2.5 million tons of coal, and a fission reactor requires 35 tons of uranium oxide produced from 210 tons of uranium ore.

10.3 Magnetic Confinement Fusion

The Lawson criterion prescribes the conditions of plasma density, temperature, and energy confinement time for net energy production in a hot, confined plasma. The leading technology to produce these conditions and enable fusion power is magnetic confinement. The principle of magnetic confinement is the use of strong magnetic fields, which confine charged particles to gyrate about lines of magnetic force, as shown in Fig. 10.3. To prevent particles from exiting the confinement chamber, the magnetic configuration is made toroidal, or doughnut-shaped, such that the field lines execute loops about the central axis.

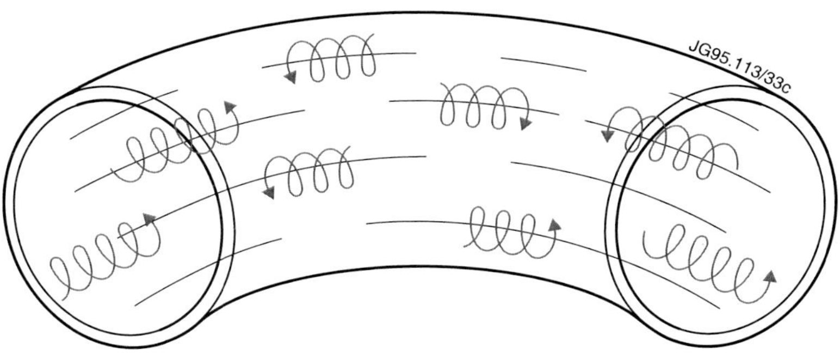

Figure 10.3 Gyration of particles about field lines. Photograph/figure supplied courtesy of JET-EFDA publications copyright Euratom.

At present, the most reactor-relevant toroidal magnetic confinement configuration is the tokamak, shown in Fig. 10.4. This is a doughnut-shaped (toroidal) plasma, with a strong toroidal magnetic field produced by external field windings. An additional magnetic field is produced by a toroidal plasma current, which is partly induced by a transformer action due to the central solenoid (see the primary transformer circuit in Fig. 10.4). Other advanced toroidal magnetic confinement concepts exist, which offer potential improvements in cost and steady-state operation. Research programs based around these concepts are important in terms of both their ability to contribute to the programmatic development of fusion power (e.g., ITER) and concept innovation and improvement.

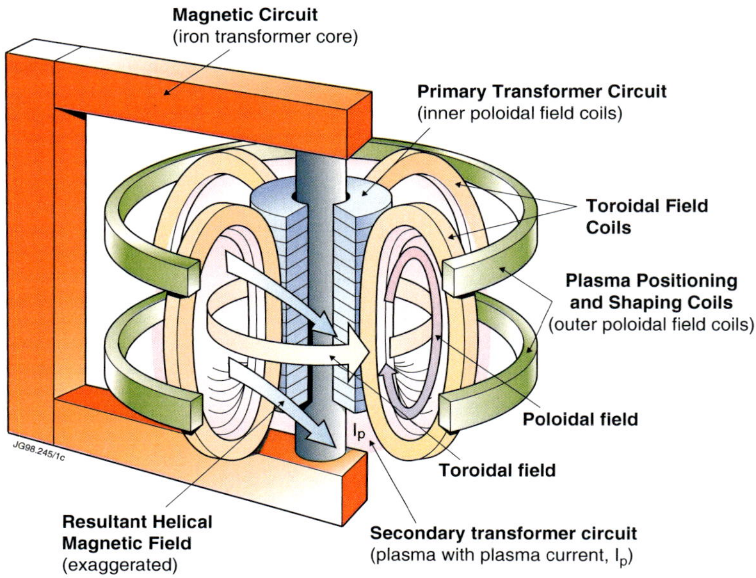

Figure 10.4 Schematic of a tokamak. Photograph/figure supplied courtesy of JET-EFDA publications copyright Euratom.

Three principle heating mechanisms are employed to heat fusion plasmas to reactor-relevant temperatures. The simplest of these is resistive heating, which operates by the same principle as an electric household radiator. Plasmas can support large electrical currents. By inducing these in the plasma, normally by transformer action, the fuel ions can be heated via electron–ion collisions—a process also known as resistive (or ohmic) heating. Another common technique for plasma heating is the use of electromagnetic waves, a process similar to cooking food in a microwave oven. In this case, antennas or waveguides couple radio frequency waves directly to the plasma and the plasma particles are heated by wave particle resonant absorption. Another technique that has yielded significant improvements in plasma performance is heating by injection of beams of energetic neutral particles. In this process, ions are accelerated electrostatically to high energy (e.g., 1 MeV for ITER), are neutralized by charge exchange with a neutraliser gas and travel in a straight line into the plasma. Once inside the plasma, they undergo a second charge exchange event with the thermalised plasma, producing an ion and a neutral particle. The high-energy ion is then confined by the magnetic field, while the low-energy neutral particle is lost to the chamber walls. Finally, via collisions, the high-energy particle loses its energy to the bulk of the plasma, heating it. Figure 10.5 shows a schematic of these heating systems.

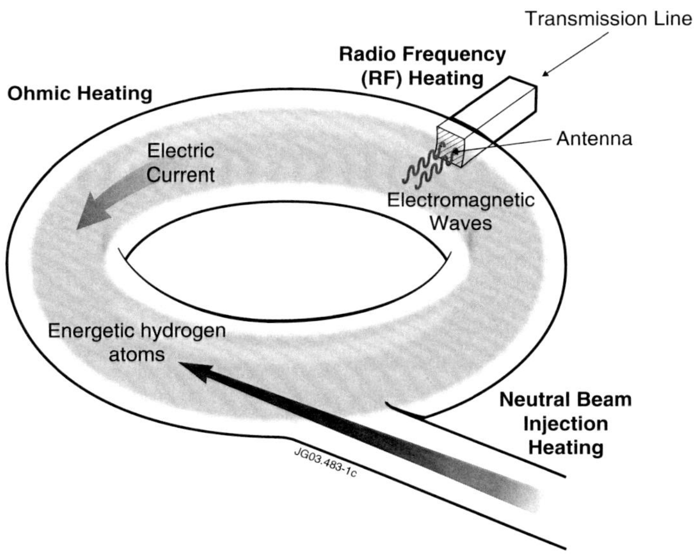

Figure 10.5 Heating mechanisms. Photograph/figure supplied courtesy of JET-EFDA publications copyright Euratom.

The use of these heating mechanisms, together with improvements in magnetic field configuration, access to improved stability and higher-energy confinement regimes, and improved control and power handling, has enabled rapid advances in fusion performance. Figure 10.6 shows the increase in the fusion triple product (left vertical axis) with an increasing central ion temperature. The right vertical axis indicates the approximate performance growth as a function of time. In 50 years of experimental research, the triple product has increased 5 orders of magnitude (i.e., by a factor of 100,000). In the same period, the plasma temperature has increased from 1 million to over 100,000,000°C. Only two experiments, the Joint European Torus (JET) and the Tokamak Fusion Test Reactor (TFTR), have been capable of handling tritium. The highest-performing discharge was in JET, with a demonstrated power gain of $Q = 0.63$ (Wesson, 2004).

$$Q = \frac{P_{\text{fusion}}}{P_{\text{H}}}$$

In most plasma experiments, however, tritium is not used. In these cases, the estimated power gain Q_{DT}, extrapolated from data, assuming the plasma to be an equal mix of deuterium and tritium, has exceeded the break-even value ($Q_{\text{DT}} > 1$). Figure 10.6 also shows the ignition barrier and the operating range of the next-step fusion experiment, ITER, which is outlined in Section 10.10.

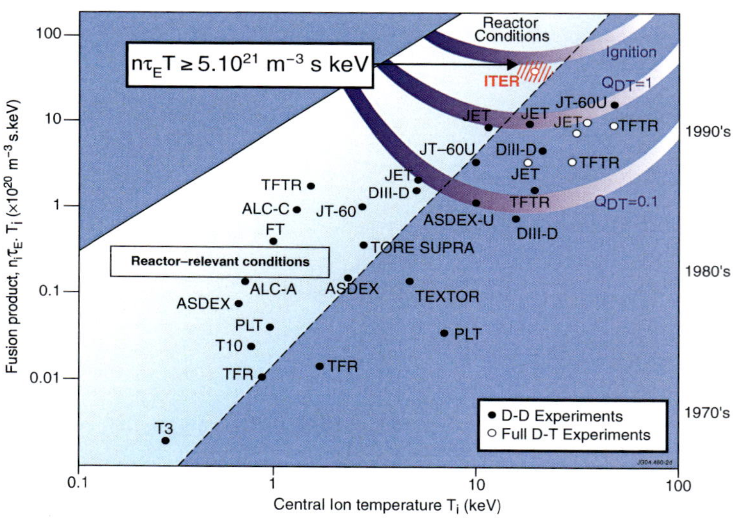

Figure 10.6 Progress in the fusion triple product, showing the ITER operating point. Photograph/figure supplied courtesy of JET-EFDA publications copyright Euratom.

10.4 Fusion Materials

Any form of energy production requires materials working at close to their physical limits to achieve optimum efficiency. In the case of fusion this means that different parts of the structure need to meet stringent requirements, which differ on the basis of the placement and role played. The relevant structures are as follows:

- First wall: High heat and radiation load and low hydrogen retention
- Diverters: Extreme heat load and low hydrogen retention
- Vacuum vessel: High radiation load, low hydrogen retention, and vacuum integrity
- Lithium blanket module: Efficient neutron capture and heat transfer
- Heat exchange system: Effective heat transfer using low neutron activation materials
- Electromagnetic coils: Superconducting material that can cope with high stress and potential neutron irradiation
- Structural materials: Long-term stability and rigidity in the environment of a moderate neutron dose and potential neutron activation

For ITER, the materials have been chosen. In the most critical heat load locations, the materials used will be tungsten, beryllium, and carbon. These are not the best choices but the best compromises at this stage. More work is needed to develop better materials to meet these demanding situations.

The most extreme conditions are experienced by the first wall and diverters exposed to a heat load of 10–100 MW/m^2 and a continuous exposure to a high flux of 14 MeV neutrons. At the same time, the first wall material must be a good thermal conductor to allow heat transfer to the heat exchanger.

The first wall has to cope with:

- a heat load of 10–100 MW m^{-2};
- 14 MeV neutron irradiation;
- 10 keV D, T, and He bombardment; and
- tritium retention.

It should also:

- be a good thermal and electrical conductor;
- have a high melting point;
- not create long-lived radioactive isotopes;
- ideally be composed of low atomic number species;

- not retain too much hydrogen;
- have a low sputtering yield;
- have good thermomechanical properties;
- have a high resistance to thermal shocks; and
- be easy to machine.

A key challenge for fusion power is controlling transport at the boundary between the hot fusion core (>10^6 K) and the low-temperature (10^3K) wall. Actively cooled components must be capable of withstanding a range of energetic (eV to keV) particle bombardment under sustained exposure to high heat loads (10 MW/m^2).

Wall materials are of critical importance because they might be emitted in some form into the core plasma, contributing to the plasma chemistry and to the energy loss processes in the plasma (Kleyn, Lopes Cardozo, and Samm, 2006). Heat and particles are transported through the edge plasma to the surrounding chamber walls or special high-heat-flux surfaces via various collisional, intermittent, and turbulent processes. The power level absorbed by the material surfaces for ITER is close to the limit materials can stand without rapid erosion. Although the plasma is relatively "cooler" at the edge of the fusion reactor, the plasma-facing components (PFCs) still receive high particle and heat fluxes.

The initial materials for PFCs of ITER have been chosen: beryllium for the first wall (due to its low Z and its ability to remove oxygen [gettering]), graphite carbon fiber composites (CFCs) for the diverter target (as carbon is a good radiator and has excellent thermal and mechanical properties), and tungsten for other areas of the diverter (due to low erosion and a long lifetime). However, each of these materials has significant shortcomings, and the mixture of three plasma-facing materials could become a critical issue after sustained operation. While beryllium is highly toxic, carbon has a significant drawback due to the complex erosion/deposition phenomena occurring through the plasma–surface interaction.

Plasma irradiation creates a dynamical surface with a changing interface and changing surface morphology and chemistry. One aspect of the material science challenge is controlling the synergistic effect of impinging helium ions and neutrons, which will be produced in the tritium and deuterium reaction. The ions can greatly affect materials and lead to significant nanostructuring of the material (Ueda, Peng, Li et al., 2012). For example, nanofuzz (tendrils of material produced on the surface) can be created on tungsten exposed to a relatively high flux of low-energy helium (10–50 eV) (Baldwin and Doerner, 2008). Figure 10.7 displays the nanofuzz that has developed on a polished tungsten target subjected to a helium ion flux of $\sim 5 \times 10^{22}$ ions $m^{-2}s^{-1}$. The nanofuzz can be up to a micrometer in length and 10–50 nm in width. In fact, the formation of this nanofuzz or indeed other subsurface structures depends on temperature and incident ion energy (Figure 10.8). Beginning at relatively low temperatures (<900 K), helium atoms can become trapped under the surface of a tungsten target. Increasing the temperature to above 900 K leads to an increase in the mobility of helium atoms and to the formation of nanometer-size bubbles, or blistering, on the surface. It is at temperatures above 1100 K that the nanofuzz structure begins to form. Above 2000 K crystallization of tungsten occurs and micrometer-size helium bubbles become apparent on the surface (Nishijima, Ye, Ohno, and Takamura, 2003, 2004).

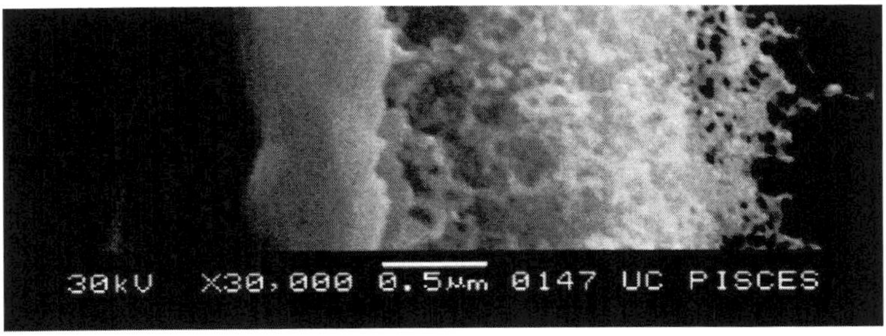

Figure 10.7 Cross-sectional SEM micrograph of a tungsten target that was exposed at 1120 K to pure He plasma for 4.3×10^3 s . Figure reprinted from Baldwin & Doerner (2008), with permission from IOP, © 2008 IAEA. *Abbreviation*: SEM, scanning electron microscope.

Nanosized structures can have significant detrimental effects on materials: they can lead to enhanced erosion, formation of dust particles, and increased tritium trapping. Over the past few years, there has been an increase in the knowledge base regarding nanostructuring of these materials under fusion-relevant conditions; however, there is still much to resolve, including decoupling of the synergistic effect of plasma and ion bombardment at the plasma–wall interface.

There is a requirement to study these materials at the nanoscale and to understand their evolution under plasma and ion irradiation. The progress in micro- and nanotechnology diagnostic tools can be applied to advance development of nanostructured materials for the plasma–material interface in fusion devices. One can use transmission electron microscopes (TEMs) and focused ion beams (FIBs) to determine the grain size, orientation, anisotropy, particle form, and distribution. Nanoindentation can be used to study the mechanical properties of the material, and positron annihilation lifetime spectroscopy (PALS) is particularly suited to probe vacancy-like defects in a nondestructive way and study the defect distribution near the surface and in the bulk after irradiation. Other valuable techniques include neutron and X-ray scattering to study structural properties of materials.

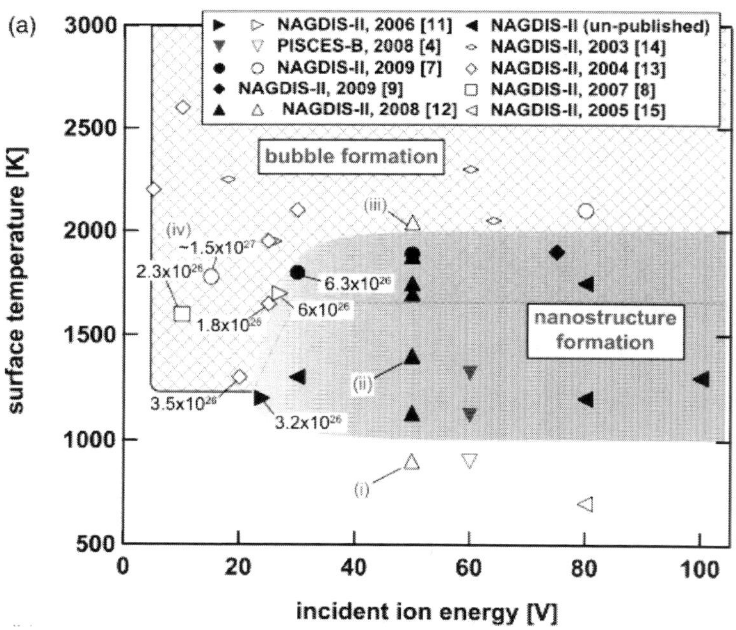

Figure 10.8 Helium irradiation experiments conducted in the diverter simulator NAGDIS-II for different surface temperatures and incident ion energy. Reprinted from Kajita et al. (2005), with permission from IOP, © 2009 IAEA.

10.5 Fusion Power Economics

Fusion power is not presently a commercial technology and is unlikely to make a contribution to the world energy mix until at least 2050. As such, economic studies of fusion power are subject to enormous uncertainties. Nevertheless, comparative studies of fusion power economics do exist, at least for Europe and the United States. A brief discussion of these reports provides insight into the commercial potential of fusion power comparative to other technologies.

Our analysis is based primarily on the work by Cook, Miller, and Ward (2002). The primary finding of this report is that using modest physics optimization and anticipated near-term materials, the internal costs of electricity (which ignores environmental impact costs) would be 50% more than fossil fuel–based electricity, a figure roughly comparable to renewables. The use of advanced materials, technology, and physics leads to an internal cost of electricity approaching fission and fossil fuels.

Cook, Miller, and Ward (2002) break down the costs of electricity into internal costs, which are the costs of constructing, fueling, operating, and disposing of power stations, and external costs, which are the "estimated" impact costs to the environment and public and worker health. Internal costs have been calculated using the mathematical model PROCESS, which encapsulates the engineering, physics, and costings of a commercial power station. Where possible, these have been independently verified against the projected costs of the next-step fusion experiment, ITER. As ITER is not a power station, the comparison is not complete. In cases where agreement was possible, agreement was within 10%. On the basis of these estimates, Cook, Miller, and Ward (2002) predicted fusion electricity between 7 cents and 13 cents per kWhr, in 1996 Euro.

Fusion economic costs compare well to other energy technologies. Figure 10.9 shows the projected internal costs of coal, gas, fission, biomass, wind, solar photovoltaic, tidal, and fusion electricity and hydroelectricity. Where necessary, storage costs of electricity have been included to make the electric power firm. For fusion, the internal costs of electricity are comparable to that of tidal power.

Due to storage costs, the costs of solar photovoltaics and wind are larger than that of fusion.

Cook, Miller, and Ward (2002) evaluate the external costs of electricity production, which are those associated with environmental damage, or adverse effects on the environment, using the ExternE method. The ExternE method was developed to evaluate the external costs of a variety of nonfusion electricity sources and assesses the entire life, fuel cycle, and death of a power station. This includes material manufacturing, construction, operation of the plant, dismantling of the plant, site restoration, and disposal of waste. At each stage factors such as hazardous chemical or radioactive emissions, road accidents, occupational accidents, accidents at the plant exposing the public to risks, and occupational exposure to hazards were considered. The adverse effects were quantified in monetary terms and summed to produce an estimate of the total external cost. The total external cost over the entire life cycle of the power station was then divided by the total electrical output to produce the external costs per kilowatt-hour.

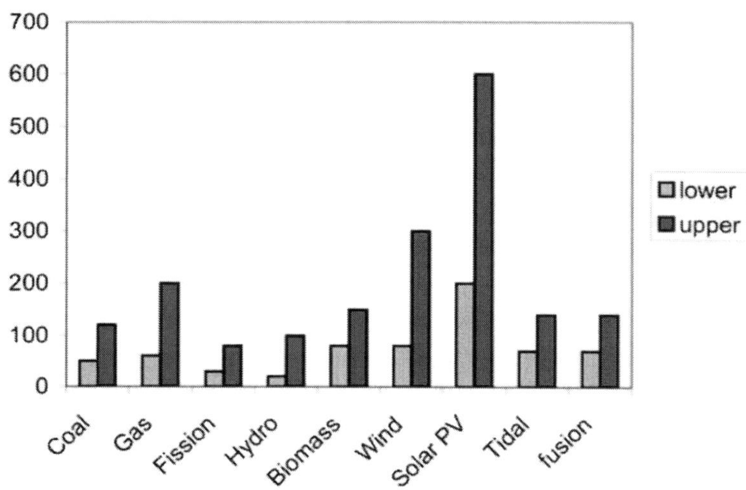

Figure 10.9 Comparative internal costs of electricity production. The left axis is in units of 1996 USm$ (i.e., USD 0.001) per kWh. Reprinted from Cook, Miller, and Ward, Copyright 2002, with permission from Elsevier.

Figure 10.10 shows the calculated external costs per kilowatt-hour of electricity generated from coal, gas, fission, biomass, photovoltaics, wind, and fusion. All the numbers shown are subject to significant uncertainty. Fusion compares well to other technologies, with external costs comparable to wind. Compared to coal, the external costs of fusion electricity are 20 times lower.

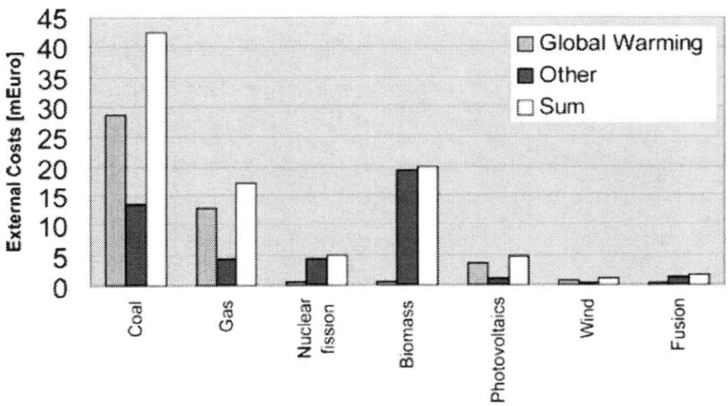

Figure 10.10 External costs of fusion power, using ExternE calculation. The left axis is in units of 2002 mEuro (i.e., 0.001 Euro). Reprinted from Cook, Miller, and Ward, Copyright 2002, with permission from Elsevier.

10.6 Greenhouse Emission Implications of Fusion Power Deployment

Nuclear energy offers a high-energy-density power supply with low greenhouse emissions. The main greenhouse emissions associated with a fusion power plant are those associated with the construction of the plant itself. As with fission, no CO_2 emissions are generated during the reaction. Table 10.1 shows the estimated relative greenhouse emissions (in tons CO_2/GWhr [electrical]). The fuel-related item is the CO_2 emission from mining, fuel processing, and transport. The processing of fuel for fusion is relatively simple, and the energy density high—hence the low CO_2 emissions per GWhr of electricity. The largest CO_2 emissions come from the operation of coal-fired power plants. Overall, fusion is predicted to have over

100 times less CO_2 emissions than coal and is the least CO_2-emitting technology assessed.

Table 10.1 Comparative CO_2 emissions from different energy technologies, in tonnes/GWhr (electrical)

Process	Coal	Natural gas	Wind	Fission	Fusion (D-T)
Fuel related	17	76	0	10	0.2
Plant materials and construction	1	1	10.2	2	5.5
Operation and maintenance	956	386	4	2	3
Decommissioning	0.2	0.02	0.4	1	0.4
Total	**974**	**464**	**15**	**15**	**9**

Source: Adapted from Meier and Kulcinski (2001)

10.7 Radioactive Waste from Fusion Power

D-T fusion produces an energetic neutron, which will be preferably captured by the lithium blanket, thereby generating more tritium to fuel the plasma. Inevitably, a small fraction of neutrons will escape capture by the lithium blanket, and some of the vessel structure enclosing the plasma will be activated by neutron capture.

Compared to fission waste, the radioactive lifetime of the neutron-activated structure is relatively short. Detailed studies of concept fusion power plant designs (Maisonnier, 2007) have shown that the radiotoxicity (the biological hazard of activated materials) of fusion power plant materials decays by a factor of 10,000 after 100 years of shutdown. All of this material, after being kept in situ for some decades, will be regarded as nonradioactive or recyclable. A small fraction (~10%) may require remote handling.

Figure 10.11 shows the projected material mass remaining after 100 years of shutdown of a concept fusion power plant (Maisonnier, 2007). About 90% of the power plant is either nonactive waste (NAW) or classified as simple recycle material (recyclable with simple radioactive-handling procedures). The remaining 10% is also recyclable but may require remote handling. The design is based on limited extrapolations in plasma physics performance and a blanket design based on near-term technology. More advanced concept designs exist, employing vanadium steel alloys and silicon carbide structures, which offer even further reduced activity.

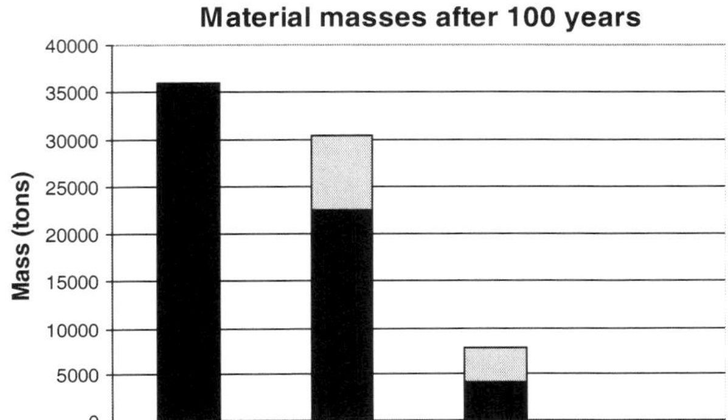

	NAW	SRM	CRM	PDW
□ Operations	0	7979	3727	0
■ Decommissioning	36004	22438	4016	0

Figure 10.11 Material masses after 100 years of shutdown of a power plant—conceptual study model B. Figure reprinted from Maisonnier et al. (2007), Power plant conceptual studies in Europe, *Nucl. Fusion,* **47**(11), 1524–1532, with permission from IOP, © 2007 IAEA. *Abbreviations*: NAW, nonactive waste; SRM, simple recycle material; CRM, complex recycle material; PDW, permanent disposal waste (nonrecyclable).

10.8 Health and Safety Implications of Fusion Power

Health and safety modeling of fusion power is an active research topic. A series of the European Fusion Development Agreement (EFDA) studies, "Safety and Environmental Assessments of Fusion Power," have explored the safety and environmental impact of fusion power (Cook, Marbach, Di Pace, Girard, and Taylor (2001). Cook, Marbach, Di Pace, Girard, and Taylor (2001) summarize fusion as offering three areas of safety and environmental advantage: zero climate-changing emissions, low consequences of worst-case accidents, and no waste management burden on future generations. This study indicates that fusion has good inherent safety qualities, with no chain reactions or production of actinides (radioactive elements with long half-lives). In addition, the radioactive fuel component tritium is both produced and consumed on-site; therefore there are no issues about transporting radioactive fuel.

Cook, Marbach, Di Pace, Girard, and Taylor (2001) identified the following key aspects of fusion reactor safety: effluents and emissions from normal operation, including planned maintenance activities; occupational safety for workers at the facility; radioactive materials and wastes generated during operation and from later decommissioning; and potential incidents and accidents. We summarize each of these in turn:

- During normal operation, the total radioactive dosage to the most exposed member of the general public (a person located at the site boundary) would by less than 1 μSv/year for gaseous leakage and less than 0.2 μSv/year for liquid leakage. In contrast, the annual dosage to an Australian due to natural background radiation is 2000 μSv/year.
- Occupational radioactive exposure to workers is estimated to be comparable to the best performance of pressurized water fission reactors.
- The majority of the radioactive materials from operation and decommissioning can be released from regulatory control in reasonable timescales. It is estimated that 60% of the material would be below International Atomic Energy Agency (IAEA) clearance levels after 30 years, growing to 80% after 100 years.
- In a worst-case accident, the total radioactive dosage to the most exposed member of the general public (a person located at the site boundary) would be comparable to the average annual natural background for a generic site. No single component failure will lead to very large consequences, and no single event can simultaneously damage the multiple confinement barriers provided in a fusion reactor design.
- Radioactive exposure to the general public due to a worst-case external event (e.g., an earthquake or a terrorist attack) is limited by the vulnerable tritium inventory, which is 1 kg. The release of 1 kg would result in a dosage to a member of the public in the plant area up to 4000 μSv—twice the yearly background radiation dose.

Maisonnier (2007) has continued to extend this study into the design of four concept commercial fusion power plant designs. Consideration is also given to improved containment concepts for the fusion reactor core, the production of activated materials during the lifetime of a fusion power plant, and their possible reduction through recycling and material optimization. Salient features of these designs include:

- Any power excursion will be self-limited to low levels by the inherent processes in the plasma. If a total loss of active cooling were to occur during the burn the plasma would switch off passively due to impurity influx deriving from temperature rises in the walls of the reaction chamber. Any further temperature increase in the structures cannot lead to melting.
- The power plant will be designed to withstand an earthquake of intensity equal to that of the most severe historical accident, increased by a safety margin.
- In case of fire, a maximum of a few grams of tritium could be released, by appropriate partitioning of the tritium inventory. At this level, evacuation would not be required.

10.9 Security and Proliferation Issues

As energy supplies dwindle, the world economy becomes increasingly vulnerable to disruption and prices can escalate. Examples of such situations are provided by the ongoing Middle East conflicts and the deep-sea-drilling accident in the Gulf of Mexico in 2010.

As proposed, fusion reactors will breed tritium for use as fuel. Tritium is useful in weapons only as an ingredient for hydrogen bombs, for which one must first have an atomic (fission) bomb, employing uranium or plutonium, as a trigger. Fusion neutrons can be used to breed plutonium, but this would involve rebuilding the reactor with uranium as the blanket, which would be complicated, costly, and highly visible. The technical challenges involved in fusion make it extremely unlikely that a rogue state would elect to develop fusion reactors rather than modify fission power reactors to facilitate weapons production.

10.10 Next Step Research and Development

The field of fusion energy science is primarily advanced by integrated international programmatic development. A key area of international R&D activity in fusion energy is the development of the next-step fusion energy project, ITER. ITER will be the first experiment to explore the "burning plasma" regime, in which the heat of the confined products is greater than the external heating. In continuous operation, ITER will yield five times more power than is required to sustain the reaction, with a total energy output of 500 MW. In pulsed mode, the power gain could be as high as 30. ITER, which is a precursor to a demonstration power plant, will determine the viability of fusion power.

The programmatic goal of ITER is "to demonstrate the scientific and technological feasibility of fusion power for peaceful purposes." The general goal is realized via a number of technical, specific aims and performance requirements, details of which can be found in the ITER technical basis (ITER Design Team, 2001). In summary, these aims are as follows:

- Produce and study inductively driven, burning plasma at $Q \geq 10$ (400–500 MW) for an "extended" time, ~400 seconds (where Q is the power amplification P_{fusion}/P_{input} defined in Section 10.2).
- Aim to produce and study "steady state," burning plasma with a noninductive drive $Q \geq 5$.
- Demonstrate the availability and integration of essential existing fusion reactor technologies, including superconducting magnets, components able to withstand high heat loads, and remote handling.
- Test components for a future reactor, including tritium-breeding module concepts (neutron power load > 0.5 MW m^{-2}; fluence > 0.3 MW year m^{-2}).

Due to a blend of physics properties and engineering constraints, these aims force the ITER design to have the specifications listed in Table 10.2.

An engineering schematic is shown in Fig. 10.12, showing the major physical systems. For scale reference, a human is shown under the reactor core. The estimated cost of the ITER project is A$20 billion, composed of A$12 billion in construction and A$8 billion in operating funds for 20 years. In fiscal terms, ITER is the world's largest science experiment. The ITER program is supported by over 30 of the world's most developed nations, representing more than half the planet's population.

Table 10.2 Key ITER operating parameters

Total fusion power	500 MW
Steady-state noninductive power gain	Q > 5
Extended pulse length (~400 s), inductively driven, burning plasma	Q > 10
Minor radius/major radius	2.0 m/6.0 m
Plasma current (I_P)	15 MA
Toroidal field @ 6.2 m	5.3 T
Plasma volume	837 m^3
Auxillary heating, current drive	73 MW
Central ion temperature	100,000,000°C

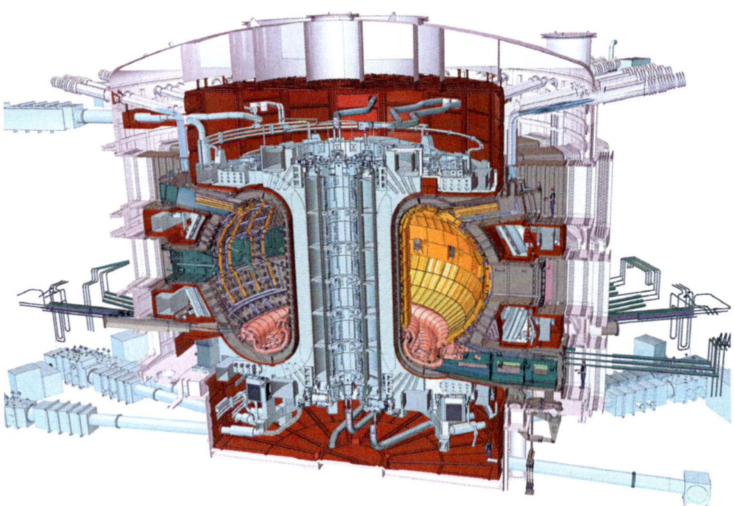

Figure 10.12 A cross-section cutaway of ITER. For scale, note the human under the reactor core. Image courtesy: ITER Organization.

10.11 Summary

In this chapter, we have done a holistic analysis of fusion power and discussed it in terms of science, engineering, and economics. Fusion is a promising technology, and significant progress has been demonstrated over the past four decades. Our current understanding of the economics, safety, security, and greenhouse emission credentials of fusion power place it as one of the most attractive energy sources in the long term. As we approach burning plasma conditions, there is a strong need to develop not only our understanding of burning plasma physics but also extreme materials to withstand the high heat and neutron flux for ITER and commercial power plants. Under such extreme conditions, nanoscience challenges arise for PFCs, such as the formation of nanofuzz and other nanostructures. Understanding gleaned from extreme materials for fusion can also help the development of science in other extreme environments, such as in fast-breeder fission reactors and high-temperature thermal power systems.

References

Baldwin, M., and Doerner, R. (2008). Helium induced nanoscopic morphology on tungsten under fusion relevant plasma conditions. *Nucl. Fusion*, **48**(2008), 035001.

Cook, I., Marbach, G., Di Pace, L., Girard, C., and Taylor, N. P. (2001). *Safety and Environmental Impact of Fusion.* European fusion development agreement.

Cook, I., Miller, R. L., and Ward, D. J. (2002). Prospects for economic fusion electricity. *Fusion Eng. Des.*, **63–64**, 25–33.

ITER Design Team (2001). *ITER Final Design Report.* Vienna: International Atomic Energy Agency.

Kajita, S., Sakaguchi, W., Ohno, N., Yoshida, N., and Saeki, T. (2009). Formation process of tungsten nanostructure by the exposure to helium plasma under fusion relevant plasma conditions. *Nucl. Fusion*, **49**, 095005.

Kleyn, A. W., Lopes Cardozo, N. J., and Samm, U. (2006). Plasma-surface interaction in the context of ITER. *Phys. Chem. Chem. Phys.*, **8**, 1761–1774.

Maisonnier et al. (2007). Power plant conceptual studies in Europe, *Nucl. Fusion*, **47**(11), 1524–1532.

McKay, W. J., McKay, A. D., Miezitis, Y., et al. (2005). *Australia's Identified Mineral Resources.* Australia: GeoSciences Australia, Australian Government.

Meier, P. J., and Kulcinski, G. L. (2001). The potential for fusion power to mitigate US greenhouse gas emissions. *Fusion Technol.*, **39**(2), 507–512.

Nishijima, D., Ye, M. Y., Ohno, N., and Takamura, S. (2003). Incident ion energy dependence of bubble formation on tungsten surface with low energy and high flux helium plasma irradiation. *J. Nucl. Mater.*, **313–316**, 97–101.

Nishijima, D., Ye, M. Y., Ohno, N., and Takamura, S. (2004). Formation mechanism of bubbles and holes on tungsten surface with low-energy and high-flux helium plasma irradiation in NAGDIS-II. *J. Nucl. Mater.*, **329–333**, 1029–1033.

Norton, G. A., and Leahy, P. P. (2006). *Mineral Commodity Summaries.* US Department of the Interior, US Geological Survey, Washington, DC: US Government Printing Office.

Oliphant, M. L. E., Harteck, P., and Rutherford, L. (1934). Transmutation effects observed with heavy hydrogen. *Proc. R. Soc. London, A*, **144**(853), 692–703.

US Department of Energy (2010). *Annual Energy Outlook 2012.* US Department of Energy.

Ueda, Y., Peng, H. Y., Lee, H. T., et al. (2012). Helium effects on tungsten surface morphology and deuterium retention. *J. Nucl. Mater.*, **442**(1–3), S267–S272.

Wesson, J. (2004). *Tokamaks,* 3rd. Oxford: Clarendon Press.

Chapter 11

Nanotechnology, Plasma, Hydrogen from Artificial Photosynthesis, and Fuel Cells: Powering the Developing World to the Sustainocene

Thomas A. Faunce[a,b] and Christine Charles[c]
[a]*ANU College of Law, Building 6, North Wing, Australian National University, Canberra, ACT 0200, Australia*
[b]*ANU College of Medicine, Biology and the Environment, Building 131, John Curtin School of Medical Research, Australian National University, Canberra, ACT 0200, Australia*
[c]Space *Plasma, Power and Propulsion Laboratory, Research School of Physics and Engineering, ANU College of Physical and Mathematical Sciences, Building 60, Oliphant Building, Mills Road, Canberra, ACT 0200, Australia*
thomas.faunce@anu.edu.au, christine.charles@anu.edu.au

To achieve a Sustainocene epoch, nanotechnology will need to assist in addressing the particular needs and requirements with regard to energy security of the developing world and remote communities. Our hypothesis is that one approach to facilitating the global deployment of artificial photosynthetic hydrogen fuel systems that support environmental sustainability involves using well-developed technologies already proven for other applications. This chapter discusses in this context the prospects for using plasma techniques

Nanotechnology Toward the Sustainocene
Edited by Thomas A. Faunce

ISBN 978-981-4463-76-8 (Hardcover), 978-981-4463-77-5 (eBook)
www.panstanford.com

to deposit nanoparticulate catalysts on low-cost substrates for use in fuel cells.

11.1 Energy Security and the Developing World

Enhancing energy security in the developing world and remote communities is critical to the achievement of the United Nations' *millennium development goals* (MDGs) (UNIDO, 2009). Africa's energy consumption per capita is low in global terms (10 and 5 times less than the United States and Europe, respectively (International Energy Agency [IEA], 2011a), and its primary energy supply is 5% of the global total (mostly from traditional biomass for cooking energy (IEA, 2011a). This disparity highlights the extent to which that continent lacks access to energy, particularly in the form of fuel. There is no clear and universally accepted definition of "energy access," but the term is often referred to in policy documents as the number of households supplied by an electricity grid system (IEA, 2011a, p. 12).

Africa, particularly its rural areas, is grossly underserved, however, by large-scale electricity grid infrastructure. Its installed generation capacity is 122.6 GW, or just 2.6% of the world total (IEA, 2011). In sub-Saharan Africa, only about 30% of the population has access to electricity (with 14% electrification in the rural areas) (IEA, 2011b). Further, those connected to the grid experience highly unstable and unreliable electricity supplies and relatively high costs per kilowatt-hour (IEA, 2011b; Sebitosi and Pillay, 2007).

Yet, the global need for an energy transition from high-carbon-emission-intensive fuels may provide an ideal driver for a technological "bypass" in developing countries and regional areas. Some of the basic preconditions are good. Solar insolation in Africa is about 1800–2850 kWh/m^2.a (World Energy Council, 2010), while its photovoltaic technical potential is estimated 33–8700 TWh/a (National Renewable Energy Laboratory, 2008). Yet, the technology, such as solar hydrogen–based fuel cells, that could most make a difference to access to "off-grid" energy in Africa is not being developed as rapidly as it could be.

The United Nations General Assembly declared 2012 the International Year of Sustainable Energy for All (United Nations,

2012), recognizing that ". . . access to modern affordable energy services in developing countries is essential for the achievement of the internationally agreed development goals, including the Millennium Development Goals, and sustainable development, which would help to reduce poverty and to improve the conditions and standard of living for the majority of the world's population." As mentioned in Chapter 1, it appointed the chairman of Bank of America and the chair of UN-Energy and director general of the UN Industrial Development Organization to cochair the secretary general's High-Level Group on "Sustainable Energy for All." With the help of the group, they were tasked with mobilizing commitments from governments, the private sector, and civil society partners to take actions that will make sustainable energy a reality for all over the next two decades (United Nations, 2012). Yet that project did not even consider the implications of globalizing nanotechnology-based artificial photosynthesis. Perhaps that technology, despite the considerable research being done in that field, seemed too speculative for them. This leads to the question of whether a more widely used nanotechnology-based approach may be what is required to spur the global deployment of artificial photosynthesis. To consider this, it is first necessary to review some of the ways in which nanotechnology intersects with the field of artificial photosynthesis.

11.2 Hydrogen from Solar-Driven Water Splitting

Over the past few years, scientists, politicians, and economists have become increasingly aware of the impending oil peak and the present global warming. A progressive transition from the hydrocarbon economy to a hydrogen economy is one of the solutions to reduce greenhouse gas emissions and to face the dwindling supplies of petroleum.

In such an economy, the energy required for motive power (automobiles or other vehicles) or electricity (stationary applications) will be derived from a reaction involving hydrogen (H_2) and oxygen. However, many improvements are needed to achieve this transition in terms of H_2 production, H_2 storage, and fuel cell cost and efficiency.

Table 11.1 United Nations' year of sustainable energy for all

Targets
1. Modern cooking appliances and fuels
2. Distributed electricity solutions: Using grid infrastructure and supply-efficient off-grid, microgrid, and minigrid
3. Large-scale renewable power: Grid connected
4. Industrial and agricultural processes
5. Transportation: Increasing fuel efficiency for all classes of vehicles; increasing the share of renewables in the fuel supply
6. Buildings and appliances: Improving efficiency through design, insulation, and retrofit of buildings and incorporating renewable self-generation options with more efficient consumer appliances and equipment
7. Energy planning and policies: Promoting direct public action and improving the legal and administrative context for successfully engaging the private sector and civil society
8. Business model and technology innovation: Overcoming barriers to sustainable energy services and technologies
9. Finance and risk management: Increasing private investment in sustainable energy through the targeted use of public and philanthropic capital
10. Capacity building and knowledge sharing for faster replication across the world

As mentioned in Chapter 1, researchers (many using nanotechnological approaches) now are striving to improve photosynthesis to achieve low-cost conversion of sunlight and water into hydrogen fuel for heating and cooking (Hurst, 2010). Some nanotechnological innovations for what is generally known as artificial photosynthesis focus on improved light capture. Such systems use mesoporous, thin-film, dye-sensitive solar cells of semiconductor nanoparticles and carbon nanotubes (CNTs) to harvest and conduct the resultant electricity (Sgobba and Guldi, 2009). Nanomaterials and hybrid organic-inorganic nanostructures are improving the solar energy conversion efficiency of existing photovoltaic units that could be used in artificial photosynthesis light capture (Guter, Chone, Philipps et al., 2009; Konstantatos and Sargent, 2010).

PSII in plants is a complex protein with 27 subunits and 32 cofactors involved in electron transfer and light harvesting. Researchers are on the threshold of making a nanotechnological mimic of this protein (maquette) that is simpler (Koder, Anderson, Solomon, Reddy, Moser, and Dutton, 2009) and incorporates designer molecules that prolong charge separation (Carmieli, Mi, Ricks, Giacobbe, Mickley, and Wasielewski, 2009). Nanotechnology is facilitating the construction of artificial photosynthetic electron pathways to this reaction center that perform a single quantum computation, sensing many states simultaneously and so enhancing the efficiency of the energy capture and transfer at physiological temperatures (Lee, Cheng, and Fleming, 2007; Engel, 2007; Ball, 2010).

Yet we are many years away from a globally widespread water catalytic system involving inexpensive and self-repairing components that operate at neutral pH with nonpure (salty or bacterially and chemically contaminated) water (Kanan and Nocera, 2008) and stable to a variety of exposure conditions in air, water, and heat (Yin, Tan, Besson et al., 2010). A major scientific challenge will be to optimize the free energy required for the overall water-splitting process (Balzani, Credi, and Venturi, 2008). Multiwalled CNTs (Toma, Sartorel, Iurlo et al., 2010) and single-walled CNTs may produce the critical breakthrough here (Sgoba and Guldi, 2009).

We are even further from perfecting the artificial photosynthesis version of the "dark reaction," in which ATP and NADPH, as well as carbon dioxide (CO_2), will be used in an enhanced version of the Calvin–Benson cycle to make locally usable food or fuel (for domestic heating, cooking, light, and transport) in the form of carbohydrate via the enzyme RuBisCO (Gray, 2009). Bioinspired self-repair strategies will ensure that this aspect survives damage from repeated cycles of thermodynamically demanding reactions (Wasielewski, 2006). New catalysts for H_2 production and methods for efficient H_2 usage (in a fuel cell to make electricity) or storage (as a fuel after cooling and concentrating) will need to be built (Magnuson, Anderlund, Johansson et al., 2009).

Some researchers consider they should primarily focus on biological approaches to artificial photosynthesis. Their focus is on genetically modifying plants and also using synthetic biology. They seek, for example, to genetically manipulate or even synthetically reproduce photosynthetic plants and bacteria to maximize their light capture and carbon reduction activities. But this approach surely has echoes of the nineteenth-century experimentalists who considered that only birds could do flight and that perhaps one day we'd find a way to make huge birds that would allow us to travel across the Atlantic or Pacific Ocean.

When we travel in aircraft across the world it is easy to see the extent to which human concrete and asphalt structures are proliferating across the face of the planet. Such structures contribute little to the ecosystems around them. They do not enrich the soil or provide oxygen or absorb carbon dioxide. Yet we are almost at the point where nanotechnology and artificial photosynthesis can be engineered into such structures so they can be made to "pay their way" in an ecosystem sense. We are reaching the point where there is an overwhelming moral obligation to undertake such research on a concerted, global scale. That point, however, is not likely to arrive for quite a few decades and the problem of efficiently using artificial photosynthesis hydrogen or ammonia fuel directly or by local conversion to electricity is a key issue for global energy security and climate change mitigation.

Ethically, one of the most immediate impacts of nanotechnology-based artificial photosynthetic technology should be the provision of fuel and electricity to families and communities in the developing world that lack access to a functional electricity grid. It is surprising then, as mentioned earlier, that the United Nations' program on sustainable energy for all made no reference to the prospects of artificial photosynthesis. Perhaps what is needed is a related bridging nanotechnology with greater immediate practical appeal. Fuel cells could be just such a technology. Plasma deposition techniques for nanoparticles in fuel cells may be suitable well-developed techniques with important sustainable energy implications (e.g., potato chip packets are 10 μm plastic films covered with 100 nm of aluminum deposited using plasma magnetrons).

11.3 Fuel Cells

11.3.1 Introduction to Proton Exchange Membrane Fuel Cells

The development of fuel cells is widely considered to be an integral part of a sustainable hydrogen economy. Fuel cells are like batteries, operating when fuel—in this case, pure hydrogen sourced from solar-driven water splitting via artificial photosynthetic techniques—is supplied to create an electrochemical reaction with oxygen (or air) that directly converts chemical energy into electrical energy.

Various types of fuel cells exist, but the one most focused on for use in automobiles relies on a proton exchange membrane

(PEM). PEM fuel cells produce water and heat as by-products of the electrochemical reaction and have high power density, high efficiency, rapid startup capability, clean operation, and flexibility in terms hydrogen fuel sources. In the generic PEM fuel cell, this membrane lies between a positively charged electrode (the cathode), which reduces oxygen, and a negatively charged electrode (the anode), which oxidizes the hydrogen fuel. A catalyst (typically platinum) is needed for both reactions to occur.

At the anode, the platinum catalyzes an ionizing reaction that breaks the hydrogen atom down into its positive ions (hydrogen protons) and negative electrons. Since only the protons can pass through the membrane (hence the name "proton exchange"), the electrons have to travel along the external electrical circuit around the PEM, generating electricity.

Anode: $2H_2 \xrightarrow{Pt} 4H^+ + 4e^-$

Cathode: $O_2 + 4H^+ + 4e^- \xrightarrow{Pt} 2H_2O$

Overall: $2H_2 + O_2 \rightarrow 2H_2O$ + electricity and heat

Optimization of a membrane-electrode assembly (MEA) requires reduction of both critical transport losses and the amount of high-cost materials used for the catalyst (e.g., platinum). Decreasing the amount of Pt while increasing the Pt utilization efficiency has been one of the major concerns during the past decade (Brault et al., 2004; Caillard, Charles and Boswell, 2007; Caillard, Charles, Boswell and Brault, 2007; Caillard et al., 2009).

A single hydrogen fuel cell delivers a low voltage (the maximum output relative to the chemical reaction is about 1 V). In practice, each cell produces around 0.7 V, about enough power to run a lightbulb, in contrast to around 300 volts needed to run a car. Thus, manufacturers "stack" fuel cells together in a series as in a dry-cell battery to increase their operating voltage. The greater the surface area of the electrodes, the greater the current.

A proton exchange membrane fuel cell (PEMFC) is a promising power source due to its high energy conversion efficiency at relatively low operating temperatures (near 80°C). PEMFC electrodes are usually fabricated by coating a carbon cloth with an ink composed of carbon particles (carbon black), clusters of platinum catalyst, and polytetrafluoroethylene (PTFE) particles. This coating is called the gas diffusion layer (GDL) and is 50 μm thick, with a platinum catalyst content of 0.35 mg cm^{-2}.

PEMFCs are being seriously considered for use on portable power sources for laptop computers and cellular phones, for transportation such as buses and cars, and for stationary applications like auxiliary power sources for domestic use in housing, possibly in conjunction with gas turbines.

PEMFCs using solid polymer membranes and operating at low temperatures (100°C) must deliver high power densities with an ultralow content of platinum catalyst to reduce cost for mass commercialization. According to the US Department of Energy (DoE), the cost of the electrocatalyst must be decreased to \$5 kW^{-1} and the maximum power density delivered by the MEA must reach 1 Wcm^{-2} (and 0.25 Wcm^{-2} at 0.8 V). However, the platinum loading in a PEMFC electrode is currently in the range 0.3–0.5 mgPt cm^{-2}, leading to an electrical performance of 0.7–0.9 Wcm^{-2} with a thin solid polymer membrane (usually Nafion® 112). Consequently, the Pt loading must be reduced by a factor of 10 (at least) in the MEA to achieve the DoE target. Therefore, both an increase in the electrical performance of PEMFCs and a decrease in the platinum loading in the MEA are of interest.

PEMFC technology has many other advantages in terms of energy sustainability, such as a low operating temperature, sustained operation at a high current density, low weight, compactness, the PEMFCs potential for a low cost and volume, long stack life, fast start-ups, and suitability for discontinuous operation. These features make PEMFCs the most promising and attractive candidates for a wide variety of power applications, ranging from portable/micropower and transportation to large-scale stationary power systems for buildings and distributed generation. For these reasons, many companies, including fuel cell technology (Ballard, UTC, Nuvera, GE-FCS, Plug Power, Intelligent Energy, NovArs, Smart Fuel Cell, Toshiba, Sanyo, and Hydorgenics), automobile (Daimler-Chrysler, Ford, Renault, Toyota, Nissan, GM, BMW, and Hyundai), and electricity (NTT, Sanyo, Samsung, and IBM), have announced various applications, new technologies, and prototype vehicles using on-board PEMFCs. In addition, many technologies utilizing PEMFCs for different applications are currently under development. There are, however, still a number of barriers that need to be overcome before this technology becomes more widespread. The main issue is cost on the membrane materials and catalysts are expensive, but ongoing research and development is constantly reducing cost and economies of scale will kick in once these cells are mass produced.

Today, PEMFC technology is promising in the mass market for automotive, stationary, and portable applications. The PEMFC offers attractive mass market applications, primarily in personal transport, aimed at reducing urban pollution. Its efficiency strongly depends on the MEA, and one of the challenges in its commercialization is the high cost of noble metals used as catalysts, for example, Pt in electrodes. As mentioned, increasing the Pt utilization efficiency in order to decrease the Pt amount in the catalytic layer has been one of the major concerns during the past decade, and this is where nanotechnology is expected to be beneficial.

11.3.2 Practical Significance of Nanotechnology and Plasmas

To summarize, a conventional, chemically produced electrode as used in a PEMFC is usually prepared by ink processes and consists of a GDL, typically carbon Vulcan and PTFE particles, and an active layer. The latter has a mixture of carbon Vulcan particles coated with platinum catalyst and a proton-conducting polymer (typically Nafion®) spread on either carbon cloth or carbon paper made of weaved or stacked carbon fibers (mean diameters of 10 μm and 5 μm, respectively). In such a structure, the Pt catalyst will be active if it is simultaneously in contact with the gas, the continuous electron-conducting media (porous carbon supports), and the continuous proton-conducting polymer.

Even with the most advanced conventional electrodes, however, this catalyst layer nanostructure does not ensure optimum Pt utilization. Firstly, the Nafion® polymer addition encapsulates Pt/C particles to form agglomerates (more or less spherical, with a diameter in the range 200–1000 nm), which consequently isolate some Pt/C particles from the electrode support by cutting off the electron pathways. Secondly, these agglomerates themselves are flooded in a low-gas-diffusivity proton polymer. As a result, hydrogen and oxygen cannot diffuse through the whole volume of these agglomerates and the Pt catalyst, far from the agglomerate surface, becomes inactive. Finally, gas diffusion, electron and proton conduction, and water removal are quite difficult in such a porous electrode due to its complex shape. There is also a need to increase fuel cell efficiency and reduce costs by improving the electrode characteristics. Attention has been focused here on the reduction of the amount of expensive noble metal catalyst (typically platinum).

CNTs or carbon nanofibers (CNFs) have been proposed for electrochemical devices to replace traditional black carbon particles in PEMFC electrodes because of their high electrical conductivity, unique surface structure, large surface area, and chemical inertia (Caillard, Charles and Boswell, 2007; Caillard, Charles, Boswell and Brault, 2007). These nanocomposites have usually been used in powders mixed with Nafion® and spread on electronic conductive supports such as carbon cloth or carbon paper. An interesting prospect is that of using fuel cell using CNTs directly grown on electrodes (i.e., using silica gel to catalyze CNT growth or a process of electrodeposition of Pt on CNTs grown by chemical vapor deposition (CVD) on carbon paper after an initial electrodeposition of Co catalyst.)

One technique here involves simply exposing the uncatalyzed GDL to a flux of platinum nanoparticles, which then diffuse into the GDL to a sufficient depth so that adequate fuel cell performance is reached. This can be called "plasma sputtering," and it has the advantage of controlling the depth profile. Typical diffusion lengths of Pt atoms into GDL are between 400 nm and 2 μm and Pt nanoclusters are formed with diameters in the range of 2–10 nm, depending on plasma deposition parameters such as argon pressure. At low current density, the electrical performance of MEAs based on such Pt-sputtered anodes is similar to a conventional electrode-based MEA but the plasma-prepared electrode has a Pt loading five times lower.

Plasmas are very hot ionized gases containing atoms, ions, and electrons and emitting photons (light). Their charged particles' motions are controlled by electromagnetic fields. Plasmas are used to manufacture propeller blades, plastics, potato chip bags, camera lenses, and synthetic diamonds, as well as semiconductor circuits, particularly computer chips. In one relevant application, an inductively coupled plasma etcher was used to fabricate titanium dioxide (TiO_2) micropillars on a transparent conducting glass substrate to increase the contact area between the TiO_2 electrode and the sensitizer dye and so develop a new dye-sensitized solar cell based on the injection of electrons of a photoexcited state of the sensitizer dye into the conduction band of the semiconductor (Sarra-Bournet, Haberl, Charles, and Boswell, 2011).

In plasma-sputtering techniques relevant to artificial photosynthesis and fuel cell electrodes, a platinum catalyst can be deposited into porous carbon materials. Platinum deposits in such conditions are observed to grow as clusters. This method can create a novel open-structure active layer as an alternative to the conventional active layer, especially for fuel cell cathodes. This structure is composed of 30 nm diameter, rod-like, electron conductors grown perpendicularly to the membrane that are coated with well-dispersed platinum nanoclusters and a 10 nm thin layer of proton conductor. In this active layer, where electron/proton conductors and gas channels are highly oriented, catalyst utilization may approach 100% (Caillard, Charles, and Boswell, 2007). These techniques allow the fabrication of the PEMFC electrode with lower platinum loading (compared to a chemically based electrode) without a decrease in efficiency.

11.4 Governance Mechanisms to Promote Fuel Cells

Assuming that plasma deposition of nanoparticles for fuel cells can assist in developing practical artificial photosynthetic systems, how can such a technology be rapidly deployed where it will be most beneficial in a public health and environment sense, that is, in the developing world? Some valuable insights may be gained here from examining approaches to supporting renewable energy by perhaps the largest funder of such projects in the developing world—the World Bank.

Since 1992, the World Bank's "solar home systems" project approved 12 projects that provide basic "energy services," such as lighting, radio, television, and operation of small appliances, to rural households that lack access to electricity grids (Wamukonya, 2007). A solar home system consists of a photovoltaic solar panel, a storage battery, a battery-charging controller, and various pieces of end-use equipment, like fluorescent lamps. Solar home systems can eliminate or reduce the need for candles, kerosene, liquid propane gas (LPG), and/or battery charging. Direct economic benefits include avoiding the costs of battery charging and LPG or kerosene purchases; other significant benefits include increased convenience and safety, improved indoor air quality, a higher quality of light than what kerosene lamps provide for reading, and reduced CO_2 emissions. Solar home systems already provide basic electricity services to more than 500,000 households in developing countries. They operate in the setting of wider development objectives, like power sector reform, rural electrification, and rural development. They are designed to overcome barriers to technology dissemination, such as lack of an established market and successful business models; unwillingness of politically and economically powerful utilities to provide off-grid electricity services; lack of consumer financing; and other policy constraints, like subsidies for competitors' archived photosynthesis fuels (i.e., oil, coal, and natural gas) (World Bank, 2006; Martinot, Cabraal, and Mathur, 2011).

The World Bank's solar home system technology program is representative of official recognition that as solar photovoltaic costs decline and commercial markets develop, while population growth outpaces the ability of electric utilities to extend rural electricity grids, opportunities are opening for off-grid electricity systems, such as thus discussed here in the context of PEMFCs.

The World Bank's solar homes program addresses various factors in developing a sustainable market, including customer satisfaction, affordability, dealer profitability, and effective supply and service chains. Such projects, thus, incorporate pilot private sector and NGO delivery models, pilot consumer credit delivery mechanisms, first-cost subsidies, support policies, codes, and standards; establish certification, testing, and enforcement institutions; and conduct consumer awareness and marketing programs (World Bank, 2006).

In 2000, one hundred eighty-nine countries agreed on a global alliance for development, the so-called Millennium Declaration, the subsequent agenda to accomplish this goal being structured in eight MDGs. The main targets of the MDGs' global alliance are governments; yet several institutions and authors have called for private sector involvement in the goals, or the targets will not be met (Nelson and Prescott, 2008).

MDG seven ("Ensure Environmental Sustainability") is the most directly related to energy and the themes of this chapter and volume. Initiatives here include managing the direct environmental impacts of companies, greening financial markets, increasing energy efficiency, and reducing carbon emissions in supply chains and supporting community-level environmental initiatives (Nelson and Prescott, 2008; Porcaro and Takada, 2004). More specifically, multinational companies are widely regarded as key agents in fostering development in southern countries, due to their global power (80% of the capital flows to southern countries comes from the private sector) (Porcaro and Takada, 2004). The involvement of the energy industry is deemed vital for achieving the MDGs, given that this industry accounts for the largest share of foreign direct investment in the region (Porcaro and Takada, 2004).

11.5 Conclusions and Future Work

Using plasmas we can now deposit nanoclusters of platinum on commercial fuel cell electrodes or we can grow CNFs onto which we can deposit such nanoclusters. This allows the creation of a large, effective surface area (thus increasing potential catalytic activity) and allows fuel feed (with CNFs, the porosity of the electrode is maintained). Similar plasma nanotechnologies can be envisaged assisting the globalization of artificial photosynthesis, particularly hydrogen fuel production and atmospheric carbon dioxide removal. The process could involve the deposition of metals (Co, Mn, Ti, and Ni) and their oxides (doped or not) on low-cost substrates (plastic) using low-cost plasma techniques.

A main advantage of such an approach is its applicability to developing nations. To facilitate the deployment of large-scale artificial photosynthesis in such societies (with all its attendant advantages in terms of environmental sustainability), relevant industries need to use well-developed technologies that are already proven for other applications. The argument advanced here is that proven plasma-spluttering techniques may provide a valuable avenue whereby nanotechnology can be used to facilitate the development of fuel cells that take advantage of hydrogen produced by artificial photosynthetic methods.

References

Allen, B. L., Kichambare, P. D., Gou, P., et al. (2008). Biodegradation of single-walled carbon nanotubes through enzymatic catalysis. *Nano Lett.*, **8**, 3899–3903.

Ball, P. (2010). Material witness: quantum leaves in fact and fiction. *Nat. Mater.*, **9**, 614–621.

Balzani, V., Credi, A., and Venturi, M. (2008). Photochemical conversion of solar energy. *ChemSusChem*, **1**(1–2), 26–33.

Brault, P., Caillard, A., Thomann, A. L., et al. (2004). Plasma sputtering deposition of platinum into porous fuel cell electrodes. *J. Phys. D: Appl. Phys.*, **37**, 3419–3423. PII: S0022-3727(04)80187-6.

Caillard, A., Charles, C., and Boswell, R. (2007). Plasma based platinum nanoaggregates deposited on carbon nanofibers improve fuel cell efficiency. *Appl. Phys. Lett.*, **90**, 223119.

Caillard, A., Charles, C., Boswell, R., and Brault, P. (2007). Integrated plasma synthesis of efficient catalytic nanostructures for fuel cell electrodes. *Nanotechnology*, **18**, 305603. doi:10.1088/0957-4484/18/30/305603.

Caillard, A., Charles, C., Ramdutt, D., Boswell, R., and Brault, P. (2009). Effect of Nafion and platinum content in a catalyst layer processed in a radio frequency helicon plasma system. *J. Phys. D: Appl. Phys.*, **42**, 045207. doi:10.1088/0022-3727/42/4/045207.

Carmieli, R., Mi, Q., Ricks, A. B., Giacobbe, E. M., Mickley, S. M., and Wasielewski, M. R. (2009). Direct measurement of photoinduced charge separation distances in donor-acceptor system for artificial photosynthesis using OOP-ESEEM. *J. Am. Chem. Soc.*, **131**, 8372–8378.

CCI Solar (n.d.). *Outreach Activities*, http://ccisolar.caltech.edu/webpage/3.

Cornell, B. A., Braach-Maksvytis, V. L. B., King, L. G., et al. (1997). A biosensor that uses ion-channel switches. *Nature*, **387**, 580–583.

Gray, H. B. (2009). Powering the planet with solar fuel. *Nat. Chem.*, **1**(1), 7–12.

Gujba, H., Thorne, S., Mulugetta, Y., Rai, K., and Sokona, Y. (2012). Financing low carbon energy access in Africa. *Energy Policy*, **47**(1), 71–78.

Guter, W., Chone, J. S., Philipps, S. P., et al. (2009). Current-matched triple-junction solar cell reaching 41.1% conversion efficiency under concentrated sunlight. *Appl. Phys. Lett.*, **94**(22), 223504.

Hilder, T. A., Gordon, D., and Chung, S. H. (2009). Boron nitride nanotubes selectively permeable to cations or anions. *Small*, **5**, 2870–2875.

Hilder, T. A., Gordon, D., and Chung, S. H. (2010). Synthetic chloride-selective carbon nanotubes examined using molecular and stochastic dynamics. *Biophys. J.*, **99**(6), 1734–1742.

Hurst, J. K. (2010). In pursuit of water oxidation catalysts for solar fuel oxidation. *Science*, **328**, 315–317.

Kalyanasundaram, K., and Graëtzel, M. (2010). Artificial photosynthesis: biomimetic approaches to solar energy conversion and storage. *Curr. Opin. Biotechnol.*, **21**(3), 298–302.

Kanan, M. W., and Nocera, D. G. (2008). In situ formation of an oxygen-evolving catalyst in neutral water containing phosphate and carbon dioxide. *Science*, **321**, 1072–1078.

Kay, A., Humphry-Baker, R., and Grätzel, M. (1994). Artificial photosynthesis. 2. Investigations on the mechanism of photosensitization of nanocrystalline TiO_2 solar cells by chlorophyll derivatives. *J. Phys. Chem.*, **98**(3), 952–959.

Koder, R. L., Anderson, J. L. R., Solomon, L. A., Reddy, K. S., Moser, C. C., and Dutton, P. L. (2009). Design and engineering of an O_2 transport protein. *Nature*, **458**, 305–311.

Konstantatos, G., and Sargent, E. H. (2010). Nanostructured materials for photon detection. *Nat. Nanotechnol.*, **5**, 391–400.

Lee, H., Cheng, Y.-C., and Fleming, G. R. (2007). Coherence dynamics in photosynthesis: protein protection of excitonic coherence. *Science*, **316**(5830), 1462–1465.

Grätzel, M. (2004). Conversion of sunlight to electric power by nanocrystalline dye-sensitized solar cells. *J. Photochem. Photobiol., A*, **164**(1–3), 3–14.

Magnuson, A., Anderlund, M., Johansson, O., et al. (2009). Biomimetic and microbial approaches to solar fuel generation. *Acc. Chem. Res.*, **42**(12), 1899–1909.

Martinot, E., Cabraal, A., and Mathur, S. (2011). *World Bank/GEF Solar Home System Projects*, http://e.nikkei.com/e/fr/tnks/Nni20110118D18JFA13.htm.

Nelson, J., and Prescott, D. (2008). *Business and the Millennium Development Goals: A Framework for Action*, 2nd ed. United Nations Development Programme.

Olah, G. A., Goeppert, A., and Surya Prakash, G. K. (2009). *Beyond Oil and Gas: The Methanol Economy*. Weinheim: Wiley VCH.

Pace, R. J. (2005). An integrated artificial photosynthesis model. In *Artificial Photosynthesis: From Basic Biology to Industrial Application* (Collings, A., and Critchley, C., ed.). Weinheim: Wiley VCH, 13.

Toma, F. M., Sartorel, A., Iurlo, M., et al. (2010). Efficient water oxidation at carbon nanotube–polyoxometalate electrocatalytic interfaces. *Nat. Chem.*, **2**, 826–831.

Porcaro, J., and Takada, M. (2004). *Achieving the Millennium Development Goals: The Role of Energy Services; Case Studies from Brazil, Mali and the Philippines*. New York: United Nations Development Programme.

Ranjit, L. R., and O'Sullivan, K. (2002). *A Sourcebook for Poverty Reduction Strategies*, vol. 2. Washington, DC: The World Bank.

Sanderson, K. (2008). Chemistry: the photon trap. *Nature*, **452**(7186), 400–402.

Sarra-Bournet, C., Haberl, B., Charles, C., and Boswell, R. (2011). Characterization of nanocrystalline nitrogen-containing titanium oxide obtained by $N_2/O_2/Ar$ low-field helicon plasma sputtering. *J. Phys. D: Appl. Phys.*, **44**(45) 455202.

Sebitosi, A. B., and Pillay, P. (2007). Modelling a sustainability yardstick in modern energisation of rural sub-Saharan Africa. *Energy Policy*, **35**(1), 548–552.

Sgobba, V., and Guldi, D. M. (2009). Carbon nanotubes: electronic/electrochemical properties and application for nanoelectronics and photonics. *Chem. Soc. Rev.*, **38**, 165–184.

Uppsala Universitat, Department of Photo Chemistry & Molecular Science (2005). *Solar-H*, http://www.fotomol.uu.se/Forskning/Biomimetics/solarh/index.shtm.

Towards Global Artificial Photosynthesis: Energy, Nanochemistry & Governance (conference proceedings), http://law.anu.edu.au/coast/tgap/conf.htm.

Tsikata, F. S., Brew-Hammond, A., and Osafo, Y. B. (2008). Increasing access to clean energy in Africa: challenges and initiatives. In *Moving Beyond Carbon: Energy Law in Transition* (Zillman, D. N., Redgwell, C., Omorogbe, Y. O., and Barrera-Hernandez, L., ed.). Oxford University Press, 163–179.

United Nations (2012). *Sustainable Energy for All*, http://sustainableenergyforall.org/objectives.

Wamukonya, N. (2007). Solar home system electrification as a viable technology option for Africa's development. *Energy Policy*, **35**(1), 6–14.

Wasielewski, M. R. (2006). Energy charge and spin transport in molecules and self-assembled nanostructures inspired by photosynthesis. *J. Org. Chem.*, **71**(14), 5051–5066.

World Bank (2006). *An Investment Framework for Clean Energy and Development: A Progress Report*. Washington, DC: The World Bank.

Yin, Q., Tan, J. M., Besson, C., et al. (2010). A fast soluble carbon-free molecular water oxidation catalyst based on abundant metals. *Science*, **328**(5976), 342–345.

Chapter 12

Nanotechnology-Based Artificial Photosynthesis: Food Security and Animal Rights in the Sustainocene

Alex Bruce[a] **and Thomas A. Faunce**[b, c]

[a]*ANU College of Law, Building 5, Fellows Road, Australian National University, Canberra, ACT 0200, Australia*

[b]*ANU College of Law, Building 6, North Wing, Australian National University, Canberra, ACT 0200, Australia*

[c]*ANU College of Medicine, Biology and the Environment, Building 131, John Curtin School of Medical Research, Australian National University, Canberra, ACT 0200, Australia*

alex.bruce@anu.edu.au, thomas.faunce@anu.edu.au

Development of nanostructured materials that facilitate the globalization of artificial photosynthesis (particularly the "food" part of that process, involving atmospheric nitrogen fixation and carbon dioxide reduction) offers unique opportunities to address some of humanity's most significant food security and climate change problems. Here we explore whether it also may have a profound impact on how our laws protect the interests of other sentient beings (such as the animals we slaughter for food) and their ecosystems. This chapter explores this concept by focusing

Nanotechnology Toward the Sustainocene
Edited by Thomas A. Faunce

ISBN 978-981-4463-76-8 (Hardcover), 978-981-4463-77-5 (eBook)
www.panstanford.com

on the role of competition and consumer law in fostering the governance framework for commercial nanotechnology products as is required to foster their marketplace transition not just towards anthropocentric energy sustainability and climate change mitigation, but environmental sustainability. To do this requires, we argue, a policy values choice—a technology-led transition away from global governance systems dominated by archived photosynthesis fuel and mass-production agribusiness corporations. Our hypothesis is that if commercial applications of nanotechnology are refocused (for example, by more effective safety and corporate regulation, as well as better options for decentralised, ethical citizen/consumer marketplace sovereignty) on artificial photosynthesis and if that does start producing distributed-source fertilizers and food on a global scale, it might take the pressure off our exploitation of nature both for large-scale commercial agriculture and "food animals" (the millions of animals, such as chickens, cows, and pigs, processed and slaughtered each day for human consumption). We examine whether this could pave the way for a long-term policy period known as the Sustainocene, where animals and ecosystems are recognized as legal "persons" capable of suing in courts (through their guardians) to protect their interests.

12.1 Governing Nanotechnology Toward Sustainability

12.1.1 Threshold Issues with Regulation of Nanotechnology in the Marketplace

More and more products sold on domestic markets incorporate nanotechnology. In the United States, for instance, the Project on Emerging Nanotechnologies Consumer Products Inventory (Wilson Institute, 2011) listed 1317 products made using nanotechnology to be sold across 30 countries to people referred to as "consumers." The largest main category was health and fitness (i.e., cosmetics and sunscreens), totaling 738 products. The most common materials mentioned were silver (313 products); carbon, including fullerenes (91 products); titanium, including titanium dioxide (59 products); silica (43 products); zinc, including zinc oxide (31 products); and gold (28 products) (Wilson Institute, 2011).

The 2009 Dutch National Institute for Public Health and the Environment (RIVM) report *Exposure to Nanomaterials in Consumer Products* identified several consumer products that could lead to high exposure to nanoparticles. These included sun cosmetics, oral hygiene, supplement and other health products, fuel for motor vehicles (after combustion), do-it-yourself (DIY) coatings, and glues and cleaning products (RIVM, 2009).

Though international governance of nanotechnology for workplace safety is an emerging reality, there seems to be little evidence that substantial efforts are being made to govern nanotechnology so that citizens are given information allowing informed choices about their domestic purchase and use of such products.

Part of the problem is the difficulty confronting both manufacturers and regulators in obtaining a comprehensive life cycle analysis about the unique risks of nanoparticles. The Organization for Economic Co-operation and Development (OECD) Working Party on Manufactured Nanomaterials (WPMN) is working toward obtaining the requisite information, particularly under Project 3, in which various nations have agreed to provide regulators with consistent toxicity data on common nanomaterials. Australia, for example, is contributing by coordinating research on nanoparticulate forms of zinc oxide (used in sunscreens), cerium oxide (used as a liquid fuel additive), and nanosilver (used as an antimicrobial). A related significant effort to harmonize international nanosafety regulation involves the International Standards Organization (ISO) Technical Committee, which has published work on nanotechnology terminology and definitions and international standards on environmental, health, and safety issues concerning nanotechnology (the ISO has proposed a definition of nanoparticle as one with all three external dimensions in the size range 1 nm to 100 nm) (Faunce, 2008).

In terms of domestic nanosafety governance, another issue is that most developed nations divide regulatory responsibilities between chemical, pharmaceutical, veterinary, and agricultural and food safety regulators, each having different requirements and approaches. To take Australia as an example, in terms of nanosafety regulation both the Australian Pesticides and Veterinary Medicines Association (APVMA) and the Therapeutic Goods Administration (TGA) regulate products rather than chemical substances. The National Industrial Chemical Notification and Assessment Scheme (NICNAS) administers the Australian Inventory of Chemical Substances (AICS), just as the TGA administers the Australian

Register of Therapeutic Goods (ARTG). These regulatory bodies, in their respective fields, require applicant manufacturers of nanoparticles to submit comprehensive characterization data sets, which include, where appropriate, particle size and characteristics. All food products supplied in Australia (including those potentially containing nanoparticles such as nanosilver) must comply with the Australia New Zealand Food Standards Code (ANZFSC) and be safe for human consumption. The use of industrial chemicals is also regulated at the state and territory levels by a range of agencies. Yet, this information is not routinely made available to people seeking to buy domestic products containing nanotechnology (i.e., through labelling) (Faunce, 2008).

Regulators of so-called consumer interests in products have shown much less interest in nanoparticles. In Australia, for example, the prevailing view of the Australian Competition and Consumer Commission (ACCC) is that the potential side effects of exposure to such nanoparticles are unknown and there is no substantial evidence about them (Australian Competition and Consumer Commission, 2012). Such an unequivocal statement doesn't correlate with the substantial scientific literature developing in the field of nanotoxicology. It does, however, reveal much about the restricted mind-set concerning nanotechnology that characterizes the consumer and competition regulators of most developed nations.

One illustrative problem for such regulators involves extrapolating to living human safety from in vitro test systems. Nanosilver, for example, is the most widely utilized consumer product application of nanotechnology, particularly in reducing bacterial spoilage of food, odors in clothes, and biofilm buildup in washing machines (Faunce and Wattal, 2010). Nanosilver, however, has been proven to have unique and deleterious dose- and time-dependent effects on cell function and viability at unusually low applied concentrations, though such in vitro data may be of limited assistance in in vivo risk assessment (Faunce and Wattal, 2010). A further concern is the difficulty of characterizing the surface chemistry of nanoparticles in biological systems when inevitably they are quickly surrounded by a protein corona (mostly albumin in humans) (Faunce, White, and Matthaei, 2008).

Mitigating the exposure of people to engineered nanoparticles used in or resulting from domestic products (such as nanosilver antiseptic sprays and diesel fuel, respectively) is likely to be extremely difficult in the absence of data suitable to create effective exposure limits. Control banding has been proposed as a precautionary tool to mitigate workplace exposures when there is not enough information for a quantitative risk assessment; but this approach is more complex for domestic uses. Similarly, existing statutory threshold "regulatory trigger" levels for the volume of manufactured nanomaterials bear little relation to the quantities in domestic products and in any event inadequately take into account the unique chemical properties and toxicities of many nanoparticles. Likewise, there is the issue of how to assign nanomaterials to domestic hazard categories and exposure scenarios and what level of domestic exposure reduction is afforded by specific engineering control levels (Faunce and Bai, 2010).

Thus, although many relevant institutional regulators are beginning to focus on manufacturers of nanoparticles (including requiring them to submit comprehensive data sets on nanoparticle size and characteristics), significant knowledge and regulatory gaps exist in all jurisdictions with respect to managing the environmental and public health implications associated with widespread domestic use of nanotechnology (Faunce, 2008).

12.1.2 Role of Citizen-Consumers in Nanotechnology Marketing

Governments in developed nations have made concerted efforts to advertise they are inquiring into the beliefs of the general public about nanotechnology (Commonwealth Government, *Approach to the Responsible Management of Nanotechnology*; Commonwealth Government, *National Enabling Technologies Strategy*). They have been less adroit in promoting mechanisms for actually informing such citizens, for example, through labeling of the risks and benefits of nanotechnologies. The asymmetric information status of nanotechnology in the marketplace derives from both a reluctance to overregulate this potentially lucrative industry and a reluctance to saddle its manufacturers with excessive burdens of responsibility for disclosure.

The international best practice in relation to efficient communication to the general public concerning nanotechnology requires certain minimum features. These include balanced policy information that accurately identifies and communicates the benefits and risks of nanotechnology, the content of information drafted at a level that is appropriate to the relevant stakeholder, the delivery form of that content varied and adapted to both content and recipient, and opportunities for stakeholder "feedback" on nanotechnologies, facilitating an ongoing dialogue of education.

Research into effective communication of nanotechnology confirms the above and stresses the need for information that is balanced, accurate, contextualized, and engaging. The particular regulatory challenge is to provide citizens who may not possess a scientific background with sufficient information enabling them to make accurate and informed decisions about nanoproducts, particularly so they can balance personal gains against social and environmental consequences (Petersen, Seear, and Bowman, 2010).

Both domestic and international studies indicate that the general public is optimistic about the potential benefits of nanotechnology; yet this optimism does not appear to be balanced against an accurate and informed awareness of the potential risks associated with nanotechnology (Petersen, Seear, and Bowman, 2010). Consequently studies about communication to citizens have recommended moving away from providing homogenous information about nanotechnology in general, toward heterogeneous information about socially valuable applications of nanotechnology, for example, in energy or medical applications (Petersen, Seear, and Bowman, 2010). Such calls have included suggestions about how regulatory systems should emphasize the precautionary principle (that regulation for public and environmental safety should proceed even in the absence of complete scientific certainty about risks) (Faunce, 2008). They also raise the issue of whether at this critical point in human history there is a broader category of information that should be provided to citizens who are purchasing nanotechnology products.

12.1.3 Existing Governance Strategies to Promote Nanotechnology for Sustainability

It is often suggested that in response to the critical contemporary challenges of overpopulation and anthropogenic climate change (particularly to human food and energy security), what is needed is more "command and control" regulation, for instance, in the form of government incentives for research into relevant new technologies by multinational corporations (MNCs). Nanotechnology, of course, should be central to such efforts if global governance exhibits a rational coalescence of human needs and technological potential. Yet, public and private sector policy is currently not oriented to encouraging nanotechnology in this important direction (Faunce, 2012a).

It might be argued that placing a price on atmospheric carbon represents one attempt to initiate a market mechanism to shift investment flows toward technological innovations (such as those based on nanotechnology) for a zero-carbon future. Considerable doubts exist, however, about whether this approach will be successful.

In Australia from July 1, 2012, for instance, the carbon price was to be A$23 per ton, rising at 2.5% per annum and transitioning to a fully flexible market price, such as that in the European Union (EU). Linked with a renewable energy target and various schemes funding research in renewable energy alternatives, this mechanism explicitly aimed to manage the transition to a low-carbon Australian economy. Yet, this Emissions Trading Scheme (ETS) received political backlash from powerful interests in the oil, coal, and liquid natural gas industries that convinced the public they would be burdened by the weight of this "tax." The conservative government elected federally in Australia in September 2013 undertook to repeal the legislation creating the carbon price and axe the A$10 billion commercially oriented Clean Energy Finance Corporation (designed to invest in renewable energy, low-pollution, and energy-efficiency technologies) and the Australian Renewable Energy Agency (ARENA), set up to administer A$3.2 billion in government support for research and development, demonstration, and commercialization of renewable energy (Liberal Party of Australia, 2013; Australian Government, Department of the Environment).

Other jurisdictions are making equally equivocal progress on carbon pricing and interrelated government support for renewable energy as a strategy for shifting corporate investment toward new technologies (such as nanotechnology) supporting sustainable and ecofriendly energy sources (Caripis, Peel, Godden, and Keenan, 2011). The EU Energy and Climate Policy, for example, aims to reduce greenhouse gas (GHG) emissions by 20% and increase renewable energy by 20% by 2020. The EU Strategic Energy Technology Plan (SET Plan) aims to accelerate development of low-carbon technologies and ensure their widespread market take-up. The European Industrial Initiative on Electricity Grid aims to enable 35% EU electricity from dispersed and concentrated renewable sources by 2020 and completely decarbonized electricity production by 2050 (EU SETIS, 2013). Yet such initiatives conflict with the financial interests of politically powerful "archived photosynthesis"-dependent (i.e., dependent on oil, coal, and natural gas) extraction and export, power generation, fuel import and distribution, transportation, and agribusiness MNCs (Sovacool, 2008). The World Energy Council, indeed, has been criticized for being a cartel of MNCs whose forecasts on renewable energy supply are in reality special interest decrees (Scheer, 2001).

In the United States, the American Clean Energy and Security Act 2009 (Waxman-Markey Bill) was passed in the federal House of Representatives but got stalled in the senate. It aimed for a 17% reduction in carbon emissions by 2020 and 80% by 2050, combined with a 20% increase in renewable energy supply by 2020. It provided extra funding for energy efficiency and renewable energy (US$90 billion by 2025) as well as enhanced resources for related basic scientific R&D (by US$20 billion). Likewise, the Save Our Climate Act (HR 3242) sought to impose a tax on carbon dioxide at oil wells, coal mines, and ports of entry for carbon-based fuels at $10/ton, rising by $10/ton per year—with the proceeds earmarked for deficit reduction. The hydrogen production roadmap produced by the US Department of Energy (DOE) and the US Council for Automotive Energies (CAE) mentions solar fuels but states that current water electrolysis units have relatively low production rates (100 kg/day) and need to be scaled up to 50,000 kg/day (United States Department of Energy and Council for Automotive Energies, 2011).

Yet, even if a global carbon price eventually is achieved, it is far from an ideal mechanism for driving research and investment toward technologies that would usher in a zero-carbon future. Carbon-intensive industries are politically and economically powerful and will not see a high price on carbon emissions in their financial interest. If the carbon price is not high enough it will not adequately incentivize the shift of investment to low- or zero-carbon technologies. Further, if a country's carbon price is linked with overseas carbon markets, then some policy makers may view this as threatening national sovereignty. Of course there is also the significant risk that some polluters may game the carbon-pricing system by utilizing exceptions, offsets, or offshore permits.

The problems with shifting investment toward zero-carbon energy solutions are exacerbated by global economic dependence on MNCs earning vast profits from carbon-intensive fuels. The Australian economy, for example, is closely tied to companies such as BHP Billiton, BG Group, Chevron, Rio Tinto, Xstrata, BP, Shell, Caltex, and Mobil (their energy exports were $69 billion in 2010–2011, one-third of total commodity exports) (Blackburn, 2013).

The capacity of such MNCs to prolong global dependence on "archived photosynthesis" fuels is enhanced by price-decreasing new resource and technology discoveries (i.e., hybrid carbon capture, shale oil, and coal seam gas extraction), strategic reductions in domestic oil refining, and the maintenance of carbon-intensive fuel subsidies, tax breaks, and strong lobbying and other levers over public policy (Blackburn, 2013). Such influence has resulted in a body of law and policy that legitimizes and incentivizes carbon-intensive industries, locking in decisions likely to freeze out frontier renewable energy technologies for decades (Sovacool, 2008).

Sovacool's (2008) influential thesis is that clean frontier technologies (such as artificial photosynthesis) symbolize a policy values choice—a transition to less anthropocentric governance systems of ownership, control, and ethics that will be considerably delayed unless five policy mechanisms are implemented: making clean power mandatory, eliminating subsidies, pricing energy appropriately, informing the public, and protecting the poor (Sovacool, 2008).

Yet in most developed nations, a centralized, carbon-intensive power generation system continues to be bolstered by subsidies, despite the vulnerability of its physical infrastructure, critical dependencies, and control systems to extreme weather events and terrorism, the likely need to source crude oil and petroleum further afield, electricity- and fuel-charging mechanisms that send inappropriate price signals to consumers, as well as inadequate antifraud and market manipulation regulation (Commonwealth Government, 2012a,b).

In the next section, we begin to consider an alternative governance approach for stimulating investment in zero- or low-carbon energy technologies, utilizing competition and consumer law. We first consider, particularly using the Australian case study, whether current nanotechnology regulation is incorporating governance mechanisms promoting a low- or zero-carbon future. We then examine the proposition that "competition and consumer" laws may facilitate citizens' choices to purchase nanotechnologies certified as assisting energy sustainability and climate change mitigation.

12.2 Competition Law, Nanotechnologies, and Food Security

12.2.1 Competition and Consumer Law in Nanotechnology Regulation?

Concerns about the safety of nanoparticles in certain products have led to calls for explicit regulation (particularly for labeling) of nanotechnology products as a matter of competition and consumer law (Faunce, 2008). What we wish to consider here is a much broader application of what is euphemistically known as "competition and consumer law." This involves how that area of law can empower citizens to drive the industry toward the safe development of technological products that create the preconditions for a sustainable world. Of particular interest to us is the capacity of such a legal mechanism to facilitate investment in technologies creating alternate food supplies to those currently dependent on the mass slaughter of "imprisoned" animals.

The so-called neoclassical school of economics makes certain assumptions about the way markets should function in order to promote efficiency and thereby enhance consumer welfare. These include the theory that markets largely free of financially failing firms will inevitably deliver optimal outcomes for producers and consumers alike. Of course, no real market has been or ever will be perfect, and deviations from this optimal competitive model occur in the form of anticompetitive conduct; information asymmetries, such as fraud; price collusion; misleading and deceptive conduct; and simple consumer irrationality (under the influence of mass media advertising).

Another key assumption is that "consumer demand" should maintain primary place as the economic driving force determining what is produced, its quantity, its price, its quality, and its social and environmental value (Johnston, 1996). Related to this assumption are the views that there are many buyers and sellers in the market, sellers produce a homogeneous product, buyers and sellers are equally informed about price, and there are no barriers to entry, meaning that firms can enter and exit the market. Finally, there is the understanding that market forces of supply and demand establish the price of the product—suppliers cannot affect the price of the product since no one firm produces more of the product than the others (Swann and McEachern, 2003).

"Neoclassical" economic theory posits that firms want to make goods or services at as low a cost as possible to themselves and, on the other hand, to sell those goods or services to consumers for as high a price as they can. In this way, firms are supposed to attempt to widen the margin between their costs of production and sale prices. The difference between the two represents the profit the firm derives from its goods or services.

On the other hand, consumers under such neoclassical economic models want to choose from among a wide variety of goods and services and also be able to buy those goods or services as cheaply as possible. The indicator of this tension is deemed to be price. Through their purchasing patterns, consumers "signal" to firms the goods or services that are preferred and the price levels they are prepared to pay for them. In this way, consumers are supposed to activate competition between corporates that will naturally distribute scarce resources to socially appropriate ends (Kovacic, 2007; Armstrong, 2008; Huffman, 2010; Schwarzkopf, 2011a). Firms respond to those signals through product innovation, improved service, and better allocation of resources as they compete with each other for customers' choice (Sylvan, 2004). This process has been described as one in which consumers vote with their "sovereign" purchasing power in the marketplace and so influence the efficient allocation of resources (Williams, 1989).

The implicit and disturbing idea emerging here, however, is that purchasing decisions may actually be viewed as a surrogate, or an alternate to political voting, particularly in a society that has privatized, that is, turned over to corporate control, the bulk of its infrastructure. Tacit acceptance of this notion could have profoundly negative consequences for the creation of strong regimes of governmental governance over markets in order to support foundational social virtues like justice and environmental sustainability.

At first glance, both competition and consumer law and artificial photosynthesis (specifically nanotechnology-based systems to make solar fuel and food) appear to have little to do with animal welfare, the rights of ecosystems, or environmental sustainability generally. Yet, perhaps the ideal purpose behind much of what is generally termed competition and consumer law is to utilize market forces in the form of signals from what is termed consumer demand (or withdrawal of it) exerting upstream market pressure on primary industry producers to implement food animal welfare initiatives (e.g., the Competition and Consumer Act 2010 [Cth]) [CCA] and Cth *Labelling Logic Report*, 2011; Bartlett, 2009).

The ACCC, for example, is tasked with determining the health risks of consumer products containing nanoscale materials (including composite materials), particularly through potential exposure to nanoparticles when the product is being used (Bruce, 2011a). The CCA, in particular under Section 18, will prohibit misleading or deceptive conduct about such "nanoproducts" in trade or commerce as that applies to corporations (as a law of the Cth) and equally to unincorporated entities (through reciprocal state and territory application legislation) (Bruce, 2009a).

In 2012, Antaria was referred to the ACCC by Friends of the Earth in a complaint about "misleading and deceptive conduct" for marketing ZinClear-IM as "nonnano," when subsequent independent tests conducted by the Australian National Measurement Institute (NMI) proved it contained a nanomaterial. Unions and the Public Health Association of Australia backed the complaint. Antaria rejected the allegations and claimed that the zinc oxide in its ZinClear-IM dispersion was of a nonnanoparticle size (Cooper, 2012).

The assumption behind laws facilitating such complaints is that such deception (should it exist) undercuts effective operation of market forces of supply and demand. The theory is that market dynamics will facilitate consumers' desires for accurate information.

We wish to go further and consider whether that information may drive an expansion of community conscience and sympathy, which may eventually encompass the relationship of nanoproducts' manufacture, marketing, and utilization to issues of energy and environmental sustainability.

Our hypothesis is that just as many citizens would be willing to pay a little extra to reduce the suffering of animals used to create the food they purchase (Leslie and Sunstein, 2007), so might they pay to purchase products deliberately targeted at creating a sustainable world. Relevant strategies could include a scheme for certifying renewable energy and food technologies integrated with climate change mitigation. We wish to explore, in other words, the relationship between nanoproducts, consumer protection legislation, consumer buying power, and planetary sustainability.

We begin by reconsidering two key aspects of consumer protection theory. The first concerns the way competition law and policy may be reorganized to empower people to make informed, moral purchasing decisions that enhance the foundational virtues of the marketplace—a governance concept we refer to as "citizen-consumer sovereignty." The assumption here is that if citizen-consumers are provided with sufficient information about renewable energy and food as well as climate change mitigation technologies, their conscience will be a stronger component in spending patterns that will signal suppliers and investors to orient technologies toward virtues such as justice, equity, and environmental sustainability. We replace the terminology "consumer" with "citizen-consumer" in order to signal the fundamental involvement, accountability, and responsibility of people through this governance process, which goes beyond defining them solely by reference to their capacity to make a limited range of purchasing decisions.

We also propose that citizen-consumers, once informed by suppliers about energy and food sustainability plus climate change mitigation issues associated with the products they are buying, will be *actually willing* to pay a price premium to promote morally appropriate technology. If there is evidence of this willingness, then a different set of questions relates to the *legal* implementation of that relationship.

It must be acknowledged that there currently is no overarching theory of consumer protection that explains market dynamics in terms of an environmental sustainability ethos. Indeed, this is reflected in the apparently incoherent profusion of citizen-consumer protection aspects in legislation and case law in differing jurisdictions (Reich, 1992). Concepts such as soft regulation, bounded rationality, neoclassical attribution, information asymmetries, biased contracting, enforced self-regulation, and shrouded attributes have been adapted from other disciplines and then applied to citizen-consumer protection issues in an attempt to explore these difficult policy issues (Bruce, 2009b; Bruce, 2011b).

Yet, the vital relationship between competition policy and citizen-consumer protection in delivering market-based benefits to society and its natural environment is increasingly being recognized at an institutional level (Kovacic, 2007; United Kingdom Office of Fair Trading, 2009). A competitive market is considered to be an "efficient" market in the sense that competition is the mechanism by which society's resources are efficiently allocated. But courts are recognizing that "in identifying the existence of competition in particular industries or markets, we must focus upon its economic role as a device for controlling the disposition of society's resources" (Re Qld, 1976).

Our claim is that when a market functions efficiently, citizen-consumers should benefit from the price competition among retailers of goods and services that are oriented to producing a sustainable world. This competitive benefit can be viewed as arising from two broad mechanisms—*interbrand* competition and *intrabrand* competition. In an efficient market, a citizen-consumer who wants to buy a solar-powered stove, for example, can visit different retailers and compare prices and certified social and environmental benefits of the product across and within different brands. All forms of anticompetitive and deceptive conduct have consequences for interbrand and intrabrand competition.

We propose that a market can properly function as a device for controlling the disposition of society's resources toward sustainability if it is working efficiently. Truly competitive markets display a number of characteristics that illustrate what is meant by the term "efficiency." Efficiency in this context may refer to technical efficiency (the effectiveness with which resources within a firm are utilized) and dynamic efficiency (the speed at which firms respond to changing problems and opportunities). When firms are unable to increase their profits through fairly marketing products characterized by worthwhile innovation, governance arrangements should ensure their pursuit of profit is channeled into increasing efficiency and better customer service rather than fraud, collusion, or other anticompetitive conduct (Economic Planning Advisory Council, 1989).

12.2.2 Citizen-Consumer Sovereignty and Sustainability

As "sovereign," rational choosers with corresponding social and familial responsibilities and obligations, consumers are viewed by neoclassical economic theory as chiefly driven by an individual utilitarian orientation to maximize personal benefits at the lowest possible cost (Kjaernes, 2011). Adam Smith is often quoted (as if thereby the idea gains additional credence) that "consumption is the sole end and purpose of all production" (Smith, 1776).

Yet, human beings also are undoubtedly motivated in large part by conscience to acquire virtue through the consistent application of universally applicable principles in the face of obstacles (Faunce, 2011). It is here we propose that it is time neoclassical economics confronted the proposition of the citizen-consumer, the marketplace purchaser who deliberately utilizes that purchasing activity to drive altruistic ends to the benefit of other sentient beings and their ecosystems.

This involves a change in the manner in which consumer protection theory is frequently said to involve the notion of "consumer sovereignty" (Schwarzkopf, 2011b; Nagarajan, 2007). William Hutt, in his 1936 book *Economics and the Public: A Study of Competition and Opinion*, argued that consumers should be aided by the modern state as one of its primary purposes in the exercise of freedom to pursue their own ends through the market (Hutt, 1936; Persky, 1993). Yet, for consumer sovereignty to work effectively, competition policy must ensure that markets present consumers (a) with a range of options and (b) the ability to select freely from among those options (Averitt and Lande, 1997; 1998). Our view is that what we term "citizen-consumer sovereignty" can thus be considered a set of governance arrangements that causes a society to shape its foundational social virtues (justice, equity, respect for human dignity, and environmental sustainability) by an overarching focus on the capacity of its individual members to consistently implement principles about buying things in response to corporate or government advertising and/or labeling, as well as the corporate and governmental responses to those signals.

Set out in such a way citizen-consumer sovereignty attempts to integrate the economic concept of consumer sovereignty with the idea of citizen sovereignty according to the rule of law under an implied or explicit social contract. Legal strategies in competition law, such as disclosing product information; prohibiting unconscionable, misleading, or deceptive conduct; and ensuring product liability, may then be viewed as providing citizen-consumers with sufficient information consistent with their sovereign status in the market and also of stimulating among them and the wider society political and normative discourse about rights and interests (Leslie and Sunstein, 2007; Bruce, 2012a,b).

Part of this shift involves citizen-consumers being willing and being provided with efficient mechanisms to express conscience-driven "ethical" and political preferences (i.e., through purchasing decisions and complaints to suppliers and regulators) in the market itself and not solely on the political forum (Korthals, 2001). Such actions, representing the theoretical and practical the transformation of consumers into citizen-consumers for governance purposes, constitute the start of a renegotiation of the global social contract to include both corporations (artificial people) and the environment.

Our citizen-consumer approach is illustrated in the area of marketplace choices about food animal products (Lea and Worsley, 2002). A large number of studies and surveys indicate that European people are concerned about the welfare of food animals, demand animal welfare–friendly products and are willing to pay a premium for them (Schroder and McEarchen, 2004; Nocella, Hubbard, and Scarpa, 2010). For example, a 2005 study indicated that 74% of European citizens believed that they could exert a positive influence on farm animal welfare through purchases of animal-friendly products and more than 60% confirmed that they were willing to pay a price premium to ensure farm animal welfare (Andersen, 2011). The principal reported obstacle involved the lack of information concerning animal welfare issues available to consumers at the time of purchase. Such research confirms, however, that there is little incentive for suppliers to satisfy consumer demand for animal welfare values if consumers are conflicted about those animal welfare values (Carlsson, and Frykblom, 2007; Mayfield, Bennet, Tranter, and Woolbridge, 2007; Tawse, 2011; Pickett, Norwood, and Lusk, 2010). We shall return to such an example again after we have considered how this citizen-consumer approach to competition law might apply to innovative products related to the crucial global issues of energy security and environmental sustainability.

12.3 Competition Law, Global Artificial Photosynthesis, and Food Security

12.3.1 Food Implications of Nanotechnology-Based Artificial Photosynthesis

It has been argued that the moral culmination of nanotechnology is global artificial photosynthesis (Faunce, 2011). The rationale for such a claim involves the capacity of this technology, once deployed in the worldwide marketplace, to allow structures such as buildings and roads to not only produce hydrogen fuel by using sunlight to split water but also make ammonia and starch through absorption and chemical reduction of atmospheric nitrogen and carbon dioxide. In this way purchases of such nanotechnology-based renewable energy products would assist in remedying the major moral dilemma facing our species: how to live peacefully and sustainably on earth without deleteriously impacting other life forms.

Central to this vision for nanotechnology contributing to a sustainable world deep into our future is creating the governance conditions in the global markets for nanotechnology-based artificial photosynthetic products to be rolled out as cheap, "off-grid," local and domestic sources of energy, fertilizer, and food. Yet, so deployed, nanotechnology-based global artificial photosynthetic products would undercut profitable market models of consumer sovereignty predicated on mass centralized production of carbon-intensive fuel and food, with suppliers showing little local community accountability or responsibility (Faunce, 2012).

A world powered by artificial photosynthesis, in which an increasing proportion of human-engineered structures operate like trees in generating hydrogen fuel and oxygen while absorbing nitrogen and carbon dioxide to make basic food and fertilizer, will be one experiencing a shift from the hopefully short-lived Anthropocene to the Sustainocene epoch. In the latter period, humanity may still control the earth and its ecosystems but this would be more as a steward, exhibiting due respect for the flourishing of all life (Faunce, 2012a).

One of the chief reasons for predicting the globalization of artificial photosynthesis is likely to be the transformative technology that ushers in the Sustainocene, is not just the amount of safe solar energy striking this planet but also its proven natural capacity to drive processes, not just of water splitting, but also of atmospheric carbon dioxide reduction. More solar energy strikes the earth's surface in one hour of each day than the energy used by all human activities in one year (Nocera, 2012). At present the average daily power consumption required to allow a citizen to flourish with a reasonable standard of living is about 125 kWh/day. Much of this power is devoted to transport (~40 kWh/day), heating (~40 kWh/day), and domestic electrical appliances (~18 kWh/day), with the remainder lost in electricity conversion and distribution (MacKay, 2009). Global energy consumption is approximately 450 EJ/year, much less than the solar energy potentially usable at ~1.0 kW/m^2 of the earth—3.9 × 10^6 EJ/year, even if we take into account the earth's tilt and diurnal and atmospheric influences on solar intensity (Pittock, 2009).

Photosynthesis as a natural process has been as important as DNA in enabling humanity to achieve its current dominant position on the earth. Photosynthesis provides the fundamental origin of our oxygen, food, and the majority of our fuels; it has been operating on the earth for 2.5 GYr (Leslie, 2009). Photosynthesis can be considered as a process of planetary respiration: it creates a global annual CO_2 flux of 124 PgC/year and an annual O_2 flux of ~10^{11} t/year (Hillier and Wydrzynski, 2008). In its present nanotechnologically unenhanced form, photosynthesis globally already traps around 4,000 EJ/year solar energy in the form of biomass (Kumar et al., 2009). The global biomass energy potential for human use from photosynthesis as it currently operates globally is approximately equal to human energy requirements (450 EJ/year) (Hoogwijk et al., 2003; Parikka, 2004; Fischer and Schrattenholzer, 2001).

Nanotechnology will be critical to achieving necessary breakthroughs in many important areas of artificial photosynthesis.

Some such nanotechnological innovations focus on improved light capture. Examples include the work by Catchpole reported in this volume. Carbon nanotubes, other nanomaterials, and hybrid organic-inorganic nanostructures can more efficiently harvest and conduct the resultant electricity (Guter et al., 2009; Konstantatos and Sargent, 2010).

Table 12.1 Energy consumption and solar power

Parameter energy/power	
Annual solar energy intercepted by the earth at ~1.37 kW/m^2	5.5 × 10^6 EJ/yr
2008 primary energy consumption	352.8 EJ
2050 primary energy consumption	590–1050 EJ
2100 primary energy consumption	880–1900 EJ
Ratio of potentially usable annual solar energy to current primary energy consumption/yr	~9000 to 1

Sources: Rogner, H.-H. (2004). Chapter 5. In *United Nations Development World Energy Assessment*. United Nations, 162.
International Energy Agency (2010). *Key World Energy Statistics*. Paris, France: International Energy Agency, 30.

Yet, of most interest as far a global food product is concerned is the much less well-researched artificial photosynthesis version of the "dark reaction." In this process ATP and NADPH, as well as carbon dioxide (CO_2), are used in an enhanced version of the Calvin–Benson cycle to make locally usable food or fuel (for domestic, heating, cooking, light, and transport) in the form of carbohydrate via the enzyme RuBisCO (Gray, 2009). So much of the focus is on solar-driven water splitting that this critical component of a fully functioning artificial photosynthetic system is overlooked. It is true that only a few plants are able (in addition to making starch from atmospheric carbon dioxide reduction) to fix atmospheric nitrogen (i.e., by nodules in their roots). But making ammonia fertilizer from atmospheric nitrogen is likely to be part of a fully functional artificial photosynthetic system.

Enhanced "dark reaction" components of artificial photosynthetic products sold cheaply to citizen-consumers across the globe could assist with reforestation of deserts and cities, crop production on marginal lands, and reduction of atmospheric CO_2 levels, as well as lowering of geopolitical and military tensions over fossil fuel, food, and water scarcity. There would be profoundly positive (in terms of sustainability) governance implications of allowing buildings and roads to create carbon-neutral fuels for domestic, community, and industrial storage as well as basic starches from atmospheric carbon dioxide or the production of ammonia fertilizer from atmospheric nitrogen.

Numerous competitively funded research teams in many nations have dedicated projects underway to develop artificial photosynthetic products for purchase by citizen-consumers (Sanderson, 2008; Faunce, 2012). A multidisciplinary international conference at Lord Howe Island in August 2011 linked senior artificial photosynthesis and global governance experts as a precursor to a macroscience global artificial photosynthesis project (Faunce, 2011). A further conference on this theme but involving potential stakeholders such as the Welcome Trust and Moore Foundation was coordinated at Chicheley Hall with the UK Royal Society in July 2014 (Royal Society, 2014).

A critical issue then becomes, what type of governance arrangement is best suited to assisting the global deployment of artificial photosynthetic nanotechnology? As will be examined in the next section, there appear to be many obstacles in current governance theory and practice to such a global transitional approach to nanotechnology. It may be that competition and consumer law has a significant role to play in circumventing some of these obstacles.

12.3.2 Governance Obstacles to the Global Deployment of Artificial Photosynthesis

One way to develop a governance framework to facilitate the global deployment of the food as well as fuel components of artificial photosynthesis is to work outward from the so-called soft law norms. These include internationally agreed targets to reduce poverty and lack of necessary fuel and food and encourage environmental sustainability and global technological cooperation, expressed in the United Nations' millennium development goals

(UN MDGs). Principles supporting similar goals (and directed to individuals, communities, and private corporations and states [Article 1]) appear in UNESCO's Universal Declaration on Bioethics and Human Rights (particularly the social responsibility principle in Article 14(b), "access to adequate nutrition and water," and Article 14(c), "improvement in living conditions and the environment." Related international law obligations arise under the international right to health (set out, for example, in Article 12 of the United Nations' International Covenant on Civil and Political Rights). Similarly, Article 15 of the United Nations' International Covenant on Civil and Political Rights (ICESCR) sets out the right to enjoy the benefit of scientific progress and its applications (REBSPA). In particular, it refers to the right of everyone to (in Article 15[1(b)]) **enjoy the benefits of scientific progress and its applications** and recognizes (in Article 15[4]) the benefits to be derived from the **encouragement and development of international contacts and co-operation in the scientific and cultural fields.**

Likewise, the United Nations Educational Scientific and Cultural Organization's (UNESCO, 1997) *Declaration on the Responsibilities of the Present Generations Towards Future Generations* states in Article 4:

> The present generations have the responsibility to bequeath to future generations an Earth which will not one day be irreversibly damaged by human activity. Each generation inheriting the Earth temporarily should take care to use natural resources reasonably and ensure that life is not prejudiced by harmful modifications of the ecosystems and that scientific and technological progress in all fields does not harm life on Earth.

Yet international trade, and more particularly international trade and investment law, resides at the heart of a system of corporate globalization driven by a much more politically and economically powerful, contradictory set of norms. Under these rules, corporate capital is permitted to take advantage of abundant natural resources (particularly timber, oil, coal, and minerals) or cheap labor to manufacture products for distribution and profitable sale throughout the world using road, rail, sea, and air freight transport; reduced tariffs; and mass-marketing techniques without necessarily having to take into account countervailing social or environmental risks (Faunce, 2012).

The World Trade Organization (WTO) is headquartered in Geneva near many of the United Nations' human rights organizations with which it normatively has so little in common. The WTO comprises a secretariat and public officials from nation states that have been involved in agreements, such as the Trade-Related Intellectual Property Rights (or TRIPS) Agreement and the General Agreement on Trade in Services (GATS), by which those states agree to not merely reduce various trade barriers but allow supranational corporations to take control of major national assets (such as intellectual property, hospital and health services, water, agriculture, power generation, and manufacturing) in a way that is very hard to undo (due to the compensation to corporate stakeholders that must be paid by taxpayers). What has been created, in other words, is a supranational corporation-controlled legal system that is pushing global governance in directions different to those of democracy-based community and civil society institutions committed to societal virtues, such as justice, equity, and, increasingly, environmental sustainability.

Tactics of particular concern as a potential obstacle to the global rollout of nanotechnology-based artificial photosynthesis products are the so-called investor-state dispute settlement (ISDS) provisions. In the 1990s civil society prevented the creation of a supranational investment protection agreement (the Multilateral Agreement on Investment, or the MAI, also known as Multilateral Investment Treaty) that would have allowed the global implementation of such provisions, but they have nonetheless proliferated in a series of bilateral and regional arrangements. Basically they allow supranational corporations to sue (before small panels of commercial arbitration lawyers with little understanding of or desire to apply international public law) other nations that have imposed governance requirements (even when in the public health and environmental interest and based on good scientific evidence) if their commercial interests are thereby impeded. Investor-state provisions surfaced in the failed MAI of the 1990s and in the 1994 North American Free Trade Agreement (NAFTA), between the United States, Canada, and Mexico. They are now part of over 2000 bilateral investment treaties (BITs) (Ranald and Southalan, 2003; Australian Fair Trade and Investment Network, 2010).

The ISDS system grants corporate investors covered by its norms a right to initiate dispute settlement proceedings (before a panel of trade lawyers known as commercial arbiters) for damages against foreign governments in their own right. Should artificial photosynthesis products begin to look as if they are likely to replace those upon which supranational corporations have substantial investments (in, say, old photosynthesis fuels or electricity distribution networks), then those corporations may well resort to investor-state mechanisms to protect their profits and inhibit the rollout by opposing governance mechanisms that work in the opposite direction.

Nanotechnology-based artificial photosynthesis research and development will also face major issues about whether intellectual monopoly privileges (IMPs) such as patents should cover its products as well as processes and functions. It is likely that in the United States the "utility" for such a patent (as is the case for DNA) will be that it must be specific, substantial, and credible. If patent ownership in this area becomes fragmented, researchers in the field may find their follow-on research hampered by the high cost and difficulty in negotiating contracts with large numbers of patent owners. Each individual patent owner, for example, without some prior licensing and sharing arrangement will have an incentive to overcharge other researchers requiring access.

Attempts could be made to overcome such governance obstacles using a model based on international law. Would Article 15(1[b]) and Article 15(4) of the ICESCR, for example, justify flexibilities to WTO, bilateral and regional trade agreements IMP provisions, or a 0.05% tax on global financial transactions to fund a coordinated global R&D project on artificial photosynthesis? Could it justify provisions requiring a wind-back of fossil fuel subsidies or preventing money laundering, transfer pricing, and other tactics of tax evasion (House of Commons, 2013; Kobetsky, 2005)? Could it justify new forms of e-governance (Williamson, 2010)?

The core component of the international legal obligations in Article 15 of ICESCR can be interpreted as aiming to protect, fulfill, and respect the scientific enterprise insofar as it contributes to achieving human rights and obligations. It aims to protect the capacity of the scientific enterprise to bring benefits to everyone through encouraging measures that permit critical analysis, honesty, and objectivity among scientific researchers and their employers and facilitate government regulatory systems based on scientific evaluation of the risks, benefits, and cost effectiveness of new technologies. For the purposes of this right it is expected that states have an obligation to ensure that science conducted within their boundaries is coherent with international human rights. Yet qualifications, exceptions, ambiguities, and reservations have been so carefully clustered around these international law obligations and their related enforcement mechanisms are so weak that the states signed on to be "bound" by them experience little impact on the residual sovereignty permitted to them by corporate multinationals.

This leads to the issue of whether there may be a more effective governance mechanism for stimulating nanotechnology R&D toward energy security and climate change mitigation technologies, such as artificial photosynthesis. We argue this approach could be competition and citizen-consumer law.

12.3.3 Competition and Citizen-Consumer Laws Facilitating Global Artificial Photosynthesis and Animal Rights

Expecting governments to satisfactorily navigate the inconsistent and imperfectly enunciated and enforced objectives that characterize nanotechnology regulation worldwide so as to reorient it toward addressing major global health and environmental issues is a process likely to confront deep-seated cultural norms, religious beliefs, and vested economic interests. Governments, it may be argued, are not supposed to allow "technologies" to help humans "play God" and reshape fundamental relationships between humans and nature. Governments are also not expected (it could be maintained) to "pick winners" among technologies and are supposed to let "market forces" dictate which technologies get researched and developed.

Instead, through properly designed competition and citizen-consumer laws, individuals and communities could drive a governance revolution sponsoring new energy security and climate change mitigation technologies through a bottom-up, or grassroots, approach. Using as a model mechanisms expanding on those developed under the CCA 2010 (Cth) and Cth Labelling Logic Report

2011, such regulatory initiatives would be driven by a mixture of governance arrangements reducing fraud and promoting competition among suppliers and informing citizen-consumers through product labeling focused not only on safety (and reducing misleading or deceptive statements) but also on public health and environmental sustainability impact (i.e., through labeled certifications by relevant nongovernmental organizations), tax concession, and other rewards for new technologies that meet renewable energy and climate change mitigation targets and initiatives to support greater citizen-consumer involvement in shaping market values.

The aim of such an approach would be no less than to facilitate investment likely to usher in a world where every building, road, and vehicle is doing photosynthesis more effectively than plants (using sunlight to split water as a source of hydrogen and oxygen while reducing atmospheric nitrogen and carbon dioxide as sources of fertilizers and starch). Such a world would be one that facilitates significant normative changes in our relationships with animals and ecosystems. Increasing information and its communication imaginatively and scientifically have already facilitated an expansion of human sympathy by which conscience that embraced slaves, women, children, and gay people as a prelude to their enhanced legal protection. Our hypothesis is that likewise, in a world where all human structures are "paying their way" in terms of sustaining our environment while providing for our basic food and energy needs, the economic imperative to exploit animals and ecosystems will fade and allow our expanding sympathies to accord them enforceable rights.

Yet some would dispute that globally distributed use of this new application of nanotechnology is likely to have such a normative consequence. Getting people to accept that their basic forms of food can be supplied by nanotechnology facilitating buildings and roads using sunlight, water, and absorbed nitrogen and CO_2 will not be easy. Western societies, for example, largely accept that meat products from slaughtered animals form a natural and indeed desirable component of human dietary requirements (Fiddes, 1991), despite the deleterious impact this process has on ecosystems (Bourdieu, 1977).

The United Nations Population Division predicts that by 2050 the world's population will reach somewhere between 8 and 11 billion people (United Nations Department of Economic and Social Affairs, 2012). Much of this population growth will occur in developing countries, where a growing middle class, with more disposable income, is expected to generate substantial demand for meat products as part of its diet (Thornton, 2010). This is particularly so in China and India, where a rapid demand for meat products is being met by mass production techniques that both degrade animals and pollute ecosystems. In satisfying this growing preference for meat products, animal farming enterprises have transitioned from small, family-owned operations to large-scale, concentrated animal farming operations (CAFOs) managed by vertically integrated corporations seeking to maximize the efficient production of animal products to satisfy both domestic and foreign demand (Hocquette and Chatellier, 2011).

Could artificial photosynthetic systems provide equivalent amounts of protein? The answer is, not directly, but they could provide staple carbohydrates and replenish degraded and presently unproductive lands. They could facilitate small-scale agriculture that involves improved relations between humans and animals. Competition and citizen-consumer law can help provide flexible and rapidly responsive means of assisting in this governance transition.

Neoclassical economic principles define corporate success in terms of profit and return on investments, where wealth is maximized through productive, allocative, and technical efficiencies accomplished through techniques of mass production of animal-based food products.

The reality for most of the animals slaughtered for their meat or farmed for their eggs in developed nations is that they do not see the sun or feel the earth, they do not socialize with other animals, and they are not able to express their natural instincts but are confined in mass factories before they are slaughtered or their eggs harvested (Ibrahim, 2007; Evans, 2006). This gulag-inspired process of factory farming involves raising animals using intensive production line methods that maximize the amount of meat produced while minimizing costs, emphasizing high stocking densities and/or close confinement, forced growth rates, high mechanization, and low labor requirements (Voiceless, 2008; Bergmann, von der Heidt, and Maller, 2010). Mass production methods for making food from animals are justified as necessary to permit the use of profit and efficiency-enhancing animal husbandry practices that would otherwise be characterized as acts of cruelty. Legislative attempts in the EU, the United Kingdom, and New Zealand to either displace or regulate the religious slaughter of animals have foundered against human rights claims of freedom of religious practice recognized by domestic *constitutions* or international human rights instruments (Silver, 2011; Australian Meat Industry Council (2009)).

Consumers are becoming increasingly sensitive to the ways in which food animals are treated and about the related issue of the impact manufacture of products has on environmental sustainability. This sensitivity is expressed both morally (i.e., in Facebook comments) and practically (i.e., through voting and purchasing decisions such as paying a price-premium for food animal products from suppliers who have implemented welfare-friendly animal husbandry practices) (Toma, Stott, Revoredo-Giha, and Kupiec-Teaha, 2012; Andersen, 2011; Naald and Cameron, 2011; McEarchen and Schroder, 2002). In response to this consumer demand, suppliers of food animal products are seeking to differentiate their products on the basis of animal welfare–friendly practices. Product labels promoting free-range, free-to-roam, organic, or cruelty-free animal husbandry practices are used by suppliers to influence consumers who, for example, "seek out free range eggs as a matter of principle, hoping to advance the cause of animal welfare by so doing." This is despite most people in Western societies not yet accepting that animals possess rights that can or should be legally protected and enforced over and above human rights claims, interests, preferences, or freedom of religious practice (Steiner, 2005; ACCC, 2010; Attorney General, 1964).

It may then be that labeling of artificial photosynthetic products that address the slaughter of animals for food and environmental sustainability would resonate with the conscience and later the purchasing decisions of such citizen-consumers. Technical problems may have to be overcome. Competition and citizen-consumer laws, for example, generally are not prescriptive of the scientific evidence required to prove misleading conduct (*Plastec Australia Pty Ltd v.*

Plumbing Solutions and Services Pty Ltd, 2012; *Olivaylle Pty Ltd v. Flottweg AG [No 4]*, 2009). Expert evidence also will be adduced as to the truth or falsity of the relevant representation and tested on the balance of probabilities, but the question is ultimately one "for the tribunal of fact and . . . not . . . for any witness to decide" (*Interlego AG v. Croner Trading Pty Ltd*, 1992).

In summary, as global deployment of artificial photosynthetic products begins to allow buildings and roads to provide for basic human food needs, may not the legal and regulatory framework that currently facilitates the efficient exploitation of animals (particularly in the chicken, pork, and beef industries) come to be as much reviled as that which perpetuated human slavery?

12.4 Conclusion

Both domestic and international studies indicate that consumers are excited and optimistic about the potential benefits of nanotechnology, though increasingly seeking accurate and readily available information (including through labeling) about its potential risks both to their own health and that of the environment.

We have here explored some normative consequences of the vision that nanotechnology's moral culmination involves, assisting artificial photosynthesis to be rolled out globally as an "off-grid" energy, fertilizer, and food product. We have suggested that in a world where human existence is fundamentally supported by such a nanotechnological application, the rights of sentient beings, such as animals and their ecosystems, could receive greater protection (Adams, 2009).

A world powered by artificial photosynthesis, in which an increasing proportion of nanotechnologically engineered structures operate like trees in using sunlight to split water to generate hydrogen fuel and oxygen while absorbing atmospheric nitrogen and carbon dioxide to make basic food and fertilizer, will be much closer to being ecologically sustainable. It might even mark a shift from the hopefully short-lived Anthropocene (or Corporatocene) to the Sustainocene. In the latter epoch, humanity may still control the earth and its ecosystems but more as a steward, exhibiting due respect for the flourishing of all life.

In these circumstances, we consider an emphasis on governance that ensures that life cycle safety for nanotechnology would not be viewed as an end in itself or primarily a means of ensuring greater profits for corporate manufacturers in the marketplace. Thus the development of the morally most valuable use of nanotechnology should be set upon a foundation of safety strategies such as the use of the precautionary principle, mandatory manufacturer premarketing notification, approval and postmarketing surveillance of human health and ecosystem impacts, and citizen notification schemes.

The next stage of nanotechnological governance, however, must aim to focus this industry on the great public health and environmental challenges of our time, including the global imperative to conserve dwindling natural resources and the need to preserve the biosphere (a field of study called "planetary medicine"). It is here we have argued that claims of government support for frontier technologies (such as artificial photosynthesis) addressing such critical aims must be linked with a fundamental shift in human ethics and lifestyle, that risks becoming platitudinous rhetoric until linked with novel practical competition law reforms that empower citizens' support of public goods in a genuinely competitive energy market.

In this context we have here advanced the proposition of the citizen-consumer, the purchaser who deliberately utilizes his or her purchasing activity in the marketplace to drive altruistic ends to the sustainable benefit of other sentient beings and their ecosystems. Our position is that this transitional regulatory approach will only work effectively if competitive and informed markets continue to place consumer interests and welfare at the center of supply and demand forces. We have briefly set out the wide-ranging types of reforms that need to take place to make markets genuinely competitive and informed.

Such measures include those aimed at reducing fraud by creating financial incentives for informants, restructuring the corporations law to ensure that maximization of shareholder profit sits alongside allocated public good as a condition of yearly registration, and restructuring capital markets to make them more responsive to social and environmental good (by imposition, for example, of a financial transactions tax).

They would also include governance measures, including those informing citizen-consumers through product labeling focused not only on safety (and reducing misleading or deceptive statements) but also on public health and environmental sustainability impact (i.e., through labeled certifications by relevant nongovernmental organizations), tax concession, and other rewards for new technologies that meet renewable energy and climate change mitigation targets and initiatives to support greater citizen-consumer involvement in shaping market values.

As we have stated, the aim of such an approach is no less than to usher in a world where every building, road, and vehicle is doing photosynthesis more effectively than plants (using sunlight to split water as a source of hydrogen and oxygen while reducing atmospheric nitrogen and carbon dioxide as sources of fertilizers and starch). Such a Sustainocene epoch would be characterized by genuinely competitive markets facilitating significant normative changes in our relationships with animals and ecosystems.

Acknowledgment

We wish to acknowledge the assistance of an Australian Research Council (ARC) Discovery Grant.

References

Adams, W. (2009). Human subjects and animal objects: animals as 'other' in law. *J. Anim. Law Ethics*, **3**, 29.

Australian Fair Trade and Investment Network (2010). *TPP Submission to Minister for Trade*, http://www.dfat.gov.au/trade/fta/tpp/subs/tpp_sub_aftinet_081103.pdf.

Andersen, L. (2011). Animal welfare and eggs: cheap talk of money on the counter? *J. Agric. Econ.*, **62**(3), 565.

Armstrong, M. (2008). Interactions between competition and consumer policy. *Competition Policy Int.*, **4**(1), 97.

Attorney General (SA) v. Bray, 111 CLR 402 (1964).

Australian Competition and Consumer Commission v. C.I & Co Pty Ltd, FCA 1511, [31] (2010).

Australian Competition and Consumer Commission (2012). *Product Safety Australia*, http://www.productsafety.gov.au/content/index.phtml/itemId/973893.

Australian Meat Industry Council (2009). *National Animal Welfare Standards for Livestock Processing Establishments: Preparing Meat for Human Consumption 2009–2010* (2nd ed.). New South Wales, Sidney, Australia.

Australian Government, Department of the Environment *Working for a Clean Energy Future*, http://www.cleanenergyfuture.gov.au/.

Averitt, N., and Lande, R. (1997). Consumer sovereignty: a unified theory of antitrust and consumer protection law. *Antitrust Law J.*, **65**, 713.

Averitt, N., and Lande, R. (1998). Consumer choice: the practical reason for both antitrust and consumer protection law. *Loyola Consumer Law Rev.*, **10**, 44–47.

Bartlett, A. (2009). Animal welfare in a commonwealth system: a commonwealth politician's approach. In *Animal Law in Australasia: A New Dialogue* (Sankoff, P., and White, S., ed.). Sydney, Australia: The Federation Press, 387.

Bergmann, I., von der Heidt, T., and Maller, C. (15–16 July 2010). *Cognitive Dissonance and Individual's Response Strategies as a Basis for Audience Segmentation to Reduce Factory Farmed Meat Consumption* (Russell-Bennett, R., and Rundle-Thiele, S., ed.; pp. 32–35), paper presented at the International Nonprofit and Social Marketing Conference, Brisbane, Australia.

Blackburn, J. (2013). *Australia's Liquid Fuel Security: A Report for NRMA Motoring and Services*. National Road and Motorists' Association.

Bourdieu, P. (1977). *Outline of a Theory of Practice*. United Kingdom: Cambridge University Press.

Bowman, M., and Redgwell, C. (1996). International law and conservation of biological diversity. *Kluwer Law Int., The Hague,* **33**, 39–40.

Bruce, A. (2009a). Animals and the trade practices act 1974 (Cth). *Australian Anim. Prot. Law J.*, **2**, 49–52.

Bruce, A. (2009b). Cognitive dissonance in the contribution of the Catholic Church to international human rights law discourse. *Adelaide Law Rev.*, **30**(1), 149.

Bruce, A. (2011a). Australia's national consumer protection regime. In *Consumer Protection Law in Australia*. Sydney, Australia: LexisNexis Butterworths, 4.

Bruce, A. (2011b). Do sacred cows make the best hamburgers? The legal regulation of the religious slaughter of animals. *Univ. of New South Wales Law J.*, **34**(1), 351.

Bruce, A. (2012a). Animals and cruelty. In *Animal Law in Australia: An Integrated Approach*. Sydney, Australia: LexisNexis Butterworths, 197.

Bruce, A. (2012b). Animals as food. In *Animal Law in Australia: An Integrated Approach.* Sydney, Australia: LexisNexis Butterworths, 221.

Carlsson, F., and Frykblom, P. (2007). Consumer willingness to pay for farm animal welfare. *Eur. Rev. of Agric. Economics,* **34**(3), 321.

Caripis, L., Peel, J., Godden, L., and Keenan, R. (2011). Australia's carbon pricing mechanism. *Climate Law,* **2**(4), 583–603.

Commonwealth Government (2012a). *Response to the Recommendations of Labelling Logic: Review of Food Labelling Law and Policy.* Commonwealth Government, 40.

Commonwealth Government (2012b). *Australian Energy White Paper.*

Commonwealth Scientific and Industrial Research Organisation (2007). *Australian Standard for the Hygienic Production and Transportation of Meat and Meat Products for Human Consumption* (FRSC Technical Report No. 3; AS 4696:2007). Collingwood Victoria, Australia: CSIRO Publishing.

Cooper, A. (2012, July 25). Sunscreen maker faces ACCC heat. *Sydney Morning Herald,* http://www.smh.com.au/national/sunscreen-maker-faces-accc-heat-20120724-22nvg.html#ixzz2fgejIH52.

Corones, S. (2010). Competition policy and economic theory. In *Competition Law in Australia,* 5th ed. Australia: Thompson Reuters, 43.

Economic Planning Advisory Council (EPAC) (1989). *Promoting Competition in Australia,* Council Paper No. 28, AGPS Canberra.

Elliot, A., and Devine, P. (1991). On the motivational nature of cognitive dissonance: dissonance as psychological discomfort. *J. Personality Social Psychol.,* **67**(3), 382.

European Commission. (2009). *Feasibility Study on Animal Welfare Labelling and Establishing a Community Reference Centre for Animal Protection and Welfare.* EU Directorate General for Health and Consumer Protection.

European Commission (2012). *European Union Strategy for the Protection and Welfare of Animals 2012–2015,* 10–11, part 3.4.

European Union SETIS (2013). *What Is the SET-Plan?* http://setis.ec.europa.eu/about-setis/set-plan-governance.

Evans, G. (2006). To what extent does wealth maximisation benefit farmed animals? *Anim. Law,* **13**, 167.

Faunce, T. A. (2008). Toxicological and public good considerations for the regulation of nanomaterial-containing medical products. *Expert Opin. Drug Safety,* **7**(2), 103–106.

Faunce, T. A. (2011a). Will international trade law promote or inhibit global artificial photosynthesis? *Asian J. WTO Int. Health Law Policy*, **6**, 313–347.

Faunce, T. A. (2011b). Governing nanotechnology for solar fuels: towards a jurisprudence of global artificial photosynthesis. *Renewable Energy Law Policy*, **2**, 163–168.

Faunce, T. A. (2012a). *Nanotechnology for a Sustainable World. Global Artificial Photosynthesis as the Moral Culmination of Nanotechnology*. Edward Elgar Publishing.

Faunce, T. A. (2012b). Future perspectives on solar fuels. In *Molecular Solar Fuels* (Book Series, *Energy*; Wydrzynski, T., and Hillier, W., ed.). Cambridge, United Kingdom: Royal Society of Chemistry, 506–528.

Faunce, T. A., and Bai, J. (2010). Biotechnology and nanotechnology regulation in Australia. In *Health Law in Australia* (White, B., McDonald, F., and Willmott, L., ed.). Sydney, Australia: Lawbook Co., 645–671.

Faunce, T. A., Lubitz, W., Rutherford, A. W., et al. (2013). Energy and environment case for a global project on artificial photosynthesis. *Energy Environ. Sci.*, **6**(3), 695–698. doi:10.1039/C3EE00063J

Faunce, T. A., and Wattal, A. (2010). Nanosilver and global public health: international regulatory issues. *Nanomedicine*, **5**(4), 617–632.

Faunce, T. A., White, J., and Matthaei, K. (2008). Integrated research into the nanoparticle proteincCorona. *Nanomedicine*, **3**(6), 859–865.

Festinger, L. (1957). *A Theory of Cognitive Dissonance*. United States: Stanford University Press.

Fiddes, N. (1991). *Meat: A Natural Symbol*. United States: Routledge Press, 5.

Food Labelling Law and Policy Review Panel (January 27, 2011). *Labelling Logic: Review of Food Labelling Law and Policy*. Commonwealth of Australia, paragraph 3.20, 47.

Hocquette, J. F., and Chatellier, V. (2011). Prospects for the European beef sector over the next 30 years. *Anim. Frontiers*, **20**.

House of Commons (2013). *Tax Avoidance: Google*. Committee of Public Accounts.

Huffman, M. (2010). Bridging the divide? Theories for integrating competition law and consumer protection. *Eur. Competition J.*, **6**(1), 7.

Hutt, W. (1936). *Economics and the Public: A Study of Competition and Opinion*. United Kingdom: Honathan Cape Publishers.

Ibrahim, D. (2007). A return to Descartes: property, profit and the corporate ownership of animals. *Law Contemp. Probl.*, **70**, 86.

Interlego AG v. Croner Trading Pty Ltd, 39 FCR 348, 387 (1992).

Johnston, C. (1996). Consumer welfare and competition policy. *Competition Consumer Law J.*, **3**, 1–3.

Kjaernes, U. (2011). Ethics and action: a relational perspective on consumer choice in the European politics of food. *J. Agric. Environ. Ethics*, **1**, 3.

Kobetsky, M. (2005). Tax treaty monitor: the aftermath of the Lamesa case; Australia's tax treaty override. *Bull. Int. Fiscal Documentation*, **59**(6), 236–248.

Korthals, M. (2001). Taking consumers seriously: two concepts of consumer sovereignty. *J. Agric. Environ. Ethics*, 201–203.

Kovacic, W. E. (2007). Competition policy, consumer protection and economic disadvantage. *J. Law Policy*, **25**, 101–114.

Lea, E., and Worsley, A. (2002). The cognitive context of beliefs about the healthiness of meat. *Public Health Nutr.*, **5**(1), 37.

Leslie, J., and Sunstein, C. (2007). Animal rights without controversy. *Law Contemp. Probl.*, **70**, 117.

Lewis, N. S., and Nocera, D. G. (2006). Powering the planet: chemical challenges in solar energy utilization. *Proc. Natl. Acad. Sci. U. S. A.*, **103**, 15729–15735.

Liberal Party of Australia (2013). *Our Plan to Abolish the Carbon Tax,* http://www.liberal.org.au/our-plan-abolish-carbon-tax.

Market Attitude Research Services (2011). *Australian Community Attitudes about Nanotechnology: Trends 2005 to 2011* (final report in August 2011).

Mayfield, L. E., Bennett, R. M., Tranter, R. B., and Woolbridge, M. J. (2007). Consumption of welfare-friendly food products in Great Britain, Italy and Sweden and how it may be influenced by consumer attitudes to and behaviour towards animal welfare attributes. *Int. J. Sociology Food Agric.*, **15**(3), 59.

McEarchen, M., and Schroder, M. (2002). The role of livestock production ethics in consumer values towards meat. *J. Agric. Environ. Ethics*, **15**, 221.

Naald, B., and Cameron, T. (2011). Willingness to pay for other species 'well-being.' *Ecol. Econ.*, **70**, 1325.

Nagarajan, V. (2007). Reconceiving regulation: finding a place for the consumer. *Competition Consumer Law J.*, **15**, 93.

Nocella, G., Hubbard, L., and Scarpa, R. (2010). Farm animal welfare, consumer willingness to pay and trust: results of a cross-national survey. *Appl. Econ. Perspect. Policy*, **32**(2), 275.

Nocera, D. G. (2012). Can we progress from solipsistic science to frugal innovation? *Daedalus*, **141**, 1–8.

Olivaylle Pty Ltd v. Flottweg AG, (No 4), 255 ALR 632, 644 (34) (2009).

Persky, J. (1993). Retrospectives: consumer sovereignty. *J. Econ. Perspect.*, **7**(1), 183 and 186.

Petersen, A., Seear, K., and Bowman, D. (2010). *Communicating with Citizens about Nanotechnologies: Views of Key Stakeholders in Australia* (final report). Monash University, 5.

Pickett, R., Norwood, F., and Lusk, J. (2010). Consumer preferences for farm animal welfare: results from a telephone survey of US households. *Anim. Welfare*, **19**, 335.

Plastec Australia Pty Ltd v. Plumbing Solutions and Services Pty Ltd, FCA 510, 2 (4) (2012).

Ranald, P., and Southalan, L. (2003). *The Australia-US Free Trade Agreement: Trading Australia Away?* Australia Fair Trade and Investment Network, www.aftinet.org.au.

Reich, N. (1992). Diverse approaches to consumer protection philosophy. *J. Consumer Policy*, **14**, 257.

Re QCMA (1976). ATPR 40-012

Royal Society (2014). *Do We Need a Global Project on Artificial Photosynthesis (Solar Fuels and Chemicals)?* Theo Murphy international scientific meeting organized at the Royal Society at Chicheley Hall, home of the Kavli Royal Society International Centre, Buckinghamshir, http://royalsociety.org/events/2014/artificial-photosynthesis-global-project/.

Scheer, H. (2001). *A Solar Manifesto* (2nd ed.). London, United Kingdom: James & James, 159.

Schroder, M., and McEarchen, M. (2004). Consumer value conflicts surrounding ethical food purchase decisions: a focus on animal welfare. *Int. J. Consumer Stud.*, **28**(2), 168.

Schwarzkopf, S. (2011). The consumer as 'voter', 'judge' and 'jury': historical origins and political consequences of a marketing myth. *J. Macromark.*, **31**(1), 8.

Schwarzkopf, S. (2011). The political theology of consumer sovereignty: towards an ontology of consumer society. *Theory, Cult. Soc.*, **28**(3), 106.

Silver, J. (2011). Understanding freedom of religion in a religious industry: kosher slaughter (shechita) and animal welfare. *Victoria Univ. Wellington Law Rev.*, **42**, 671.

Smith, A. (1937). *The Wealth of Nations* (original work published in 1776). New York: Modern Library, 625.

Sovacool, B. (2008). *The Dirty Energy Dilemma: What's Blocking Clean Power in the United States*. Praeger.

Steiner, G. (2005). *Anthropocentrism and Its Discontents: The Moral Status of Animals in the History of Western Philosophy*. US: The University of Pittsburgh Press, 6–7.

Swann, M., and McEachern, W. (2003). *Microeconomics: A Contemporary Introduction*, 2nd ed. Melbourne, Australia: Thomson/Nelson.

Sylvan, L. (2004). Activating competition: the consumer-competition interface. *Competition Consumer Law J.*, **12**, 191–192.

Tawse, J. (2010). Consumer attitudes towards farm animals and their welfare: a pig production case study. *Biosci. Horizons.*, **3**(2), 156–157.

Thornton, P. (2010). Livestock production: recent trends, future prospects. *Philos. Trans. R. Soc.*, 2854–2855.

Toma, L., Stott, A. W., Revoredo-Giha, C., and Kupiec-Teaha, B. (2012). Consumer and animal welfare: a comparison between European Union countries. *Appetite*, **58**, 597.

United Kingdom Office of Fair Trading (2009). *Joining Up Competition and Consumer Policy: The OFT's Approach to Building an Integrated Agency*.

United Nations Conference on Trade and Development (2003). *Dispute Settlement: Investor-State*. United Nations Publication, http://www.unctad.org/en/docs/iteiit30_en.pdf.

United Nations Educational Scientific and Cultural Organization (1997). *Declaration on the Responsibilities of the Present Generations Towards Future Generations*, http://portal.unesco.org/en/ev.php-URL_ID=13178&URL_DO=DO_TOPIC&URL_SECTION=201.html.

United Nations Educational Scientific and Cultural Organization (1972). *Convention Concerning the Protection of the World Cultural and Natural Heritage*, http://whc.unesco.org/en/conventiontext/.

United Nations (December 10, 1982). *Convention on the Law of the Sea*. Art. 1, para. 1, 1833 U.N.T.S. 397.

United Nations Department of Economic and Social Affairs (2012). *World Population Prospects: The 2010 Revision*, http://www.un.org/popin/.

United Nations Food and Agricultural Organisation (1983). *International Undertaking on Plant Genetic Resources*. Art 1. Res 8/83.

United Nations (December 17, 1979). *Agreement Governing Activities of States on the Moon and Other Celestial Bodies*. Art. 1, 18 I.L.M. 1434.

United Nations Framework Convention on Climate Change (UNFCCC) (7–18 December 2009). Draft decision -/CP.15 CONFERENCE OF THE PARTIES. Fifteenth Session, Copenhagen, FCCC/CP/2009/L.7 18 December 2009.

United Nations (1966). *International Covenant on Economic, Social and Cultural Rights.* Adopted and opened for signature, ratification and accession by General Assembly resolution 2200A (XXI) of 16 December 1966 entry into force 3 January 1976, in accordance with Article 27, http://www.ohchr.org/EN/ProfessionalInterest/Pages/CESCR.aspx.

United Nations (December 6, 1988). *Legal Status of the Atmosphere.* Para 1. UN Res 43/53.

United Nations (2012). *Millennium Development Goals,* http://www.un.org/millenniumgoals/

United Nations (January 27, 1967). *Treaty on Principles Governing the Activities of States in the Exploration and Use of Outer Space, Including the Moon and Other Celestial Bodies.* Art. 1, 18 U.S.T. 2410, 610 U.N.T.S. 205.

United States Patent and Trademark Office (2001). Utility examination guidelines. *Fed. Regist.,* **66**(4), http://www.ustpo.gov/web/offices/com/sol/notices/utilexmguide.pdf

United States Department of Energy and Council for Automotive Energies (2011). *Hydrogen Production Roadmap: Technology Pathways to the Future. Freedom CAR & Fuel Partnership Hydrogen Production Technical Team*

Voiceless: The Animal Protection Institute (November 2008). *From Nest to Nugget: An Expose of Australia's Chicken Factories.*

White, S. (2007). Regulation of animal welfare in Australia and the emergent Commonwealth: entrenching the traditional approach of the states and territories or laying the ground for reform? *Fed. Law Rev.,* **35**(3), 347.

Wilson Institute (2011). *The Project on Emerging Nanotechnologies,* http://www.nanotechproject.org/inventories/consumer/analysis_draft/.

Williams, P. (1989). Why regulate for competition? In *Regulating for Competition: Trade Practices Policy in a Changing Economy* (James, M., ed.). Sidney, Australia: Centre for Independent Studies, 13.

Williamson, A. (2010). Using mixed methods to discover emergent patterns of local eDemocracy. *AI Soc.,* **25**, 321–333.

Chapter 13

Toward the Sustainocene with Global Artificial Photosynthesis

Thomas A. Faunce,[a,b] **Alex Bruce,**[c] **and Angus M. Donohoo**[a]
[a]*ANU College of Law, Building 6, North Wing, Australian National University, Canberra, ACT 0200, Australia*
[b]*ANU College of Medicine, Biology and the Environment, Building 131, John Curtin School of Medical Research, Australian National University, Canberra, ACT 0200, Australia*
[c]*ANU College of Law, Building 5, Fellows Road, Australian National University, Canberra, ACT 0200, Australia*
thomas.faunce@anu.edu.au

The rapid global deployment of artificial photosynthesis (AP) (as a zero-carbon source of hydrogen fuel, ammonia, and starch) has been proposed as a moral culmination of nanotechnology in this era of anthropogenic (or more accurately corporatogenic) climate change, loss of biodiversity, environmental degradation, and energy and food insecurity (GAP, 2011; Faunce 2012a). Successful AP technology is generally envisioned in this context as a nanotechnology device that derives energy and food through not only replicating but also improving upon the photosynthetic process by which plants store and produce energy. The fundamental hypothesis this chapter will

Nanotechnology Toward the Sustainocene
Edited by Thomas A. Faunce

ISBN 978-981-4463-76-8 (Hardcover), 978-981-4463-77-5 (eBook)
www.panstanford.com

test is this: If a technology such as AP can decentralize global energy production and extinguish reliance on fossil fuels within a realistic time frame as well as provide local families and communities with cheap fertilizer and starch, then a long-term period of sustainable human flourishing in harmony with nature (the Sustainocene) can develop from that foundation. This hypothesis requires the testing of four core assumptions:

- The likelihood of a viable, high-quality, easily deployable energy generation and storage technology that does not rely on the combustion of fossil fuels or create serious long-term health and environmental consequences
- The rapid, global adoption of such a technology
- The adoption of such a technology having the effect of reversing or mitigating climate change and environmental degradation and spurring a long-term period of stable, sustainable human flourishing in harmony with nature
- The transition from global agribusiness characterised by corporate-centric neoliberal market structures to localised community-oriented enterprises. This will become particularly important as a means of curbing the use by large corporations of their market power to hinder or destroy community-based AP initiatives.

This chapter will begin with a glimpse of the social framework and governance structures of the Sustainocene. It will consider why globalizing AP is a reasonable approach to get us there. This will be followed by an overview of AP and of the scientific and technical features that make this technology so enticing as the basic energy and food source for the Sustainocene. It will then analyze the likely obstacles to rapid deployment of such technology, principally, significant corporate investment in maintaining the status quo reliance on fossil fuels as humanity's central fuel source. Corporate resistance to the introduction of new technologies that threaten not only established economies of scale, but also future lucrative income streams dependent on archived fossil fuel sources will be examined by studying the inhibited progress of a categorically similar innovation—the electric car. It will then consider a development of a macroscience global artificial photosynthesis (GAP) project as a means overcoming some of such problems and how this can be encouraged by the general public and policymakers.

13.1 Artificial Photosynthesis Toward a Sustainocene

13.1.1 Can Humans and the Earth Flourish Forever?

13.1.1.1 A path to the Sustainocene

Contemporary physics indicates that precisely predicting the future is impossible, since below the nanoscale nature is not deterministic, and is non-causal, and thus constant minute variations in initial conditions create significant changes that make the behavior of even fully deterministic systems unpredictable over time (Poincaré, 1908, 28; See generally Chapter 4; Heisenberg, 1935, 7–8, 17–19; Bell, 1964, 403, 407). Yet, the capacity to plan to achieve idealistic visions and anticipate probable and possible outcomes is one of the intellectual and cultural traits that has provided *Homo sapiens* with an evolutionary edge and is a distinguishing characteristic of high-achieving individuals (Osvath and Osvath, 2008; Faunce, 2012b).

This book is intended to present a plausible vision of how nanotechnology may assist in delivering a better future for humans and their environment. Developing such a vision is critically important because humanity is rapidly approaching tipping points for the safe, long-term occupation of the earth, arising from overpopulation, pollution, environmental destruction, and energy and food insecurity (Schellenhuber, 2009). Apocalyptic conclusions extrapolated from such problems become more reasonable when considering the precarious toehold of life in a geological timescale. The earth has experienced at least six major extinction events in its history, during which anywhere between 95% and 75% of all species have been wiped out (Newitz, 2013). Any plausible vision of a sustainable human future on this planet must strive toward species existence in harmony with its natural and manufactured environment on a million year timescale.

The Sustainocene is one attempt to embody such a vision (Faunce, 2012b). It is a period where new information nanotechnologies allow citizens to have a more direct say in governance, not solely through the easily corruptible mechanism of electing a political representative. It is a period where registered species and ecosystems can enforce rights through human guardians. It is a period no longer characterized by corporate globalization as a dominant economic and social model. The transition to a Sustainocene will particularly necessitate a re-evaluation of the neoliberal principles of State–market relationships so that, for example, corporate agribusinesses are not in a position to misuse their established market power to eliminate or damage emerging businesses built upon AP technology. Through these emerging community-based agribusinesses, each family and community can gain adequate food, energy, and water through the built structures of their immediate environment. It is a period when people readily strive for virtue because the connection of conscience, of established virtue, with an ecocentric consciousness is a dominant component of education systems. It is a period of human stewardship over the earth designed to approximate the legacy (and so repay the ethical debt) that the evolution of life has given humanity. It is an era when nanotechnology has provided the technological capacity to indulge in such implicit idealism and expand humanity's circle of sympathy and compassion.

In some ways it is possible to see this transition as a shift from an emphasis of human to planetary medicine, moving from the "diagnostic" phase (such as that characterizing human medicine in the nineteenth century) toward the development and deployment of a planetary therapeutic (Faunce, 2012b). This transition is also consistent with a move away from the self-centred and exploitative corporate imperatives that characterise neoliberal market structures, to a model of trade and exchange that recognises the explicit interdependence of markets, businesses, consumers and the environment (Sullivan 2012–2013).

In this book a quiver of different nanotechnological approaches have been proposed for targeting many of humanity's fundamental problems. A spectrum of other technologies undoubtedly will play their part in such a transition (National Academy of Sciences [NAS], 2009). We now turn to what we hypothesize is the most likely candidate for a planetary therapeutic and the moral culmination of nanotechnology: global artificial photosynthesis (GAP).

13.1.2 A Viable Technology: Can Artificial Photosynthesis take us Forward?

Sol rains upon us into more convenient forms (chemical bonds) [that] will power a new economy of abundance. Evolution in its patient blindness managed to develop photosynthesis; with mindful insight, we will do better (Wilczek, 2009).

As Nobel laureate Frank Wilczek observes, the sun is a prodigious source of energy. Indeed an hour's worth of solar output strikes the earth with more energy than an entire year's worth of human consumption (Nocera, 2006). It seems obvious that once properly tapped, solar power will indeed drive new economies that are characterized not just by abundance but by abundance that is as equally distributed across the face of the earth as natural photosynthesis is now.

Organisms began photosynthesizing approximately 3.4 billion years ago and commenced adding appreciable levels of oxygen to the atmosphere via this process 3 billion years ago (Crowe et al., 2013). The basic process involves converting light energy into chemical energy by splitting the molecular bonds in water and producing carbohydrates through the synthesis of water and atmospheric carbon dioxide. However, the process itself is far from basic in terms of chemical science, and improving upon it in a manner congruent with equitable global deployment presents a major technical challenge.

The photosynthetic process as it has evolved in plants and some other organisms (such as cyanobacteria) is—in the terms of the energy-hungry demands of humanity—relatively inefficient. Although the quantum efficiency in plants (the number of photons transferred to the reaction centre) is nearly 100% (Lambert, Chen, Cheng et al., 2012), if calculated in terms of the energy content of the biomass that can be harvested, the efficiency of crop plants typically does not exceed 1%. Blankenship et al. (2011) placed a limit on AP efficiencies of ~24% for a photoelectrolysis cell (that produces hydrogen fuel from water—discussed below). However, this was subsequently criticized on the basis that their calculation incurred many losses related to metabolic processes that can be avoided in synthetic systems, such as the dark reactions described later (Hammarström, Winkler, Gray et al., 2011).

Accurately calculating efficiencies with these technologies and processes is complex and difficult. The *conversion* efficiency (the ratio between the useful output and the input of energy) for example ignores the integrated social, economic, and environmental costs of the system (such as the costs of transport, waste disposal, land, water, capital, climate change, etc.) (Blankenship et al., 2011).

It is also far from clear whether efficiency should be the primary concern when determining which technologies humanity should rely on. For instance, is a clean, renewable technology with 5% efficiency more valuable than a dirty technology with 80% efficiency, particularly if the 5% technology can provide *sufficient* power? Such considerations ought to be essential and indeed, if we are to move toward a Sustainocene, *must* be essential.

Stephens, Ross, Mussgnug et al. (2010b) note that if a technology were 100% efficient (mass conversion efficiency) and could capture average incident solar energy at 170 W/m^2, a mere 0.017% of the surface area of the earth could be used to supply global energy demands. Taking into account more reasonable parameters, they calculated that with 3% efficiency, and an incident power of 350—650 W/m^2 (average irradiation in temperate and tropical regions, respectively), a microalgae biotechnology could provide current human energy consumption by taking up a mere ~1.5%–2.7% of nonarable land (Stephens, Ross, Mussgnug et al., 2010b). In the face of the rapidly increasing global population, and the potential diminution in arable land due to climate change, clean distributed fuel and food technologies driven by sunlight, water, and atmospheric carbon dioxide (plus atmospheric nitrogen) provide perhaps the most plausible path toward a Sustainocene.

13.1.3 Photosynthesis: The Technical Challenge

Photosynthesis can be divided into four distinct phases of energy storage and conversion, each of which would need to be mastered to create a "complete" AP technology. However, viable technologies could be developed that would incorporate only some of these phases or that would use distinct technical components for each. As described by Robert Blankenship the energy storage phases of photosynthesis are (1) light absorption and energy delivery by antenna systems, (2) primary electron transfer in reaction centers, (3) energy stabilization by secondary processes, and (4) the synthesis and export of stable products. These first three are sometimes referred to as light reactions, while the fourth is often referred to as dark reactions (because they are light independent) (Blankenship, 2013).

Mastery of all of these phases requires major innovation coupled with breakthroughs in light harvesting, charge separation, catalysis, semiconductors, nanotechnology, modeling from synthetic biology and genetic engineering, photochemistry and photophysics, photoelectrochemistry, catalysis, reaction mechanisms, and device engineering (Faunce, Styring, Wasielewski et al., 2013). At present, to make a viable AP technology suitable for the marketplace, researcher Daniel Gamelin has described three essential prongs that it must deliver. AP must be (1) efficient, (2) cheap, and (3) robust. Although any two of these criteria is currently available, a technology incorporating all three at once has proved elusive (Jones, 2012).

In human efforts to replicate this system, there have been at least two major directions of development, and the majority of AP systems can be roughly divided into two categories, photosynthetic organism and photocatalytic organism systems. Photosynthetic organism systems utilize forms of natural and synthetic algae and bacteria to produce biofuels like ethanol via "natural" photosynthetic means. Photocatalytic systems generally use synthetic nanoscale antennae for capturing photons (light particles), before using that photon energy in catalytic chemical reactions that drive the splitting of water and the generation of hydrogen fuel, as well as (potentially) carbon dioxide reduction, which converts and stores the greenhouse gas, usually within carbohydrates or metal hydrides. This chapter will avoid going into depth about the extraordinary variety of research that is being done in this field, yet in addition to these broad categories there are a variety of innovative technologies being developed, including, for example, solar thermochemical reactors (heat engines driven by sunlight [Ermanoski, Siegal, and Stechel, 2013]) and powdered photocatalysts (powders that can be added to water to split it and produce hydrogen directly [Kudo and Miseki, 2009]). For a good overview of the technology generally see Barber (2009).

As Wilczek asserted, there are good reasons to be optimistic and to expect that human ingenuity should be up to the task of creating a viable technology with efficiencies considerably greater than that accomplished by nature. This optimism cuts straight to one of the assumptions that we raised in our initial hypothesis: the future globally distributed deployment of a technology that could support a Sustainocene. There seems to be sufficient evidence to suggest the existence of such a renewable energy technology is indeed plausible and that there may be beneficial consequences if policymakers begin to embrace a vision of it powering the earth.

13.1.4 Hard Realities: Impediments to Delivering a Viable Technology

The bulk of present global energy needs are met through what may be termed "archived photosynthesis" sources, or "fossil fuels," particularly oil, coal, and natural gas. These fuel sources are the outcomes of an ancient biological process that stored solar energy in chemical bonds. They are accessed via centralized global distribution networks that require decades-long turnaround and intense levels of private investment for the controlling multinational corporations, leading to yearly profits in excess of those of the income of many nations. The concern of such corporations to prolong this revenue stream is directly contributing to critical problems such as greenhouse gas emissions, ocean acidification, and geopolitical instability, not to mention significant localized environmental degradation and destruction at the points of extraction, the processing locations, and even potentially during transport (Faunce et al., 2013; Stern, 2007). These corporate concerns are simply a reflection of the governance structures that have developed in the roll-out of neoliberal policies implemented by States since the Second World War (Peck, 2001). As States have progressively withdrawn from direct governance agendas, responsibility has shifted to privatized forms of industry governance, particularly in the agribusiness and supermarket sectors (Lockie and Higgins, 2007).

Where democratically elected governments were once responsible for ensuring corporate compliance with safety and consumer standards, neoliberalism has resulted in the corporations themselves assuming responsibility for their own market compliance (Busch, 2011). The shift from state regulation of market behaviour to self-regulation has resulted in a proliferation of private forms of market-based regulatory mechanisms that are administered by the corporations themselves (Richards et al, 2013).

Most evident in the supermarket and agricultural industries, this transition has resulted in small family-based suppliers being forced to comply with private regulatory standards imposed by monopsonist buyers; standards that frequently shift compliance costs onto the suppliers or the consumers.

In any roll-out of AP technology and the distribution of AP outputs (fuel, starch), particular attention needs to be paid to the role and governance of corporations as either providers of AP technology or as large buyers of AP technology or outputs. Blindly implementing neoliberal policies of self-regulation may result in large corporations imposing private forms of regulation that have the effect of excluding the family and community oriented producers of AP outputs from future markets (Richards et al, 2013).

The political infiltration of such corporations is powerful, and has been evident in their efforts in undermining other revolutionary technologies like the electric car (discussed below). As the fossil-fuel industry is a trillion dollar enterprise, they are well resourced to maintain the status quo. These corporations recognise that radically changing the nature of global energy generation would involve a massive undercutting of their profits. This would have deleterious impacts on the global economy as it is currently styled: a complex system run by wealthy corporations that are legally required to subvert components of the social contract for the maximisation of shareholder profit, and which are abstracted as entities from most social, personal or environmental connections (Faunce, 2012a).

Economic cost has long presented a major governance challenge to suitable investment in alternative power sources to fossil fuels. A focus on the cost per unit of fuel has been the dominant factor underwriting policy decisions. Until pollution is properly priced, truly innovative technologies will struggle to find a purchase in the market place (Stephens, Ross, King et al., 2010a).

The current regime of intellectual property also presents a major governance obstruction in the widespread adoption of novel technologies (Boldrin and Levine, 2010). The systems of information disclosure and the monopoly rights embedded in the patent regime are likely to have a chilling effect on any serious efforts to bring artificial photosynthesis (AP) to the marketplace, given their effect on other novel technologies (Rees and Opeskin, 2011). A technology as complex as AP will involve a multitude of innovations that will require the synthesis of many different pieces of intellectual property.

The way that research is undertaken may also be an impediment. In most developed nations the funding model for academic research has been shifting toward exploiting monopoly rights for profit, even when the research is initially undertaken via competitive grants schemes funded by taxpayers (Australian Research Council [ARC], 2001). This "privatized," corporate-driven research structure could act as a significant barrier to developing complex, new distributed energy and food technologies likely to support the Sustainocene (such as AP). Significant work on developing AP's governance framework will be required to foster a functional regime for effective, prompt sharing and expansion of the relevant critical knowledge. In the recent past, inadequate governance controls over competitor technologies have hindered the development of similar innovations such as the electric car. We now propose to interrogate that case study.

In the case of the electric car, California introduced legislation in 1990 that contained provisions forcing automotive manufacturers that sold more than 35,000 vehicles to ration automobile sales, depending upon the vehicle's emission standards. It contained mandatory credit-linked quotas for sale percentages of cars with different classes of emissions and tied credits to the sale of low-emission and zero-emission automobiles. The credit scheme was designed to provide a protected market for the manufacturers that would encourage radical innovation and enable them to nurture new technologies. Manufacturers responded to the regulations in two different ways—either by "systemic substitution," involving incremental improvements of existing technology like hybrid vehicles, or through "disruptive innovation," with major research and development (R&D) investment into fully electric vehicles (Pilkington and Dyerson, 2006). Clearly AP falls predominantly into the latter category.

Yet such rapid disruptive innovation proved to be a bridge too far for the major petrol car manufacturers, who, in concert with major oil corporations, applied serious political pressure that resulted in the detrimental weakening of the regulatory incentives for electric cars over several years. The manufacturers of truly market disruptive fully electric vehicles were forced to seek technology suppliers beyond their normal networks, which produced a high risk of new competitors entering into their commercial field. Such regulatory strategies ended up putting the manufacturers onto a defensive footing rather than encouraging them to embrace new opportunities. According to Pilkington and Dyerson, "Complex system issues rather than technological barriers . . . prevented the widespread adoption of alternative fuels" (Pilkington and Dyerson, 2006).

Multiple lessons arise from this case study with respect to the prospects of a disruptive technology like AP being used to drive a technological revolution toward a Sustainocene. A primary lesson is that legislation can be effectively used to incentivize and force major technological innovation. A system of mandatory investment in "clean" technologies from "dirty" industries that are of a sufficient size could provide a major source of investment in disruptive clean energy technology. However, another primary lesson from this example is that entrenched power threatened by legislation will fight change tooth and nail. Although the legislation appeared to initially function well, subsequent governments were too susceptible to powerful interest groups, which were highly effective in dismantling the progress that had been made.

This poses a serious problem for a GAP project since AP is a technology geared to providing humanity a future on geological timescales, not merely on the decades of current life expectancy or on the months of the next electoral cycle. The fossil fuel interests that successfully fought against zero-emission vehicles are the very same players that are most threatened by AP technology. A secondary lesson that can be taken from the electric car example is that companies will be very reluctant to outsource sections of their technology and therein weaken their monopoly positions. This is an interesting point to keep in mind as we consider a GAP project.

13.2 A Global Artificial Photosynthesis Project Toward a Sustainocene

13.2.1 A GAP Project

A GAP project is envisioned as a macroscience project on an equivalent scale to the Human Genome Project (Faunce, 2012a; Faunce, T.A., Lubitz, W., Rutherford, A.W., et al., 2013). As discussed earlier, AP is an extraordinary technical challenge, requiring major scientific innovation in a variety of fields and disciplines. As has been stressed, however, the prize is enormous, and the prospect of clean, renewable energy that subtracts carbon dioxide from the atmosphere is exceptionally alluring. For this sort of technology to develop, there needs to be new and coherent governance structures that enhance and facilitate cooperation, as well as novel legal avenues that direct wealth and incentive away from resource destruction and toward sustainable outcomes. A GAP project needs to be a global collaboration on an enormous scale, involving scientists, industry, policymakers, and a public that is willing to adopt the technological shift.

In light of the political and industrial difficulties already discussed, it may well be the public reaction that proves crucial in decentralizing global energy markets through novel technologies. It will require considerable public enthusiasm to move humanity away from the centralized extraction, processing, and combustion of finite resources into localized renewable energy production. A driving vision behind GAP sees a Sustainocene powered by AP technologies integrated into vast numbers of human structures, such as roads, paths, buildings, and roofs, therein localizing energy production into any sufficient human structures exposed to sunlight, water, atmospheric nitrogen and carbon dioxide. These integrated AP technologies would capture solar energy and either convert it into fertilizers or, through the use of nanotechnology fuel cells with high surface areas, convert it to hydrogen fuel (Faunce, 2012a). This sort of development may perhaps occur only in the face of considerable public enthusiasm toward the adoption of this sort of technology.

The complexity of perfecting artificial of photosynthesis demands an interdisciplinary response. For a GAP project to be effective, it is envisioned that it will require cooperation in an array of commercial, technical, and scientific fields. However, other macroscience projects have blazed a similar trail and provide signposts and precedents for cooperative efforts on an equivalent scale (Faunce, 2012a; Faunce, T.A., Lubitz, W., Rutherford, A.W., et al., 2013). Examples include the Human Genome Project, the Large Hadron Collider, and the International Thermonuclear Experimental Reactor. More industrially focused corollaries can be found in projects like the Boeing 747—an aerospace project that brought together a vast swathe of disparate companies and industries. In projects of this nature, it is not merely the technical challenges that are confronting, and standard engineering becomes a form of untidy "social organization" (Arthur, 2009).

Globally, investment in AP technologies is considerable, though still miniscule compared to the investment in other technologies such as fossil fuels. At this stage the largest single investment is US$122 million over five years through the Joint Center for Artificial Photosynthesis, run by the US Department of Energy and based at the California Institute of Technology and the Lawrence Berkeley National Laboratory. Globally, all developed nations have "intense and advanced" research projects into AP (Faunce, 2011). The Solar Fuels Institute (SFI) has formed links between some large projects involved in this part of AP in the United States and in Europe and seeks to leverage private capital. It is anticipating AP-style solar fuel technologies to enter the marketplace by the 2020s (SFI, 2013). A GAP project would work toward getting a product that is efficient, cheap, and robust into the marketplace as soon as possible.

The chief commissioner of the now defunct Australian Climate Commission, Tim Flannery, recently suggested that "there will be no decisive moment in the climate battle. Rather, it's a war of attrition, being won one solar panel, one wind turbine, and one electric car at a time" (Flannery, 2013). But what of historical technology transition? Does this hypothesis need to be accepted, or can a technology like AP so rapidly alter the environment and so severely undercut the profits of major fossil fuel companies that it could seriously erode their business model within a short space of time?

Radical technological changes can dramatically alter the balance of human affairs. Mechanized agriculture, automotive transport, and telecommunications are just a few examples of truly transformative technologies, and although these transformations took place over the span of centuries and decades, the exponential curve of technological progress (as characterized, for example, in Moore's law) suggests that the time span for similar breakthroughs in the future need not be so long. As this book has made clear, humanity appears to be on the brink of—or is indeed well beyond the threshold of—a nanotechnology revolution. It seems plausible that new energy technologies like AP, driven by nanotechnological breakthroughs, could play a strong part in the next major technical revolution. Visions of a nanotechnology future take a variety of forms—some bright and positive, others dark and pessimistic (Reynolds, 2004). The Sustainocene shines as a brilliantly optimistic aspiration, offering an image of a future in which humanity acts as a steward rather than a plunderer and as a creator rather than a destroyer.

13.2.2 Mere Utopia: Is a GAP-Driven Sustainocene an Impossible Dream?

Utopian speculations . . . must come back into fashion. They are a way of affirming faith in the possibility of solving problems that seem at the moment insoluble. Today even the survival of humanity is a utopian hope (Brown, 1985).

A Sustainocene is, in many ways, a utopic vision of the future, for it is founded on an idea of a peaceful, reciprocal relationship with nature—on a symbiotic coexistence rather than one that is more akin to a parasite feeding on its host. A Sustainocene can only be realistically conceived of as emerging from a peaceful state of coexistence in which human conflict has diminished. It is sometimes observed in popular discourse (though is long since disputed; see, for example, Smith, 1776) that capitalism performs as a zero-sum game in which losers and winners must always be created by the nature of the system (*Wall Street*, 1987). While conceptually, sustainability does not preclude the possibility of competition across any number of spheres of human activity, it does suggest a need for there to be a greater even-handedness in the distribution of access to energy and energy generation technology than currently exists (Lawrence, Liu, Yakovenko, 2013). It is evident that such even-handedness flies in the face of these conventional, inequitable economic realities.

However, reassuringly, conventions can change. Much of the allure of magic-bullet technologies lies in their capacity to not so much eradicate economic realities but re-envision them. A reasonable vision of energy generation that does not step too far from the present reality might consider a world where clean power and electricity are as free and easy to come by as safe drinking water is in the vast majority of the developed world (World Bank, 2013). AP goes even further than this and not only offers the possibility of environmentally sustainable, cheap, and readily available fuel but also has the capacity to produce clean fertilizers as a by-product and remove climate-affecting greenhouse gases from the air (Faunce, Styring and Wasielewski et al., 2013).

A vision of a Sustainocene may well be a utopic vision, but this does not detract from its underlying seriousness, and although "utopia" translates as "no place," the efforts of Thomas More (or more likely his friend Erasmus) in creating the original work *Utopia* in 1516 were not mere jest. Whoever was its author, *Utopia* was a vision not of heaven or of a flawless fantasy of a world where every whim was catered for. It was rather a conception of an ideally functioning state made up of an ideally principled and regulated populace. While there were certainly elements of jest and parody, there is still a rigid iron purpose underlying much of the work. It is a conceivable dream of how the world ought to be. So too is the Sustainocene. In the same way *Utopia*'s author envisioned a better life evolving from the local application of various foreign Utopian laws and principles, this chapter envisions a better world evolving from a successful GAP project (More, 1516; Turner, 2003).

13.2.3 How a Sustainocene May Evolve from New Ways of Thinking

A shift to the common adoption of conscientious thinking could be interwoven with the ascent of revolutionary new technologies and ideas like AP. In the context of the Sustainocene, implied in this aspirational policy goal is a movement away from resource exploitation by state and corporate multinationals and from the conflict that resource competition engenders between communities and nation states. For a Sustainocene to exist, mineral resource extraction must transition toward major recycling programs. Readily available energy from technologies such as AP, coupled with responsible recycling policies, may be instrumental in facilitating a transition of this nature.

Some schools of thought point to resource control as being the key incentive for the majority of conflict for centuries (Lenin, 1916; Bakeless, 1921; Klare, 2001; Acemoğlu et al., 2012). The current market-oriented concept of resource control through intensification of corporate power only serves to exacerbate resource scarcity through market manipulation, monopoly power, misuse of buyer power and information asymmetries (Faunce and Bruce, 2014). The unthinking roll out of neoliberal governance policies, characterised by private forms of market compliance compound resource scarcity by acting as potential exclusionary tactics, foreclosing smaller community-oriented producers from the market. The market of the Sustainocene must necessarily evolve away from neoliberal governance strategies and toward more inclusive economic pathways to the roll out of AP technology and its products.

It seems evident that the Sustainocene can only truly arise in a context where war over resources has diminished to the point of irrelevance—indeed this could perhaps be the point at which such an epoch could be identified as arising, that is, at the time at which sustainability supplants conflict. A peaceful epoch of human coexistence is more easily envisioned in a world of postscarcity, where competition for energy and resources is no longer a dominant driver of human conflict, which tends to fracture along tribal, religious, cultural, corporate profit, or jingoistic lines. While there can be little doubt that resource competition would continue irrespective of abundant renewable energy, it is logical that conflict generally would decline if energy generation was removed as a potential flashpoint.

If it is to truly be an epoch on a geological scale, the Sustainocene cannot merely embrace the wealthiest elite of humanity but must be a species-wide enablement of access to clean, renewable, sustainable technologies. Thus the relevant issues must be global issues, with agreement and regulation that cross the borders of nation states and that involve all people irrespective of their ethnicity, culture, religion, or employment status. By providing clean energy and reducing the effect of greenhouse gases, AP could play a major role in forging a technological epoch where environmental degradation diminishes purely as a function of a lack of need for exploitation. In that context, the Sustainocene could be perceived as an emergent property of more evolved human interactions.

With regard to this sort of long-term sustainability, there is little incentive within the current system of corporate production for businesses to utilize any type of sustainable development or to use novel technologies toward this agenda. The vast majority of companies will not put themselves at a short-term disadvantage by developing or using technologies like AP in a way that provides the greatest advantage to future sustainability. This is particularly so if companies can shift the cost of production onto either suppliers or consumers. By uncritically adopting neoliberal governance policies, companies can potentially exclude smaller entities that offer competing outputs based on AP technology by imposing privatised compliance or certification schemes. Rehabilitating corporate behavior in the Sustainocene will require significant re-evaluation of present theories of market regulation and competition policies. Unfortunately, as discussed before, in the current system of political power, truly effective movement toward sustainability may struggle to be driven by regulation.

While we have discussed benevolent visions of AP, another more unfortunate future can be anticipated where a complex technology like AP ends up polluting the environment of the poor and enriching the wealthy, as has happened with many modern high technologies in the Third World (Nnorom and Osibanjo, 2008; Puckett, Byster, Westervelt et al., 2003). A vision such as this evidently runs counter to the Sustainocene ideal and exposes much of the value in keeping such a grand vision in the picture. Embodied in the notion of the Sustainocene is the idea of an equitable distribution of technology and the recycling and reusing of goods—not merely putting them out of sight and out of mind. While the authors are optimistic that AP will develop along sustainable lines, the potential for AP to become a driver of conflict and resource depletion should not be overlooked and may be dependent on the materials that are used to develop and construct the technology.

The fundamental economic structure of supply and demand creates a considerable problem for the role of any technology in its part in the realization of a Sustainocene. All technologies will require materials for their construction, and while abundances of different resources vary, it is not clear at this stage whether certain AP technologies—particularly photocatalytic nanotechnologies—can be constructed to the requisite standard (as demanded in the first proposition of the hypothesis of this chapter) using only affordable, earth-abundant materials (Hurst, 2010). This is one of the reasons that the Sustainocene ideal is so crucial as a guiding aspiration in a GAP project. AP needs to not only be a fossil-fuel breaking technology, it also needs to be sustainable itself.

At its heart, the Sustainocene can be seen as a response to the great public health and environmental challenges of our time. In the face of major ecological destruction, the Sustainocene arises as a logical response to the essential need for survival. Thomas Hobbes postulated that in a governed society people surrender their rights to the Leviathan (the commonwealth or monarch) to ensure their protection from death at the hands of other people (Hobbes, 1651). Human beings in the Sustainocene are likely to flourish within a governance framework that recognizes that the survival of humanity is contingent upon surrendering certain rights to the environment. This is already being achieved in certain legal systems that are instilling components of the ecosystem with legal personality, as has happened now in several countries. A river in New Zealand and even "nature" itself in Ecuador have been granted legal privileges similar to those that corporations enjoy with their status as "artificial" persons (Vines, Bruce, and Faunce, 2013). In such initiatives we can see the light of the Sustainocene beginning to beckon us from the darkness of a world turned over to governance by those artificial people subverting every better human impulse to the maximization of profit.

The long-term survival of humanity can be seen as emerging from a balanced relationship that grants the environment and ecosystems rights that ensure their protection in perpetuity. In this light the security of the planet as well as that of humanity is contingent upon human endeavor pushing us toward the Sustainocene. Our capacity to imagine a world sustainably dedicated to realizing the unity of consciousness operates, like the foundational concepts of space and time, as a precondition to our mental experience of existence. Given the transcendental importance of this synthetic approach to knowledge, perhaps we should refer to a Global Synthetic Photosynthesis Project. Such usage, highlights that nanotechnology's destiny is not merely to be found in the development of synthetic photosynthesis but in the wisdom-rich world that will flow from its global deployment. An important governance step toward this would be a UNESCO Declaration declaring photosynthesis as common heritage under international law not just of humanity but of the earth.

References

Acemoğlu, D., Golosov, M., Tsyvinski, A., et al. (2012). A dynamic theory of resource wars. *Q. J. Econ.*, **127**(1), 283–331.

Arthur, W. (2009). *The Nature of Technology: What It Is and How It Evolves.* (New York: Simon and Schuster), 94.

Australian Research Council (2001). *National Principles of Intellectual Property Management for Publicly Funded Research.* Canberra, 6.

Bain & Company, Inc., (2011). The Great Eight: Trillion Dollar Growth Trends to 2020. Available at: http://www.bain.com/Images/BAIN_BRIEF_8MacroTrends.pdf (last accessed 13 Feb 2014).

Barber, J. (2009). Photosynthetic energy conversion: natural and artificial. *Chem. Soc. Rev.*, **38**, 185–196.

Bakeless, J. E. (1921). *The Economic Causes of Modern War: A Study of the Period: 1878–1918.* New York: Mofatt, Yard.

Bell, J. (1964). On the Einstein Podolsky Rosen paradox. *Physics*, **1**(3), 195–200.

Blankenship, R. (2013). *Molecular Mechanisms of Photosynthesis.* Oxford: Wiley.

Blankenship, R., Tiede, D., Barber, J., et al. (2011). Comparing photosynthetic and photovoltaic efficiencies and recognizing the potential for improvement. *Science*, **332**(13 May 2011), 805.

Boldrin, M., and Levine, D. (2010). *Against Intellectual Monopoly.* Cambridge: Cambridge University Press.

Brown, N. (1985). *Life against Death: The Psychoanalytical Meaning of History.* Middletown CT: Wesleyan University Press.

Busch, L. (2011) The private governance of food: equitable exchange or bizarre bazaar? *Agricultural Human Values*, 345.

Crowe et al. (2013). Atmospheric oxygenation three billion years ago. *Nature*, **501**, 535.

Ermanoski, I., Siegal, N., and Stechel, E. (2013). A new reactor concept for efficient solar-thermochemical fuel production. *J. Sol. Energy*, **135**(3), 031002.

Faunce, T. A. (2011). Will international trade law promote or inhibit global artificial photosynthesis? *Asian J. WTO Int. Health Law Policy*, **6**, 313–347,

Faunce, T. A. (2012a). *Nanotechnology for a Sustainable World.* Cheltenham: Edward Elgar.

Faunce T. A. (2012b). Towards a global solar fuels project: artificial photosynthesis and the transition from Anthropocene to Sustainocene. *Procedia Eng.*, **49**(2012), 348–356.

Faunce, T. A., and Bruce, A. (2014) Governing the global food system towards the sustainocene with artificial photosynthesis, in Steier G and Patel K (eds), *International Food Law and Policy*. Springer (forthcoming).

Faunce, T. A., Lubitz, W., Rutherford, A. W., et al. (2013) Energy and environment policy case for a global project on artificial photosynthesis. *Energy Environ. Sci.*, **6**(3) 695–698.

Faunce, T. A., Styring, S., Wasielewski, M., et al. (2013). Artificial photosynthesis as a frontier technology for energy sustainability. *Energy Environ. Sci.*, **6**, 1074–1076.

Flannery, T. (2013). Cleaner energy: a revolution that ordinary people can drive. *The Guardian*, 22 November 2013.

GAP (2011). Towards global artificial photosynthesis (GAP): energy, nanochemistry and governance (conference proceedings), http://law.anu.edu.au/coast/tgap/conf.htm (last accessed 28 October 2013).

Hammarström, L., Winkler, J., Gray, H., et al. (2011). Shedding light on solar fuel efficiencies. *Science*, **333**(15 July 2011), 288.

Heisenberg, W. (1935). *Is a Deterministic Completion of Quantum Mechanics Possible?* (Crull, E., and Bacciagaluppi, G., trans. 2011), http://people.isy.liu.se/jalar/kurser/QF/assignments/Heisenberg1935.pdf (last accessed 10 January 2014).

Hobbes, T. (1651). *Leviathan* (Martinich, A. P., ed). 2002 edition. UK: Broadview.

Hurst, J. (2010). In pursuit of water oxidation catalysts for solar fuel production. *Science*, **328**, 315–316.

Jones, N. (2012). Turn a new leaf. *New Sci.*, 14 April 2012, 30.

Klare, M. (2001). *Resource Wars: The New Landscape of Global Conflict*. New York: Henry Holt.

Kudo, A., and Miseki, Y. (2009). Heterogeneous photocatalyst materials for water splitting. *Chem. Soc. Rev.*, **38**, 253–278.

Lambert, N., Chen, Y., Cheng, Y., et al. (2012). Quantum biology. *Nat. Phys.*, **9**(January 2013).

Lenin, V. I. (1916). Imperialism, the highest stage of capitalism. X, http://www.marxists.org/archive/lenin/works/1916/imp-hsc/ch10.htm (last accessed 15 October 2013).

Lawrence, S., Liu, Q., Yakovenko, V. (2013). Global inequality in energy consumption from 1980 to 2010. *Entropy*, **15**, 5565–5579.

Lockie, S. and Higgins, V. (2007) Roll-out neoliberalism and hybrid practices of regulation in Australian agri-environmental governance. *Journal of Rural Studies*, **23**, 1.

More, T. (1516). *Utopia* (Turner, P., trans. 2003). London: Penguin.

Nocera, D. (2006). On the future of global energy. *Daedalus*, **135**(4; Fall 2006), 112.

Nnorom, I. C., and Osibanjo, O. (2008). Overview of electronic waste (e-waste) management practices and legislations, and their poor applications in the developing countries. *Resour., Conserv. Recy.*, **52**, 843–858.

US National Academy of Sciences (2009). *America's Energy Future: Technology and Transformation; Summary Edition*, https://www.princeton.edu/mae/people/faculty/socolow/americas-energy-future.pdf (last accessed 7 January 2014).

Newitz, A. (2013). *Scatter, Adapt and Remember*. Collingwood: Black.

Osvath, M., and Osvath, H. (2008). Chimpanzee (Pan troglodytes) and orangutan (Pongo abelii) forethought: self-control and pre-experience in the face of future tool use. *Anim. Cognit.*, **11**(4), 661–674.

Pilkington, A. and Dyerson, R. (2006) Innovation in Disruptive Regulatory Environments: A Patent Study of Electric Vehicle Technology Development, 79–91. In *European Journal of Innovation Management*, **9**(1).

Poincaré, H. (1908). Science and Method (Maitland, F., trans. 1969). New York: Dover.

Puckett, J., Byster, L., Westervelt, S., et al. (2003). *Exporting Harm: The High-Tech Trashing of Asia. Basel Action Network and Silicon Valley Toxics Coalition*, http://www.ban.org/E-waste/technotrashfinalcomp.pdf (last accessed 16 October 2013).

Rees, A., and Opeskin B. (2011). Current issues in gene patenting. In *Disputes and Dilemmas in Health Law* (Freckelton, I., and Petersen, K., eds.). Sydney: The Federation Press, 277–295.

Reynolds, A. (2004). *Century Rain. Ace.* This is a science fiction work that envisions a "Nanocaust"—a decimation of all life on Earth from out-of-control nanotechnology.

Richards, C., Bjorkhaug, H., Lawrence, G., Hickman, E. (2013). Retailer-driven agricultural restructuring—Australia, the UK and Norway in Comparison Agricultural Human Values, **30**, 235.

Schellenhuber, H. J. (2009). Tipping points in the earth system. *Proc. Natl. Acad. Sci. U S A*, **106**(49), 20561–20563.

Solar Fuels Institute (2013). About. *Trends Plant Sci.*, **15**(10), 554–564, http://www.solar-fuels.org/ (last accessed 9 January 2014).

Stephens, E., Ross, I. L., King, Z., et al. (2010a). An economic and technical evaluation of microalgal biofuels. *Nat. Biotechnol.*, **28**(2), 126–128.

Stephens, E., Ross, I. L., Mussgnug, J. H., et al. (2010b). Future prospects of microalgal biofuel production systems. *Trends Plant Sci.*, **15**(10), 554–564.

Stern, N. (2007). *The Economics of Climate Change: The Stern Review.* Cambridge: Cambridge University Press.

Smith, A. (1776). *The Wealth of Nations. 2009 Edition.* Blacksburg, VA: Thrifty.

Turner, P. (2003). Introduction. In *Utopia* (1516). London: Penguin.

Vines, T., Bruce, A., and Faunce, T. A. (2013). Planetary medicine and the Waitangi Tribunal Whanganui River report: global health law embracing ecosystems as patients. *J. Law Med.*, **20**, 528–541.

Wall Street (1987). Directed by Oliver Stone. 20th Century Fox. Dialogue by Gordon Gekko: "It's a zero-sum game: somebody wins, somebody loses. Money itself isn't lost or made, it's simply transferred."

Wilczek, F. (2009). Edge.org. World Question Center. Question: What will change everything? Answer: Homesteading in Hilbert Space. http://www.edge.org/q2009/q09_12.html (last accessed 7 January 2014).

World Bank (2013). *Improved Water Source (% of Population with Access).* Water supply data, http://data.worldbank.org/indicator/SH.H2O.SAFE.ZS (last accessed 10 January 2014).

Index